PLANNING FOR GOOD OR ILL

PLANNING FOR GOOD OR ILL

A Review of the UK Planning System:
1944 to the present day

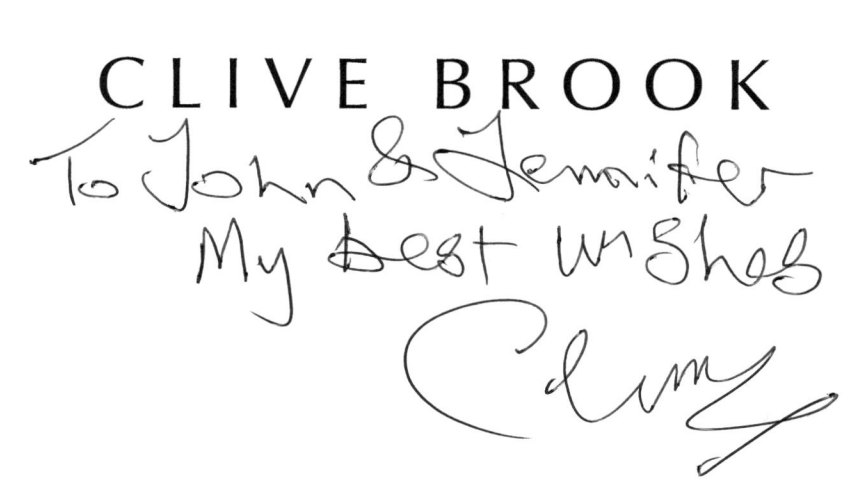

CLIVE BROOK

*To John & Jennifer
My best wishes
Clive*

BROWN
DOG
BOOKS

First published 2023

Copyright © Clive Brook 2023

The right of Clive Brook to be identified as the author of this work has been asserted in accordance with the Copyright, Designs & Patents Act 1988.

All rights reserved. No part of this book may be reproduced, stored in a retrieval system, or transmitted in any form or by any means, electronic, electrostatic, magnetic tape, mechanical, photocopying, recording or otherwise, without the written permission of the copyright holder.

Published under licence by Brown Dog Books and
The Self-Publishing Partnership Ltd, 10b Greenway Farm, Bath Rd,
Wick, nr. Bath BS30 5RL

www.selfpublishingpartnership.co.uk

ISBN printed book: 978-1-83952-613-8
ISBN e-book: 978-1-83952-614-5

Cover design by Kevin Rylands
Internal design by Mac Style

Printed and bound in the UK

This book is printed on FSC® certified paper

*To Jill for her love and support,
Claire, Paul and Richard (professionals in the built
& natural environments)
And for our (Alex, Hayden, Florence and Rory) and your next
generation in the hope that they inherit a healthier planet*

Contents

List of Illustrations		viii
List of Abbreviations		x
Introduction		xii
Chapter 1	Awakening and Reconstruction 1944–1966	1
Chapter 2	Governance, Politics and Public Service 1966–1984	21
Chapter 3	Health and Wellbeing – A Renewed Focus for Planning in the Twenty-first Century	61
Chapter 4	Consultancy and the Private Sector 1984–2020	90
Chapter 5	Planning, Design and Beauty	114
Chapter 6	Development Management and the Scope of Planning Control	138
Chapter 7	Green Belts – The Need For a New Approach	172
Chapter 8	Planning and Risk Management – Consideration of Public Anxiety and Fear	210
Chapter 9	Regional Planning – Its Rise and Fall and the Case for a Resurgence	226
Chapter 10	Planning and Transport – What Kind of Relationship?	247
Chapter 11	Homes for Everyone?	270
Chapter 12	Climate Change – The Role of Town and Country Planning	297
Chapter 13	Planning and the Environment – Achieving Enhancement and the Mitigation of Harm	319
Chapter 14	Conclusions – Learning from the Past and Meeting Future Challenges	341
Notes		366
Select Bibliography		380
Acknowledgements		383
Index		386

List of Illustrations

1. Salt's Mill: Saltaire World Heritage Site — 198
2. Saltaire worker's housing: remains highly marketable today — 198
3. Three Cliffs Bay, the Gower Peninsula: The first Area of Outstanding Natural Beauty (AONB) — 199
4. Beamsley Beacon, North Yorkshire: The Yorkshire Dales National Park meets the Nidderdale AONB — 199
5. Ilkley – Panorama of the town environs — 200
6. Ilkley – Primary environmental constraints (GIS mapping) — 200
7. Ilkley – Secondary environmental constraints (GIS mapping) — 202
8. Ilkley – North and South Pennine Moors Special Areas of Protection (SPA) & Special Areas of Conservation (SAC) – habitat mitigation zones — 202
9. Ilkley – Landscape Strategy Plan — 203
10. Ilkley – Areas selected as potential development allocations in the Bradford Local Plan — 204
11. 3D model of Ilkley Roman Fort – scheduled ancient monument plus assemblage of listed buildings — 205
12. All Saints Church Ilkley development project – external panorama — 205
13. & 14. Internal photographs of the refurbished Church — 206
15. Multi award-winning sustainable residential development by Citu – alongside the River Aire Leeds City centre — 207
16. Marmalade Lane, Cambridge – multi award-winning cohousing community — 207
17. The Wheel of Sustainable Development & the three interlocking challenges — 208
18. Competition for land-use, solar farm near Bentham, North Yorkshire — 209
19. The Leeds-Liverpool Canal south of Skipton: access to heritage, nature, recreation, and symbolising hope for the future — 209

Image Credits
Richard Brook (5, 6, 7, 8, 9, 10, 12, 19)
Author (1, 2, 3, 4, 13, 14, 15)
Bradford University Department of Archaeology (11)
National Design Guide MHCLG January 2021 (16)
Image 17 developed by the author and Richard Brook based on a similar image in the National Design Guide
Image 18 courtesy of geography.org.uk under creative commons licence

List of Abbreviations

ALC	Agricultural Land Classification
AMA	Association of Metropolitan Authorities- was wound up in 1997 and replaced by the LGA
AONB	Area of Outstanding Natural Beauty
CCC	Committee for Climate Change
CIC	Community Interest Company
CIL	Community Infrastructure Levy
CPRE	Council for the Preservation of Rural England
CSE	Centre for Sustainable Energy
DEFRA	Department for the Environment, Food and Rural Affairs
DETR	Department of the Environment, Transport and the Regions- created in 1997 and dissolved in 2001 and replaced by three separate Departments- MHCLG, DEFRA & DfT
DfT	Department for Transport
DLUHC	Department for Levelling-Up, Housing & Communities- replaced MHCLG in September 2021
EIA	Environmental Impact Assessment
ELMS	Environmental Land Management Schemes
EOR	Environmental Outcome Report- proposed as a replacement for EIA in the Levelling-Up & Regeneration Bill
HFA	Health for All- WHO Healthy Cities Project.
HIA	Health Impact Assessment
HWB	Health and Wellbeing Board – established under the 2012 Health and Social Care Act
ICNIRP	The International Commission on Non-Ionising Radiation Protection
JNCC	Joint Nature Conservation Committee
LGA	Local Government Association
LNRS	Local Nature Recovery Strategy
LPA	Local Planning Authority
LVC	Land Value Capture

List of Abbreviations

MAFF	Ministry of Agriculture, Fisheries and Food – now part of DEFRA
MHCLG	Ministry of Housing, Communities and Local Government- replaced by DLUHC
NCC	Natural Capital Committee
NHS	National Health Service
NPPF	National Planning Policy Framework
NPPG	National Planning Practice Guidance
NRN	Nature Recovery Network – as proposed by Natural England
ONS	Office for National Statistics
PAS	Planning Advisory Service
PPS	Planning Policy Statements – superseded by the NPPF in England
PSB	Public Service Board (Wales)
RPG	Regional Planning Guidance – introduced in 1986
RSS	Regional Spatial Strategy – progressively replaced RPGs from 2004 onwards
RTPI	Royal Town Planning Institute
SAC & SPA	Special Area of Conservation & Special Protection Area- introduced by the European Habitat Regulations
STP	Sustainability and Transformation Partnership – NHS and local councils in England working together on health planning and outcomes
TCPA	Town and Country Planning Association
TfN	Transport for the North
WHO	World Health Organisation

Introduction

My central aim in writing this book has been to evaluate the extent to which the modern UK town planning system, introduced in the years immediately following the Second World War, has been a force for good in people's lives. Post 1997 devolution to Scotland, Wales and Northern Ireland transferred powers, including planning, to new government arrangements. Most of my references in this book are to the planning system as practised in England with one or two highlights examining notable examples of advances in Welsh planning. There have been many criticisms of the planning system and its outcomes, and of those professionals involved in its delivery in both the public and private sectors.

While the origins of the planning of towns can be traced back over thousands of years to the period when early civilisations built towns and cities, often as fortified settlements, during the first agricultural revolution, the genesis of the modern UK planning system has most of its foundations in the adverse health, social and welfare impacts produced by rapid population growth, industrial expansion, and the pace of urbanisation in the nineteenth and early twentieth centuries. However, the realisation of a complete operational system did not occur until almost five decades of the twentieth century had passed, when the impacts of the Second World War and a very determined post-war government combined to provide the necessary impetus.

Along the way, several social and industrial philanthropists sought to improve the living conditions and wellbeing of the mid to late Victorian populace, with a number of them establishing model communities to house their workers and others in much better conditions (George Cadbury at Bournville, Titus Salt at Saltaire and Joseph Rowntree at New Earswick, being notable examples).

This book is essentially a combination of memoir, recent history, assessment, and analysis. My approach has been strongly based on my personal experiences as a town planner over five and a half decades (1966 to the present), initially in various local government posts and subsequently, over the last 37 years, as a planning consultant. This version of my review of our town and country planning system concludes in factual and analysis terms in January 2023 as further reforms to the planning system in England are under consideration.

In writing this book I have essentially followed a twin focus, which covers the relationship between the health and wellbeing of the population, both individually and collectively, and the various impacts of the planning system. The second and integral theme considers the general success and failures of the system, both perceived and actual, and how they have been addressed. These perceived successes and failures are considered from the points of view of individuals and sections of the public when directly and indirectly affected by planning policies, decisions and outcomes, and the views of planning professionals, commentators, developers, and politicians.

My analytical review centres on the following core question: does town and country planning, as devised, practiced, and amended, from 1947 to the present day, constitute a force for good, with largely benign outcomes, or has it been a restrictive and misused system, which has negative impacts on people's livelihoods and the environment in which they live? We need to consider these matters because they are of great importance for the future wellbeing of society and the qualities of the environment we occupy as well as the future role and direction of town and country planning.

The health and wellbeing of UK citizens is now one of the core objectives of our town planning system, enshrined in current national planning policy and those local development plans that are up to date. The concern of some leading politicians and philanthropists for the basic health and wellbeing of the population of the UK in the last half of the nineteenth century and the early part of the twentieth century was a product of widespread poor living and working conditions in the new urban concentrations with unsanitary conditions, the spread of disease and limited life expectancy. Rural areas were also relatively deprived of good basic services but benefitted by degrees from cleaner air to breathe.

At the end of the Second World War, the Attlee Government of 1945–50 displayed a deep combined concern for the health and wellbeing of their fellow countrymen, which resulted in a comprehensive health and welfare system of which the new Town and Country Planning legislation was a part, albeit at this stage junior in status. The health and wellbeing of citizens had expanded to include social, economic and some environmental dimensions. The aftermath of the war required new jobs and housing for returning soldiers, rebuilding and infrastructure development to replace war-damaged areas, and comprehensive economic renewal.

The English National Planning Policy Framework (Ministry of Housing, Communities and Local Government), July 2021, states that the overriding purpose of the planning system 'is to contribute to the achievement of sustainable development'. The definition of sustainable development, which is still used in this central policy document, is that adopted in Resolution 42/187 of the United Nations General Assembly:

> 'At a very high level, the objective of sustainable development can be summarised as meeting the needs of the present without compromising the ability of future generations to meet their own needs.'

In order to achieve sustainable development, the English Government set out three overarching and interdependent objectives, 'which need to be pursued in mutually supportive ways (so that opportunities can be taken to secure net gains across each of the different objectives)'. The three objectives are:

"a) **an economic objective** – to help build a strong, responsive and competitive economy, by ensuring that sufficient land of the right types is available in the right places and at the right time to support growth, innovation and improved productivity; and by identifying and coordinating the provision of infrastructure.
b) **a social objective** – to support strong, vibrant and healthy communities, by ensuring that a sufficient number and range of homes can be provided to meet the needs of present and future generations; and by fostering well-designed, beautiful and safe places, with accessible services and open spaces that reflect current and future needs and support communities' health, social and cultural wellbeing; and
c) **an environmental objective** – to protect and enhance our natural, built, and historic environment, including making effective use of land, improving biodiversity, using natural resources prudently, minimising waste and pollution, and mitigating and adapting to climate change, including moving to a low carbon economy"

This National Planning Policy Framework (NPPF), which was first introduced in March 2012, is the core document providing the policy basis for the preparation of development plans and the taking of planning decisions by individual local planning authorities. To ensure that sustainable development is pursued in a positive way, at the heart of the Framework is a **presumption in favour of sustainable development** (see paragraph 11 of the Framework for the ways in which this central presumption is to be applied in plan-making and decision-taking).

This approach has set a very high bar for local planning authorities and their planning departments, for developers and their planning consultants and indeed for the government itself via the Secretary of State's decisions and those of the Planning Inspectorate.

For decades, preparing plans and making planning decisions on applications has been about applying a planning balance through the assessment of various impact criteria associated with development proposals and their location and siting. Nathaniel Litchfield, one of the leading planners of his generation, introduced the concept of the Planning Balance Sheet for decision-making. There is some tension between the achievement of the first of the three overarching objectives and the other two. While some approaches to local development planning and planning applications take the required holistic approach to delivering truly sustainable development, this is far from being an easy task and has presented the current planning system with a major challenge.

The health and wellbeing of our people are at the heart of the second of the three objectives, which collectively comprise sustainable development, but the assessed outcomes from the first and third objectives are also relevant to health and wellbeing. In later sections of this book, I examine the extent to which the three objectives and sustainable development outcomes are being achieved and how they might be better achieved in the future by putting the health and wellbeing of the population at the very heart of the planning system.

While this is not a textbook, I hope that students of town planning of all ages in the UK and in those countries whose town planning systems have been based on, or strongly influenced by, the UK system will find my approach helpful in their studies. Members of the public and politicians who become involved in the system via individual and perhaps controversial planning applications in their locality, or as activists seeking to influence development plan outcomes, are also part of my target audience. Finally, I am seeking to provide support and inspiration to fellow practitioners as they face the many challenges of their chosen profession.

The complexity of the relationships played out within the planning system and impacting upon it have increased over each of the last five decades despite successive governments setting out with the aim of simplifying the system introduced by their predecessors. The result is that modern UK Town Planning now has a much wider remit with far more statutes, regulations, development plan documents, guidance, and procedures.

The very concept of planning and control is alien to many individuals. The system and the professionals involved in public service, and more recently in consultancy, have received much criticism from the media and politicians in local and national government. Sections of the public have also been vociferous opponents at public meetings, in consultations and on a one-to-one basis, though they are often not truly representative of the community in which they live.

These criticisms, combined with the many, often unhelpful, changes to the system, resulted for a long time in a loss of confidence and purpose amongst planners, particularly those engaged in the public sector. A further aim in writing this book has been to communicate my passion for the important work in which planners are involved and for its central relevance to resolving some of the biggest issues that we need to confront now and, in the decades, to come.

In the following chapters I demonstrate how the town and country planning system can and does provide benefits to individuals and communities, which in turn provide positive contributions to our health and wellbeing. While making this case I recognise that these benefits are not universally available to all sections of the population, particularly those living in our deprived communities. There is much work to be done in regenerating and retrofitting these urban and rural areas to spread the economic, social, and environmental benefits of good planning.

Chapter 1

Awakening and Reconstruction 1944–1966

Huddersfield, West Yorkshire, 26 February 1944. I enter a world torn apart by more than four years of war, which had spread onto many fronts. My birthplace is an industrial town famed for its prowess in engineering and textiles manufacture, which lies at the centre of an island of resistors who have by this time experienced the evacuation of 330,000 men of the British Expeditionary Force from Dunkirk (27 May–4 June 1940), the aerial Battle of Britain (10 July–31 October 1940) and the heavy bombing of the Blitz (7 September 1940–11 May 1941).

Twelve months prior to my birth, the course of the war had begun to turn in favour of the Allies. The victory of the Red Army at the Battle of Stalingrad in February 1943 and the earlier German defeats at El Alamein in 1942, which led to the end of the North Africa campaign in May of that year, meant that Germany was already beginning to lose the war when the D-Day landings occurred on the beaches of Normandy on 6 June 1944. These landings were the long-awaited beginning of the last chapter of the war, and the unconditional surrender of the German forces took place on 8 May 1945, known as Victory in Europe Day, or VE Day.

I knew nothing of the war and the peace that followed, and in subsequent years there was little family discussion of what had happened. One of my few surviving memories is of the shell of a German incendiary bomb 'stored' in our cellar, which I recall my father saying had fallen in our front yard without exploding. I was aged three or four and the bomb was nearly as big as I was.

My father had not been in the forces during the war as he was working in an essential industry and as a part-time fireman during German bombing raids. My hometown had suffered its first bombing raids on Friday 14 March 1941, as part of a series of targets focussed in and around the City of Leeds. These targets included the Avro factory at RAF Yeadon (now Leeds-Bradford International Airport), where Lancaster bombers were being made, and the munitions factories at Barnbow and Thorp Arch to the east and north-east of the city. The Huddersfield target was the David Brown factory to the south-west of the town, where my father worked as a gauge engineer in the gear manufacturing division. During the war, this factory produced parts for the

2 Planning for Good or Ill

Supermarine Spitfire and then, in peacetime, tractors and parts for the Aston Martin marque. David Brown's initials were used for the series of luxury sports cars, subsequently starring in most of the James Bond films. Other towns and cities, including London, Coventry, and Glasgow, suffered more frequent raids, much greater damage, and higher loss of life.

The engineering skills developed by Huddersfield-based companies were further featured in the war effort. Thomas Broadbent and Sons primarily made textile machinery, but during the war they manufactured four mini submarines. One of these, the X20 Exemplar, was towed to within 60 miles of the French coast on 2nd June 1944 to guide in those parts of the main D-Day invasion fleet designated for landings on Juno Beach of 6 June.

It is only in recent years that I have gained a fuller understanding of the effects of the Second World War on the social and economic fabric and physical infrastructure of the UK, and an awareness of the extent of the sacrifices made during and after the war. As a young boy and teenager growing up during this post-war period, I had a limited appreciation of the extent of the sacrifices during the war, in terms of lives given fighting for freedom from Nazi oppression and the economic cost borne by the UK, which would take years to address. A happy working-class childhood in a close-knit neighbourhood, where families looked out for each other, shielded me from the aftermath of the war, including food rationing.

It is important to understand the enormity of the war and its aftermath, as they were key influences on the emergence of the post-1945 welfare state, of which the modern town planning system was a part. Reading Anthony Beevor's excellent history of the Second World War, supplemented by documentaries and newsreels viewed over several years, has dramatically brought home the extent and impacts of the conflict.

Britain had stood alone against the might of the German armed forces following the invasion and occupation of France and Belgium in 1940. Things had gone badly, with the British Expeditionary Force hemmed in at Dunkirk. The evacuation of some 330,000 Allied troops by the Royal Navy and an armada of small craft in May 1940 averted disaster and demonstrated the spirit of the country in extreme adversity. This was further demonstrated later that year in the aerial Battle of Britain. The fear of invasion was strong, but Hitler did not progress his sea/land-based attack plan (Operation Sealion). Despite this, and the RAF's success against the Luftwaffe, great threats continued through sustained German U-boat attacks on Atlantic convoys delivering food and other supplies from North America, and the continued bombing of our industrial heartlands and main cities. Following the bombing of the American fleet at Pearl Harbour in December 1941, the United States entered the war, and this significantly helped to turn the tide towards victory in 1945.

The United States had introduced Lend-Lease aid in 1941 to assist Britain in purchasing American military equipment; just in time, as UK reserves approached exhaustion. This aid was abruptly withdrawn on 2 September 1945, which dealt a severe blow to the incoming Government. In Spring 1945, the Labour Party had withdrawn from the wartime coalition Government led by Winston Churchill and forced a general election. Labour won a landslide victory, taking more than 60 per cent of the seats in the House of Commons. Labour leader, Clement Attlee, became the new prime minister on 26 July 1945.

The introduction of an Anglo-American loan on 15 July 1946 went some way towards restoring economic stability, though the loan was primarily made to support British overseas expenditure in the early post-war years. This provided no assistance to the new Government's ambitious policies and plans for welfare reform, nationalisation, and rebuilding.

The war and its aftermath brought major social change. All classes of British society had 'pulled together' during the privations and the direct and indirect threats of the war years. The country was united in support of their armed forces as they fought on many fronts. Women had for the first time been employed in a large range of jobs, many of which had previously only been undertaken by men.

The immediate post-war years (1945–51) have been appropriately called 'Austerity Britain'.[1] This title is the first of a two-book volume of David Kynaston's penetrative socio-political history covering the first six post-war years, and forms part of a projected sequence of volumes covering the period 1945 to 1979, when Margaret Thatcher came to power with a determination to dismantle much of the post-war settlement.

The rationing of bread, the most staple of foods, existed from 1946 to 1948, and meat was rationed until 1954. Andrew Marr records how rationing was introduced during the war and continued post-1945 with the issue of 44 million ration books, the establishment of the Ministry of Food, regional offices and 1,400 food control committees.[2]

The 1945 election victory by the Labour Party was without doubt a seismic shift in politics following the coalition Government of the war period and the strong leadership of Churchill. Clement Attlee and his ministers came to power facing a set of major challenges. Jonathan Schneer summarises these in the final chapter of his book *Ministers at War*.[3]

Andrew Marr records the shock of Churchill's election defeat when, in many respects, he was at the peak of his powers.[4]

Looking back, this was collectively a truly remarkable set of achievements, which included the production of the first comprehensive town and country planning system, embodied in the 1947 Town and Country Planning Act and related statutes. Two sets of roots led to the creation and growth of this modern system of planning. The first set I examine were those of the Second World War and its aftermath to 1950. The second, deeper set has its origins in the Victorian slums created as a result of the Industrial Revolution and the rapid, largely un-planned, expansion of towns and cities, particularly in the second half of the nineteenth century. However, there is a strong common theme, which has fed these roots and continued to nurture the organism that is modern town planning at key periods of its growth. This is the health and wellbeing of the population.

Four key reports were produced in the war years that had a major influence on the 1947 Town and Country Planning Act. Three of these, the Barlow, Scott, and Uthwatt Reports, were directly related to lead planning issues of the period. The fourth, the Beveridge Report, proposed a wide-reaching system of social insurance and welfare, which would be universal and make provision against the loss or interruption of earnings due to unemployment, sickness, or old age, as well as the provision of family allowances, all paid for by national insurance contributions rather than taxes.

The Barlow Report (January 1940) was the product of the Royal Commission on the Distribution of the Industrial Population, which had been established in 1938 to analyse the problems of urbanisation and the inadequacies of the pre-war planning system. There was particular concern over uncoordinated growth in the south-east and, to a lesser extent, around other cities such as Birmingham, Glasgow and Manchester, and the parallel decline and depression in other parts of the country. The Report did not question the general economic value of urban concentration itself, but strongly concluded that there came a point when the drawbacks of further growth outweighed the advantages. Peter Self[5] emphasises that the novel part of the Report was to emphasise the economic as well as the social drawbacks of continued growth. The social problems, in particular pre-Second World War housing and living conditions, were well recognised. The Barlow Report continued to have a significant influence on national planning policy for a couple of decades. The terms of reference of the Commission included the consideration of what social, economic, or strategic disadvantages resulted from the concentration of industries and the industrial population in large towns or in particular areas of the country.[6] The final report of the Raynsford Review of Planning in England, 'Planning 2020', considers the Barlow Report as an important part of the 'impressive' technical case for planning, including its advocacy of a national plan.[7]

The Scott Report (1942) – the Report of the Committee on Land Utilisation in Rural Areas – also supported the idea of a strong central planning authority in its examination of the loss of agricultural land to urban development, including the development requirements. The Uthwatt Report (1941) – Final Report of the Expert Committee on Compensation and Betterment – influenced the content of the 1947 Act and, together with the Scott Report, advocated the appointment of a 'Super-Minister' armed with strong powers of coordination to achieve integrated planning on a number of fronts across key departments including transport, agriculture, housing, and the Board of Trade. Peter Self recognised in 1961 that this was a big ask that would never be realised in the British system of government unless the departments concerned could be closely integrated.[5] Over succeeding decades various attempts at integration have been introduced at national and local government levels.

The fourth influential document was the Beveridge Report (1942) – Social Insurance and Allied Services.[8] This major report has been widely recognised as part of the case for effective planning. Raynsford concludes that effective planning 'was a mainstream part of the wider construction of the welfare state' and 'reflected the wider consensus on the case for planning based on welfare economics'.[9] In his Report, Sir William states that the plan for a social security system 'is put forward as part of a general programme of social policy'. He considered that it was one part only of an attack on five great evils. These evils he summarises as physical want, disease, ignorance, squalor, and idleness.

Until I began some of my research for this book, I had not realised that a Town and Country Planning Act was given royal assent during the year of my birth, though apparently this is not regarded as a significant piece of planning legislation in the context of the subsequent 1947 Act, nor did it achieve much in terms of addressing the many weaknesses contained in the earlier 1932 Act. Of much greater note is the publication of the White Paper, 'The Control of Land Use',[10] which together with the four reports already mentioned, strongly influenced the content of the 1947 Act system and its proposed operation. The following extract from the White Paper appropriately summarises the need for comprehensive land use planning as perceived by the war-time coalition Government:

'Provision for the right use of land, in accordance with a considered policy, is an essential requirement of the government's programme of post-war reconstruction. New houses, whether of permanent or emergency construction; the new layout of areas devastated by enemy action or blighted by reason of age or bad living conditions; the new schools which will be required under the Education Bill now before

Parliament; the balanced distribution of industry which the government's recently published proposals for maintaining employment envisage; the requirements of sound nutrition and of a healthy and well-balanced agriculture; the preservation of land for national parks and forests, and the assurance to the people of enjoyment of the sea and the countryside in times of leisure; a new and safer highway system better adapted to modern industrial and other needs; the proper provision of airfields – all these related parts of a single reconstruction programme involve the use of land, and it is essential that their various claims on land should be so harmonized as to ensure for the people of this country the greatest possible measure of individual well-being and national prosperity.'

This summary statement not only provides a cogent argument for a comprehensive national system of town planning as the war drew to its close, but it also continues to have a resonance today, with its emphasis on the health and wellbeing of the population echoed in current national policy. It is significant, as Andrew Marr reminds us,[11] that both Beveridge and Clement Attlee had become social workers in the east end of London before the start of the First World War, which undoubtedly influenced their future approaches to the health and wellbeing of their fellow man.

This is an appropriate point at which to briefly review the earlier foundations of the modern town planning system, covering the period from 1850 to 1940. The current UK town planning system, centred around the 1947 Act and its successor legislation, is to a significant extent the result of the rapid and largely unplanned growth of towns in the nineteenth and the first part of the twentieth century. The need for a composite solution to this rapid and unplanned urban expansion and the consequent problems which burdened town and city dwellers was increasingly recognised in the latter half of the nineteenth century. Essentially, rapid, unplanned urbanisation at high densities, combined with poor quality housing and an unthinking juxtaposition of homes and industry, led to a series of adverse health, social and economic outcomes for many households. As well as poor quality housing and pollution from nearby industry, the lack of, or poor quality of, infrastructure compounded these problems. This was particularly true of drainage, water supply and waste disposal systems. The personal impacts of these adverse outcomes were all too often cumulative in nature with, for example, poor health, leading to unemployment and increasing poverty.

It is also significant that town planning has a few deeper roots, which predate the Victorian era, and include positive examples of relatively well-planned urban settlements. In a number of cases, across the centuries, aesthetics, good design, infrastructure provision and environmental considerations have played a part in creating communities that were healthier and did give degrees of consideration to the wellbeing of their citizens. The Greek and Roman Empires provide early examples of good sanitation and other infrastructure, as well as considered urban design within their town plans, which were beneficial to the wellbeing of their citizens, though Rome apparently had its areas of slum housing. The European Renaissance of the fifteenth and sixteenth centuries produced a limited number of completed planned towns. Notable examples are Palmanova, Scamozzi's nonagonal town regarded by Pevsner as 'the most famous example' of town planning of the period;[12] Pienza in Italy and Zamosc in Poland, the latter being designed from scratch by an Italian engineer, Giulio Savorgnano. There were, however, many examples of plans and realised examples of urban design for improved living in parts of towns. These were drawn up by leading architects, many of whom were of Italian origin. The substantial re-planning of parts of London came about much later under the creativity of John Nash. With the help of his patron, George IV, he created high quality terraces, squares, and crescents with the inclusion of mews housing to the rear, in the early part of the nineteenth century. He was followed by the builder Thomas Cubitt who initially built solid middle class residences in Highbury and Islington, followed by new housing on a large part of the Duke of Bedford's estate in central London in the mid-1820s and subsequently by more expensive housing in Belgravia, Pimlico and the 250-acre estate of Clapham Park. Of final note in this context of historic examples of good town planning and urban design is the work of Baron Hausmann in Paris, who as prefect of the Seine was commissioned in 1853 by Emperor Napoleon III to carry out a massive urban renewal programme to give large parts of the city air, open space, better connectivity and to make it more beautiful.

The significance of this diversion to mention some notable examples of good town planning is to question why these examples and others were not more influential in creating a stronger and more successful drive for an earlier universal system of town planning. Two immediate answers emerge but only provide a partial explanation. The first is that the examples of well-designed Georgian and Regency residential developments in London were backed by wealthy landowners and patrons and also by the well-off middle and upper classes, who were seeking quality residences together with a good living environment. The second reason is that the high-density small houses for the poorer working class were built cheaply in tight groupings, often by speculative developers with no interest in providing wider streets, good service infrastructure and integral open spaces. The term jerry-builder was coined in the nineteenth century, but its specific origin is unknown. The Concise Oxford English Dictionary states that it is 'sometimes said to be from the name of a firm of builders in Liverpool or to allude to the walls of Jericho, which fell at the sound of Joshua's trumpets'.

I have found William Ashworth's book, *The Genesis of Modern British Town Planning*,[13] to provide the best coverage and insight into those key influences on the eventual emergence of the 1947-based system prior to the commencement of the Second World War. He was born at Todmorden, then part of Yorkshire, in 1920, educated at the local Secondary School and then at the London School of Economics where he gained his BSc in Economics in 1946 and then his PhD. His post-graduate thesis became the basis of his book. He lectured at LSE in Economic History and was subsequently Professor of Economic and Social History at the University of Bristol.

In the introduction to his book, Ashworth briefly reflects on the nature of town planning and refers to a common perception regarding it as primarily an aesthetic subject, an extension of the art of architecture. On page 1, he goes on to agree that:

> 'Any formal history of town planning would be bound to give much attention to its aesthetic aspect. But this book is not a history of the planning of towns that has been proposed or accomplished in the last century and a half. Its aim is rather different. It is to try to discover why there has gradually arisen a widespread public demand that town planning should be adopted as one of the normal functions of public activity.'

He considers that it is unlikely that this public demand can 'be attributed merely to an increased sensitivity to visual beauty among many members of the community, to a wish that they themselves should promote its extension and a demand that they themselves should be enabled to live in the midst of it'. He concludes that: 'A powerful feeling of this kind has not manifested itself to any great extent in other activities.' He identifies that the objectives that modern town planning has sought to promote go well beyond aesthetics and relate to health and social and economic wellbeing. He makes the crucial link between these wider objectives of town planning and the results which have been claimed for it and the growth of the town planning movement and in the interest of the wider public:

> 'There was no town planning movement worthy of the name until a sufficient body of people was convinced that town planning could

make a unique and necessary contribution to the happiness, welfare, and prosperity, particularly of townspeople, but ultimately of the whole nation. The principal origin of the movement must be sought in the slow creation of that belief.'

The following chapters of his book chart this slow process of development of a strong public belief in town planning. It is notable that Ashworth refers to happiness, welfare, and prosperity as three key attributes sought by the general public and leading advocates of town planning throughout the long period of emergence of 'the movement'. I have identified these as a focal part of my book, and we will see how these have re-surfaced into national planning policy and current development planning and decision making. However, the significance of current and future town planning to these key attributes is not widely understood or appreciated by the population of the UK.

Ashworth goes on to identify town planning as a specific part of the full range of social policy that arose to address the nature and scale of the problems resulting from rapid urban growth. He describes his work as being 'essentially a study of the generation of a new social movement by the interaction of human thought with economic and social conditions'. He also identifies at the outset a basic and important truth that the idea of town planning 'was not an end in itself but only a means to practical changes and these involve a cost which someone has to bear'. People needed to be convinced that town planning 'was sufficiently desirable to justify the spending of money on it'.

Ashworth's thesis indicates that the emergence of contemporary town planning was part of a caring response to the adverse living and working conditions experienced by the majority of human beings in nineteenth and early twentieth century Britain. Population growth and rapid, frequently unplanned, urbanisation placing many houses very close to polluting industries, was the central cause of adverse health, social and economic impacts. It became apparent to a few leading medical and other professionals that the spread of diseases such as cholera and typhoid was directly related to overcrowding and the lack of basic sanitary provision. Ashworth refers to the Public Health Reports produced by Sir John Simon, a government medical officer and one of the leaders for sanitary reform in the latter half of the nineteenth century. Ashworth quotes a specific example of the cholera epidemic of 1854, documented by Sir John, where he compares the 2,000 deaths in Newcastle-on-Tyne and Gateshead with only four deaths in Tynemouth, some eight miles away, where sanitary regulations and drainage had been significantly improved.

A number of writers, including Ashworth, have charted the influence of a few Victorian industrial philanthropists who found good reasons to vastly improve the living and working conditions of their employees by building new settlements associated with a relocation of their mills or factories away from the squalor of the major urbanisations. In 1851 Titus Salt, the owner of several woollen mills in Bradford, embarked on a major project to replace his existing mills with a single new mill and to build round it a new town for his workforce. The new mill, opened in 1853, covered 6.5 acres and employed 3,000 people, with the workers initially being conveyed by rail from Bradford. Salt's plans for the complete community were clear prior to the opening of the mill, and Ashworth provides the following quote from the Manchester Guardian of 21 September 1853:

'The architects are expressly enjoined to use every precaution to prevent the pollution of the air by smoke, or the water by want of sewerage, or other impurity… Wide streets, spacious squares, with gardens attached, ground for recreation, a large dining hall and kitchens, baths and washhouses, a covered market, schools and a church; each combining every improvement that modern art and science have brought to light, are ordered to be proceeded with by the gentleman who has originated this undertaking.'

A series of social and other infrastructure was provided in succeeding years, and by 1871 Saltaire, as planned by its founder, was completed with the opening of a 14-acre public park. In common, as I suspect, with William Ashworth, I am a proud Yorkshireman and continue to marvel at the quality of the buildings, streets, and environment in what is now the designated World Heritage Site of Saltaire, which I frequently visit. However, this is not a museum piece, as the mills in particular have been restored and transformed into a thriving and eclectic mix of industrial, office, retail, and leisure facilities, including dedicated art galleries for the works of the world-renowned and Bradford-born artist David Hockney. Alongside the refurbished mills, the houses and streets are fully occupied by a vibrant community, which provides testimony to the brilliance and longevity of the original project.

A few other leading industrial philanthropists are worthy of mention, but there is a large time gap before the emergence of the next notable example at Bournville. This is a further indication of the slow emergence of the town planning movement, despite the model of Saltaire, which was widely acknowledged at the time for its qualities. As Ashworth states, it is not a model that was copied: 'Outside the large cities, efforts to improve the physical environment were, with this one exception of Saltaire, not only few but feeble.'

Bournville commenced in 1879 with the building of a new chocolate factory, but it was not until 1895 that George Cadbury began to build the wider model village. In 1900 his philanthropy continued when he transferred all his financial interests in the project to the Bournville Village Trust on terms that ensured that all profits were ploughed back into further extensions and improvements to the new settlement.

The foundation of Bournville was followed by another notable example of a model village at Port Sunlight, which was begun in 1888. This was a different form of philanthropy driven by W H Lever and centred on devising a means of profit sharing to benefit his employees. He applied the workers' share of profits to the building of houses to be let at a reduced rental. His central theme was 'prosperity sharing'. The scheme was carefully planned and included many amenities for the residents.

The next key event and personality to emerge in the slow progression of the movement for town planning was the publication of a book by Ebenezer Howard, entitled *Tomorrow: A Peaceful Path to Real Reform*, in 1898. He was the founding father of the Garden City Association in 1899, which became the Town and Country Planning Association (TCPA) in 1941, of which we will hear more in various parts of this book. As Ashworth explains, the term garden city was not new at this point in time, but it was Howard's holistic approach to the purpose, planning and realisation of garden city communities that set him and his ideas apart and became a blueprint for a composite town planning solution in this and other countries. While the totality of his ideas has not yet been realised, significant parts have been and continue to be a major influence on the planning of new towns and garden settlements.

Howard's book was reprinted in 1902 and retitled *Garden Cities of tomorrow* It was re-printed again in 1946, nearly 50 years after the original version, with a preface by FJ Osborn. The copy in my library is the 1989 re-print by Attic Books[14] and contains a Foreword by David Hall, then Director of the TCPA. In this he quotes from Osborn's Preface to the 1946 edition:

'Let it be remembered that in reading this book we are studying a blueprint nearly 50 years old. What is astonishing is not that it has faded on the edges, but that its centre remains so clear and bright.'

In his Foreword to the 1989 edition, Hall refers to his delight that Howard's 'great work has been republished because the fundamental principles that underlie the Garden City idea continue to be of great relevance today'. More than 100 years on from first publication, Howard's principles and ideas remain very much alive. Ashworth does not consider the individual ideas contained in Howard's composite proposals to be particularly original, but it was their holistic nature, his personality and enthusiasm, and his understanding and deep feeling for the condition of mankind within the burgeoning cities, which enabled them to be actively progressed soon after the initial publication. After the formation of the Garden City Association in 1899 there followed the establishment of the Garden City Pioneer Company, with sufficient capital by 1902 to purchase an estate and prepare a plan for the first garden city at Letchworth. In 1919 Howard was responsible for the purchase of an estate at Welwyn, which formed the major part of the site on which Welwyn Garden City was built.

In a new introduction included in the 1989 edition of *Garden Cities of Tomorrow*, Ray Thomas, a former Director of the New Towns Study Unit at the Open University, provides a review of what he calls 'Howard's Neglected Ideas'. Prior to summarising these it is helpful to state the essence of Howard's blueprint. His 'invention' was the principle of decanting population from the large conurbations, creating the scope for urban renewal within the towns and cities, via groups of 'slumless, smokeless' garden cities, with a central city population of 58,000 and a number of satellite cities of circa 32,000 people. This garden cities system was to be connected by rapid rail links, with the intervening countryside providing for agriculture, woodland, and recreation. Land reform is the first of Howard's neglected ideas. He envisaged that new towns would be built under municipal ownership, with the local authority acquiring the land required at agricultural values and, through continued ownership of the land, 'would capture for the community the urban property values which construction of the town would create. Residents and other occupiers would pay a rate-rent to the municipality for their use of the land.'[15] The 'rate-rent' was a combination of a nominal ground rent, and the annual rates bill payable to the new town authority.

The second of Howard's neglected ideas was what he called 'The Unearned Increment', more often known as 'betterment' in the compensation and betterment conundrum, which has dogged successive decades of town planning. Howard was able to demonstrate how, in the new town developments at Letchworth and Welwyn, the unearned increment could be captured, whereas in the existing cities and towns the increase in property values resulting from infrastructure and other municipal investments largely accrued to landlords. The Uthwatt Report approach to compensation and betterment was preferred by the government when drafting the 1947 planning system, but as we will see, this system of nationalising development rights failed within just a few years. Howard's alternative, which sought to replace landlords in both the new towns and the old cities so that development gain could accrue to municipalities, has never gained real support. The third of the neglected ideas was the establishment of development corporations to run the new towns. Howard envisaged the establishment of trusts, and his early ideas were implemented at Letchworth. The trust or development corporation would run the new town estate on behalf of the municipal authority who would receive the surplus from rents after interest charges had been paid. However, the 1946 New Towns Act established the new town development corporations and made them responsible to central government. There was an allowance made for the potential future transfer of the assets of the new towns to the local authorities at a later stage, but the 1959 New Towns Act established the New Towns Commission to take over the assets of new towns. By the 1980s the development corporations were instructed to dispose of all properties with a market value and they were eventually wound up.

Howard's idea that local authorities should own the land assets of new towns and thereby obtain the financial benefits from the use of the land has been almost completely extinguished in his own country. The fourth idea was that of Social Cities, a system of garden cities with a control of the optimum population size, the central city at 58,000 and the satellites at 32,000. This pattern of growth was quite novel and very different from the way in which the major conurbations were continuing to grow, with their large, focal, central business district to which people commute via increasing distances along numerous radial routes. As Ray Thomas says in his Introduction, 'the debate on the value of these ideas is unresolved'.[16] I have spent rather more space summarising Howard's ideas as they remain influential in this and other countries, they are unique in their comprehensive nature and Howard, with what a number of commentators refer to as 'his invention', has made a quite extraordinary impact on the emergence of modern town planning.

The first town planning legislation emerged in 1909 (The Housing, Town Planning, &c. Act 1909) and both Ashworth and the Raynsford Review[17] attribute this to ongoing concerns regarding the unhealthy condition of towns, a product of the concern over basic living standards and a growing interest in public amenity and higher quality placemaking. Further planning legislation followed in 1919, 1923 and 1932, but in common with the 1909 Act there were two crucial flaws. The 'requirements for plan-making were voluntary' so many authorities did not prepare plans and those who did 'had no way of effectively enforcing their plans because there was no need for landowners to apply for planning permission'.[18] In order to prevent the development of land a local authority was required by the Act to pay the landowner full compensation. Winston Churchill also pointed out in 1909 that Howard's 'unearned increment' was a key omission, as local authorities were not provided with a way of recouping any of the increase in land values created by the provision of infrastructure that resulted from any of those plans and development activities that did emerge. Ashworth sums up the period 1909 to 1939 as follows:

> 'Altogether, statutory town planning was marked by the most drastic limitations. Its application was an option exercisable only in the face of deterrents; it applied in any case only to a small proportion of land and a narrow range of conditions; and it was concerned only with the physical layout of land and buildings; the social considerations which might guide that layout were left either to be disregarded or to be sought empirically for every separate scheme. Statutory town planning was no more than a halting advance, not a conclusive victory.'[19]

Raynsford concludes that the town planning system in the inter-war period 'was weak and fragmented and could not deal with the legacy of chronically poor housing conditions'.[20]

In conclusion, the movement for town planning between 1850 and 1939 was sporadic, slow to gather any momentum until the beginning of the twentieth century, and notable examples of model towns and the later garden cities were not universally accepted or copied. This slow build up adds greatly to the significance of what was then achieved in both the war years and the short post-war period up to 1950.

Prior to leaving the early twentieth century, there is one more significant landmark to mention, which is the establishment of the Town Planning Institute in 1914. The idea of town planning as a new and distinctive area of expertise began to be seriously formulated when a number of established professionals in surveying, civil engineering, architecture and the law began working together on local government planning schemes post the 1909 Act, though I think Ebenezer Howard can rightly claim some influence, although he was not a member of any of these professions. The Institute was formally established on 4 September 1914, when its articles of association were signed. The first three of those articles summarise the purpose of the new Institute:

> 'To advance the study of town planning, civic design, and kindred subjects, and of the arts and sciences as applied to those subjects.
>
> To promote the artistic and scientific development of towns and cities.
>
> To secure the association and to promote the general interests of those engaged or interested in the practice of town planning.'

Architects had been particularly influential in the establishment of the new institute as they saw this as a means to advancing their aesthetic design skills to the planning of whole communities. It is therefore notable that emphasis was placed from the outset on the fusion of science and art in the work of the professional planner.

I have identified two sets of roots that were the main influences in establishing the modern system of town planning centred around the 1947 Act. It is important to briefly review the scope of the central planning act and the associated statutes that make up 'the 1947 planning system' to appreciate their landmark status, their longevity and their range of influence. Writing shortly after the 1947 Act was passed by parliament, S A de Smith[21] opens his legal review article by stating that the Act 'has been called a skeleton Act; but it is a singularly robust-looking skeleton. Its 120 ample sections and eleven formidable schedules dwarf the more modest New Towns Act, its companion in the new planning code. Nevertheless, so extensive are the powers of the delegated legislation conferred by the Act that it may be some years before its full implications become clear.'

As well as the central 1947 Act the system also included:

- The Distribution of Industry Act 1945, which produced powers for the restriction and positive promotion of industry.
- The New Towns Act 1946 provided the Minister of Town and Country Planning with the powers to make orders designating an area as the site of a new town and to establish a Development Corporation to plan and deliver all aspects of each of these planned settlements.
- The National Parks and Access to the Countryside Act 1949 provided for the establishment of national parks in England and Wales and set up a framework for their administration, protection, and enhancement; powers for the Nature Conservancy and local authorities to establish nature reserves; rationalisation of the laws and procedures governing public footpaths and securing public access to open country.

The central Planning Act comprised several key and novel components, including the requirement that all the newly designated planning authorities (the county boroughs and county councils) had to prepare a development plan for their area. In 1932 there were 1,441 planning authorities, as a result of the new system there were 145. The planning powers given predominantly to these local authorities, as opposed to a centralised government agency (at this time the Central Land Board), 'forever welded the fate of planning to the wider fate of local government powers, finances, and boundary changes'.[22] The planning authorities were given three years to prepare their comprehensive plan and then submit it to the Minister of Town and Country planning for approval. The second main component was bringing all land, with the exceptions of agriculture and forestry use, under a system of development control whereby the local planning authorities granted planning permissions, which could be subject to conditions regulating the use or development of any of the applicant's land. The Act contained a section defining development, and provision was made for applicants to apply to the local planning authority to determine whether a proposed use of land constituted development within the meaning of the Act. Unlike the American zonal system of planning, which was also used in some European countries, the UK system of development plans and development control, using those plans as the basis for decision making on applications, also gave considerable discretion to decision makers as the plan did not determine the final outcome. Planners and politicians were able to balance the provisions of the plan with other material considerations in order to reach a decision. This is a fundamental component of the UK system, which has essentially remained intact in the current legislation. As we shall see, what constitutes a material planning consideration has been widely debated by planners and politicians, but also, crucially, by the courts, and there have been several landmark decisions that have extended the scope of the planning system. In addition, the concept of the planning balance has become a focal component of the decision-making process and planning practice.

The 1947 Act introduced the nationalisation of development rights and the associated development values. Landowners lost the absolute right to develop their land but could continue to enjoy the existing use and could, if they wished, apply for planning permission themselves or via a contractual agreement with a developer. The landowners consequently only owned the existing (1947) use rights and values in their land. The Act also established a system of comprehensive land taxation, whereby the increase in land values resulting from the grant of planning permission was taxed at 100% with the income going to the Central Land Board to be generally used for housing and infrastructure development. As S A de Smith clearly states, 'The potential "development value" of land is thus extinguished, and land will therefore change hands at a price roughly corresponding to its "existing use value". To compensate the owners of interests in land which are consequently depreciated in value, payments will be made from the Exchequer. This will remove the heaviest burdens of compensation from the shoulders of local authorities. If the Act also ends speculation in land values, it may perhaps prove the most significant measure passed by the present Government.'[22]

The 1947 Act was accompanied by a new government department with the new Secretary of State having extensive reserve powers over the planning system. For most town planning issues this created an uncomfortable relationship over the extent to which central government should intervene in matters of policy and practice. This position has essentially continued to the present day with different degrees of emphasis.

The Raynsford Review,[23] in its examination of the performance and problems associated with the current planning system, concludes that the 1947 Act and its associated legislation, complex though this now seems, created a system capable of fulfilling the social, environmental, and economic objectives of reconstruction and long-term land management. Even though the record of delivery was soon to be challenged, there was a logic and clarity to the structure of the system that has never been matched. Reviews of the 1947 system of town planning are generally favourable. The fact that the central 1947 Planning Act was not repealed in full until 1990 underlines its strengths, as does the presence of several elements of the planning process in the current statutory system.

My good friend Roger Suddards[24] and his colleague Sir Desmond Heap, leading planning lawyers of their time, were responsible for drafting the town planning legislation of several Commonwealth countries using the framework and contents of the 1947 Act.

In the opening section of this Chapter I briefly referred to my very limited appreciation of the extent of the human, economic and social sacrifices that had been necessary to sustain Britain's central role in the war effort and the physical and economic reconstruction which followed. It is appropriate to end with a short personal reflection of what it was like for one young boy to live during this 'Age of Austerity', which also featured the successful creation of a comprehensive welfare state.

The geography and landscape of my world seemed extensive at the time, and the freedom granted by my parents and those of my circle of friends ('our gang') to play out added to this sense of extent and opportunity. Our house was a two- up two- down terraced unit that faced into a courtyard made up of three terraces linked in an irregular form and a fourth side formed by the wall of the outdoor toilet block, with two households sharing a WC. This was part of an inner urban community of Huddersfield, a solid northern town. The central yard area was our immediate play area while the back lane formed our football and cricket pitch. Given the proximity of the other dwellings we sometimes annoyed our neighbours with our activities within the yard, though on other occasions it was transformed, with various props and 'dressing- up', into our theatre, generally causing amusement for the adults. The rear 'sports pitch' was contained by the walls of our immediate terrace and, beyond this, a row of garages. At the far end was a low wall, and at a lower level a large woodyard and mill building. A vivid memory of mine is climbing the wall and descending the reverse side to retrieve our football from beneath a pile of wooden planks and being chased up the wall at speed by a large rat! I recall the local Methodist minister on his way to and from church joining in our games of cricket, usually as a good bowler.

Within a mile of our house was a park and recreation ground and, beyond that, woodland. I begged the older boys playing football in the 'rec' to let me join in, which they initially did under sufferance. In the wood we 'discovered' an old Roman kiln and did our bit for amateur archaeology by uncovering ancient tile and brick fragments. Most of the wood was private land but this did not inhibit our enjoyment and imagination. A valley below the wood was occupied by a stream that became another playground and favourite picnic spot. We had a great sense of shared adventure and a right to roam. I personally felt a huge sense of freedom and variety as we walked, ran, and later cycled through our local landscape. Occasionally in our early teenage years we ventured beyond familiar territory. With the help of early map reading skills, we 'found' Robin Hood's grave in Kirklees Park, part of the grounds of the medieval Priory, where legend has it that Robin Hood shot his last arrow as he lay dying from a poisoned wound. Several years passed before we discovered that the grave was a Victorian invention.

I grew up in a close and tightly knit urban community about two miles from the centre of town. There were weekly visits to town, the Monday market, and shops with my mother, either walking or catching the trolley bus. Huddersfield was a village of little consequence prior to the Industrial Revolution but grew rapidly as a centre of textiles and engineering during the nineteenth and early twentieth century. Both of my parents had left school at the age of 13 to contribute to family incomes, where there was insufficient money to send them on to secondary school. Their experience and some innate love of learning led them to channel some of their hopes and missed opportunities into my future education. My father was initially an agricultural worker and then trainee blacksmith in the hamlet of Upper Heaton to the north-east of Huddersfield, before later joining David Brown Industries as a gauge inspector. Amongst the very few books in our house I later discovered police training manuals and learned that he aspired to pass the entrance exams to become a PC.

My mother trained as a tailoress and, during my early childhood and school years, worked from home making suits and other garments from lengths of material, including the fine quality worsted cloth for which my home town was world renowned, 'brought-in' by a local entrepreneur. The dining table in our small living room/kitchen was cleared to form her workspace as the pattern was pinned carefully to the cloth prior to cutting with large tailor's shears and then sown together, originally on a treadle Singer sewing machine. I recall the beautifully tailored school blazer for my first day at secondary school and the later, high-quality, worsted wedding suit, the cloth length carefully selected from the warehouse. She was a relatively rare 1950s example of an earlier cottage industry.

The 1944 Education Act introduced the eleven-plus selection exam to determine which pupils would be selected for grammar school and which would go to secondary modern. I recall that my parents had taken a quiet interest in the nature of the exam and had ordered some published examples of the type of questions that might arise, my mother taking an interest in English and writing and my father in mathematics. At the time I did not put much weight on this influence but was nonetheless keen to show my teachers at the new junior school I had been moved to the previous year, as a result of catchment re-organisation, that I was brighter than my results in the entrance tests, which had downgraded me to the 'B' form. I well remember the day when I got my results and had passed to go to the top choice grammar school, Huddersfield College. I can see the headmaster of my junior school (we likened him to the character of Toad in *Wind in the Willows*) coming down the steps towards me in the entrance and expressing his shock at my result, and then, that afternoon, my mother's joy as she went to inform the neighbours, which was strange because she wasn't usually so outwardly demonstrative.

In these formative, youthful years and up to the point where I scraped into university, via the newly introduced clearing-house system, I never gave much thought to the career I would follow. Three years at Swansea University, from 1963 to 1966, culminated in the award of a below average honours geography degree and a new girlfriend of three months, who I was to marry 15 months later. Of my fellow 32 honours geography graduates, a number went to American universities to study for a higher degree/earn money as an associate professor, several into teaching, a few into commerce, and I became the sole graduate trainee town planner. The factors influencing my choice appeared almost at the last minute, when the University careers officer said that if I didn't want to teach or cross the Atlantic, I might think of town planning. I had, in my third-year studies, been most inspired by Gerald Manners,[25] who introduced me to various strands of urban and rural economic geography, including the work of Peter Self, Michael Chisholm and Peter Hall, his contemporaries at St Catherine's College, Cambridge, and this may have tipped the balance of my career choice. My degree thesis was a poor examination of the purposes and performance of the West Yorkshire Green Belt, which although a key planning topic, contained little planning content and analysis.

Looking back now to this time of leaving university, job applications and entering the world of salaried work it is tempting to try and find direct connections between the birth of the modern town planning system during my formative years and my own early life experiences and eventual career choice. I think this may be a rather fanciful approach to be quickly abandoned and replaced by a realisation that there are strong influences of a northern working-class youth, values instilled by parents, community, and friends, by access to and love of the countryside, but also urban culture and townscapes, and above all fairness and opportunities for all. I now appreciate some common threads and influences with those who helped to bring our modern town planning system into existence.

Chapter 2

Governance, Politics and Public Service 1966–1984

This chapter covers my personal experiences as a town planner in English and Welsh local government and blends this with a review of the rising influence of politics and the importance of public opinion in the process and outcomes of the town planning system. I examine the ways in which the professionalism and public service role of a planner in local government during this period interacted with political influences at central and local government levels, and how there was a growing realisation that the involvement of the public was essential to securing outcomes that met local requirements alongside national and other policy objectives. It is also relevant in this context to look at the purpose of planning, the ethics that govern the involvement of professionals and politicians, and the extent to which the system was operating for the public good.

Subsequent chapters, while they are more topic based, will continue to follow some of the sub-themes introduced here, while also pursuing the twin overlapping themes of the book.

I entered my local government career as a graduate trainee with Essex County Council in September 1966 having attended various interviews at West Yorkshire and Lancashire County Councils and the Greater London Council. I declined the offer of jobs at the Barnsley and Pontefract area offices of West Yorkshire largely on the basis that they were too close to home. The County Planning Officer of Lancashire, Udolphus Aylmer Coates, had fallen asleep during my afternoon interview at County Hall in Preston, either being bored by my answers to questions from the panel of three officers on the raised dais above me or, I like to think, having had too large a lunch. Hertfordshire County Council had declined to move the time of my interview, which clashed precisely with that at County Hall Chelmsford, so Essex was my choice. I was one of a group of six social science graduates appointed to a pioneering 'sandwich course' made up of a programme of 24 months split into 6 monthly alternating sessions at the Chelmer Institute of Higher Education and within teams at County Hall. This new planning course led to the award of a postgraduate diploma at the end of the two years of concentrated study and work experience. A problem soon emerged when we met with our course leader, Dale Robinson, an architect planner with an Essex based practice. He informed us that the Royal Town Planning Institute (RTPI) review board had only granted the course partial recognition meaning that our postgraduate diploma would not provide the automatic gateway to professional membership of the RTPI. Consequently, we would have to sit all of the external exams of the professional body as well as the internal college exams. The external exams were divided into parts I and II, each comprising four papers, and had to be taken in London during the early summer of 1967 and 1968.

While neither I nor my colleague students followed up the report of the RTPI's review panel, I later became aware that their attitude to planning education was centred around the few university courses that existed at the time and the strong influence within the RTPI of leading academics. Roy Kantorowicz at Manchester University was one of the leading proponents of undergraduate town planning courses to the detriment of post graduate courses including my own at the Chelmer Institute of Higher Education. The debate within the RTPI on this issue was quite intense.

The topics we were to cover in our internal and external studies had a wide range and a strong practical basis. They included the theory and practice of town planning, planning history, civil engineering (especially drainage and highways), planning law, architecture and the economics of property and development. A major design project completed part II of the external exams. We were soon to find that the breadth and intensity of the study required far exceeded that of our university honours courses. We had two full-time lecturers, Dale, and John Gyford,[1] and a number of part-time lecturers covering specialist topics very often based on their main professional occupation.

A further shock awaited all six of our close-knit band when, at the end of the first year, we learned that we had all failed the external history of planning exam. This seemed rather odd considering that most of us had studied history as a subsidiary subject at university. Following further investigation, it transpired that the academic marking the exam paper was in fact the author of the key course textbook and, while our reading and lectures on the topic had a very good range, our essay answers had apparently given insufficient consideration to his views and interpretation. This failure in one of the four exams making up part I of the external exams required all of us to re-sit the totality of this part alongside part II the following summer.

Despite these setbacks, Dale and John were determined, as we were, that the course would be a success and that we would collectively demonstrate to the RTPI and wider academia that this was a very appropriate model for educating practicing planners. Part II of the external exam culminated in a master planning exercise where the candidates were required to arrive at the central London exam hall on the appointed day to receive the project brief in a large white envelope at around 9 am, prior to boarding one of the awaiting coaches ready to transport us to the chosen site. Having read the brief on the coach we then had from 10 am to 1 pm to carry out our personal survey of the site and prepare notes and sketches. We returned to the steps of the exam hall, having grabbed a quick sandwich, to face a further three-hour session in the hall, by which time we had to complete our master plan design and a short presentation report.

The selected site we had visited was that of the former Battersea Fun Fair, and we were tasked with accommodating an international convention and business centre with supporting hotels, car parking, services, and landscaping. Each week in college over the two-year period we had been prepared for this daunting test. Thursdays at college were the designated design project day and each week Dale provided us with a new project to work on and produce our master plans. I recall our first workshop when we were to produce a master plan for a local shopping centre in Basildon new town. I can see Dale now, with a bright beam on his round face, fingering his bowtie and opening this first morning session by expressing his view that individually and collectively we had no design or presentation skills and that, as social scientists, we probably couldn't draw a straight line. He quickly followed this with a promise that over the weeks ahead he would equip us with the graphic design and other presentation skills to get us all through this exam and that, while we might not make professional master-planners, this would be a key part of our planning armoury in our future careers.

The Thursday afternoon session was devoted to a group critique, which Dale chaired with a gentle touch, great positivity, and good humour. Initially, the criticisms of our colleagues' work were crude and negative, but in time we recognised that constructive criticism and mutual support was more rewarding and likely to contribute to our collective success. Dale advised that my first effort had resulted in the design of an uninviting 'wind tunnel'.

In the afternoon session of the master plan exam, I was seated opposite a young man who I presumed was an architectural student, as he had equipped himself with various drawing instruments, and after more than two hours had passed, was still working on a precise form for the main building complex. Meanwhile with my newly acquired presentation skills, based around cartoon techniques, I was able to progress to the incorporation of the scheme's hard and soft landscaping zones, leaving about half an hour to complete the required presentation report. My ability to measure time expended when writing essay or report format answers on other exams was not so effective.

Experience has taught me that I owe a considerable debt of gratitude to those who have made large contributions to equipping me with work and life skills. We were all successful in the external and internal exams crammed into the final term of 1968. Dale and John were friendly and inspiring lecturers who gave our early careers quite a boost by instilling confidence and a practical 'can-do' approach as well as the acquired planning knowledge. This first two years of intensive study and work was lightened by seeing my girlfriend from university most weekends, usually meeting up in central London, where Jill was working as an administrator at the Royal College of Nursing and staying at the YWCA close to the Kings Road in Chelsea. We became engaged in November 1966 and were married ten months later at the parish church in Easthamstead, Berkshire, close to the RAF staff college where both her parents worked. Jill had moved to a new job teaching languages at a large comprehensive school in Chelmsford. Weekday evenings were a continuation of work for both of us, Jill preparing lessons and marking homework and I working towards those second-year exams. On Saturdays we both played first team sport for Chelmsford Hockey and Rugby Clubs, usually meeting up afterwards with friends at the Rugby Club disco. Sundays were definitely a day to chill out, with irregular visits to the Chelmsford Baptist Church for Sunday evening worship.

My four periods of work experience within the 300-employee County Planning Department were divided between the teams dealing with town centres, research, and information systems. This meant that I did not gain direct experience in the preparation of development plans or in dealing with planning applications, the two core areas of the 1947 Act system. I spent the first two weeks in town centres learning to colour wash town plans and confessed to the team leader that I had hoped for more challenging tasks that would help build my practical experience. He brought me down a peg or two by stating this was a core job of the planning technicians and it was important to both learn what other employees did and the colour coding then in use. He then allowed me to progress onto various town centre surveys of retail and other commercial land uses, car parking, traffic routes, etc. Towards the end of this first six months, I was allowed to make some inputs into the Wickford Town Centre Plan, the first positive planning contribution of my career.

I was pleasantly surprised to find that my move to the research team involved a variety of different planning challenges and that the leader of this small three-person team encouraged a much more open and participative approach. He was friendly and encouraging and I was invited round to dinner on two or three occasions. Project involvement included work on investigating changes in commuter patterns within and across the County boundary, and research, with adjoining counties, on the Third London Airport project. I have subsequently reviewed the long and tangled political and planning history that led to the eventual selection of Stansted as the preferred site.

* * *

The Third London Airport – Site Selection Process

Stansted and Foulness in the Thames estuary were two of the shortlisted sites in Essex for which our team prepared socio-economic research and data. David McKie, a political correspondent with the Guardian, had his book *A Sadly Mismanaged Affair – A Political History of the Third London Airport*[2] published in 1973. This book concentrates on the period 1961 to 1973. Yet the 'affair' was by no means over. A short review of the decision-making process surrounding the search for London's third airport is worthwhile at this point, because it illustrates the awkward interaction of planning and politics at the national and local levels and the continuing difficulties experienced by successive English governments in handling major infrastructure projects in the absence of a national plan, or statutory regional plans, and an appropriate and efficient governance system.

Stansted had originally been held in reserve as a potential third London airport since the 1950s, and commercial operations had begun in 1966 after it was placed under BAA control, being used by holiday charter companies who wished to escape the higher costs associated with operating from Heathrow and Gatwick. However, after a public inquiry at Chelmsford, held in 1966–67, the Government set up the Roskill Commission (the Commission for the Third London Airport 1968–71). At the time this was the longest sitting public planning inquiry, taking seventy-four days in all, and examining 160 witnesses (including colleagues from Essex County Council). Mr Justice Roskill was armed with a 23-strong research team and six fellow commissioners, and a great deal of preparatory work was carried out prior to examining witnesses. Remarkably, Stansted was not selected as one of the four shortlisted sites by the Commission, and they recommended Cublington in Buckinghamshire. However, the Conservative Government at the time, under Prime Minister Edward Heath, agreed with the minority recommendation of Colin Buchanan[3] that Foulness should be selected. In 1974, Harold Wilson's Labour government cancelled the Foulness project due to the national economic difficulties. Stansted then re-emerged as an option in the report of the Advisory Committee on Airports Policy (ACAP) and the Study Group on South East Airports (SGSEA) and was eventually selected by the Conservative Government in December 1979 from a shortlist of six sites. A new terminal and associated infrastructure for Stansted was considered at the Airports Inquiries of 1981–83, followed in 1984 by the Inspector's report and the White Paper decision of 1985. The new terminal building was finally opened in 1991!

David McKie[4] considers the selection process for the Third London Airport to be 'the biggest single planning decision in British history' and that 'the vacillation of successive governments in deciding the site has exposed the fundamental weaknesses in the whole process of decision making in government'. He regards the history of deciding where the airport should be located as 'a classic instance of an attempt in a democratic society to find an acceptable balance between conflicting interests'. Given that he reached his conclusions in 1973, the subsequent continuation of the saga underlines that successive governments in the 1970s and 1980s did not fully learn the lessons of the largely independent Roskill Commission process, the internal organisation of national government to deal with such landmark planning decisions, and the impact that local pressure groups could bring to bear on the decision-making process.

Peter Hall[5] is mentioned in McKie's book as a leading geographer who, at the time of the Roskill Commission, referred to it as 'the most rational dispassionate procedure that good minds could devise'. He was not alone in his view when the Times called the Commission 'the grandest exercise in pure reason which any British government has yet mounted'. Hall was subsequently to emerge alongside others as critics of the Commission, referring firstly to the cost-benefit model developed specifically for the Roskill Inquiry and providing the example of the high value put on air traveller's time. Hall also refers to the third London Airport planning decision as being political, in common with most major planning decisions. He goes on to observe, in this context, that the public vote for national and local politicians, generally on five- and three-year cycles, based on a confusing mix of policies where planning issues are very rarely present or apparent. Public pressure groups form around particular issues as they did in defence of Cublington and other short- listed inland sites for the third London airport. The pressure group against Cublington had spent a great deal of money in getting the Commission's decision overturned.

* * *

My final team posting was into the small and embryonic Information Systems unit, along with a promotion to Senior Planning Assistant. The central task of this team of two, working closely with the Research Team and with other South East county planning authorities, was to identify and test potential information sources for the creation of demographic and wider socio-economic computer models. If successful, these would provide strong supporting evidence to assist the preparation of development plans, the determination of planning applications and the planning and implementation of public infrastructure projects. The Essex Planning Department already had a powerful team producing large-scale land- use transportation models, which required considerable computer power at this point in time (1968/69), and consequently the work was shared with Hertfordshire and Kent County Councils. Hall[9] summarises the early work on combined urban growth and travel models in the 1960s, which were first developed in the United States and taken up in the UK by a small number of pioneering authorities.

The Information Systems team therefore had strong links to the land-use transportation modelling work. The creation of a reliable demographic database utilising locally derived information, rather than the national census produced every ten years, was a core target. Local medical records and associated transfer of individual records to a new GP surgery could, if accessed on a guaranteed confidential basis, provide regular basic demographic data, including in and out-migration statistics. There were also downsides to be investigated. For example, not everyone registers with a doctor, or leaves their current GP practice when moving their place of residence. Major utility companies, including electricity and water authorities, presented other possibilities.

My interest in this work was short-lived and, combined with our wish to leave the South East, led to a move to Nottinghamshire in early 1970. The graduate training course sponsored by Essex County Council proved to be a boost to my career, but otherwise the experience of working in a large planning department of circa 300 staff had its limitations. The communication between teams was relatively poor and the County Planning Officer, Douglas Jennings-Smith, had what could only be described as an aloof management style, which permeated down through parts of the department. I had made several friends through social events and sport. Due to my Yorkshire background, I was pressed into captaining the cricket team, despite my playing limitations with bat and ball. A short leaving ceremony took place and colleagues informed me afterwards that I was just a few places short of being the hundredth person to leave in just over three years. I still cannot believe that this rate of staff turnover was so high, but nonetheless it reflected some problems that needed resolution.

* * *

Public Participation – The Skeffington Report, 1969

Towards the end of my time at Essex County Council in 1969, an event of national significance to the future of the planning system occurred with the publication of the Skeffington Report: The Report of the Committee on Public Participation in Planning (January 1969). The Committee was chaired by Arthur Skeffington, MP for Hayes and Harlington and Joint Parliamentary Secretary to the Ministry of Housing and Local Government. There were 26 members of the Committee with a domination of white, middle-class, and middle-aged men and only four women members. Many recent commentators agree that this was, in several respects, a landmark report, which still has relevance and resonance today. The ability to distinguish it from other national planning reviews of the period 1965–1975, which dealt with key aspects of planning procedures and were concerned primarily with speeding up the planning system and administrative efficiency, is a differentiating factor. Its core aim, content, recommendations, and rather novel presentation, using several cartoon drawings to get key messages across, also sets it apart. The major aim was to engage the public earlier and more effectively in the planning process through the preparation of development plans.

The need for and purpose of the Commission was motivated by the sense that local planning processes were being disrupted by tensions and misunderstandings between the public and the planning profession. The following words of Arthur Greenwood, the Minister for Housing and Local Government, bluntly summarise the position:

> 'Attitudes have got to change; we have got to get rid of the idea that the planners and the planned are on different sides of the fence and we must study ways of getting them talking together.'[6]

Public confidence in the planning system had declined in the latter half of the 1950s and the 1960s. This was primarily due to the belief that the 1947 Act system had created a largely top-down and bureaucratic approach to large-scale development planning, which was not accessible to the public in terms of them being able to understand the plans, and larger-scale development proposals (education) and the lack of consultation (participation). In addition, there was a growth of protest movements objecting to major road schemes, loss of heritage and local character in comprehensive redevelopment schemes, and loss of countryside and landscape character.

The Skeffington Committee was set up following the publication of the Planning Advisory Group (PAG) Report on 'The future of Development

Plans' and the 1968 Town and Country Planning Act, which introduced a new system of development plans, comprising structure and local plans, together with the statutory requirements for publicity and participation in their preparation. Skeffington and his Committee were appointed in March 1968, following the adoption of the 1968 Act two months earlier. They were charged with devising a system and methodology for publicity and participation, which would facilitate the much needed early public involvement in and understanding of the new development plans.

It was largely political problems that motivated government interest in participation. If an efficient and effective planning system was to be delivered, the current Labour Government under Harold Wilson, and the successor Tory Government under Edward Heath, who came to power in the 1970 election, had to ensure that the public were involved and in a meaningful way, that there was better understanding and cooperation between the public and planners, and that the growing public concerns with the pace and nature of development and the ensuing loss of townscape and landscape character in a number of areas was addressed.

Despite certain failings, the landmark significance of the Skeffington Report is further demonstrated by two recent reviews, the first in 2014, when the entire Report was re-published by Routledge as part of their series 'Studies in International Planning History',[7] and in 2019, when a series of articles in the journal Planning Theory and Practice, also published by Routledge (Taylor and Francis Group),[8] were produced to mark the fiftieth anniversary of the original report.

I consider that the Report has had, and to a large extent retains, a significance to the wider planning system in that it marks the emergence of a four-way relationship in planning between governance, planning, politics, and public involvement. The language used is similar to that which we would use today, indeed I think it has a particular resonance as there has not been an attempt by government to produce a similar review, even though the arguments used are more powerful today than they were in 1969. The first of these powerful arguments includes the complexity and spread of the issues with which the planning system must deal. The second is the extent to which the public, in growing numbers, through greater understanding, lack of confidence in government and greater concern regarding the issues faced by mankind, are now demanding greater and more meaningful participation. Climate change and the decline in biodiversity are at the top of the list of these concerns, and the planning system has a key role to play in the resolution and mitigation of these issues (see Chapter 12). The following is the context quote from page 3 of the Skeffington Report:

'It may be that the evolution of the structures of representative government which has concerned western nations for the last century and a half is now entering a new phase. There is a growing demand by many groups for more opportunity to contribute and for more say in the working out of policies which affect people not merely at election time, but continuously as proposals are being hammered out and, certainly, as they are being implemented. Life, so the argument runs, is becoming more and more complex and one cannot leave all the problems to one's representatives. They need some help in reaching the right decision, and opportunity should be provided for discussions with all those involved.'

Skeffington and his Committee were charged with providing examples and guidance on the ways in which publicity and participation could be developed locally in practice.

Given the contents of the government White Paper 'Planning for the Future' (August 2020), the Raynsford Review (November 2018), RTPI reports and the Building Better Building Beautiful Commission (January 2020), the approach to public participation in planning may be about to change.

The interest of national government in participation, post the Skeffington Report, has been limited, though it can be argued that, following the passing of the Localism Act, this is increasing. The performance of local government, while much better than their political colleagues at national level, remains patchy and lacking in full commitment and purpose. There are, however, several examples of professional planners, in local council positions, consultancy and academia, advancing the use of new and tested means of public engagement in the planning process. This has expanded now to the point where there are a few planning consultancies who work almost exclusively on community participation projects.

Finally on this topic, I refer to Skeffington's suggestion that, at the outset of the development plan making process, planners should approach local communities to provide information and ideas as well as their aspirations. The appointment of a community liaison officer is recommended. Prior to the introduction of Neighbourhood Development Plans, planning authorities have relied on the production of issues and options reports with consultation on these internally prepared drafts being the first stage of public participation in the plan-making process. This has generally denied the public the opportunity, at an early stage, to contribute their ideas and aspirations.

Government, at national and local levels, planners and to some extent the public themselves, have failed to adequately recognise and implement many of the important recommendations of Skeffington. In 1969 the role of the private sector developer in the planning system was limited and much of the 1947 Act system was predicated on the assumption that government would be the main player in producing and implementing plans and projects. This is no longer the case and, as we shall see, the interests of the market have become much more significant. As a planner at this time I was over-influenced by process, procedures and end results, giving too little thought to the ways in which we could truly engage with communities and plan for the public good.

* * *

Nottinghamshire County Council, 1969–1972

I was appointed to the post of Economic Development Officer for the County early in 1970. Although this position was in the County Planning Department it did include a wider corporate role involving other departments, and there were key links with the constituent district councils and, in particular, their directors of engineering and technical services who were responsible for physical infrastructure and industrial developments. I was initially and nominally throughout my appointment directly responsible to the County Planning Officer, Jack Lowe, a town planner, and landscape architect. He was an inspiring leader and had a somewhat flamboyant and entrepreneurial approach for a chief planning officer of this time. My position was organisationally in the Research team of the Development Plan Group, though in practice my role was varied and involved cooperative working with several teams. I was encouraged by Jack to take initiatives and make recommendations. This approach started from the first week of my appointment, when he took me along to the County Industrial Development Committee. I had never attended a local government committee before, but such was his confidence in me and his need to attend another meeting, that having introduced me and dealt with a couple of items he left the meeting, leaving me as the sole departmental representative. It was readily clear that members of this committee, and later the wider Council membership, had a very high regard for my chief officer, both as a person and a professional advisor.

Via this initial committee meeting and early discussions with Jack, I was briefed on the role of the Joint Industrial Officer for Nottinghamshire and Derbyshire, Jack Holmes, who was based at County Hall in Matlock. His role and responsibility covered the attraction of inward investment to both counties, and he had significant previous experience in this field. However, key Nottinghamshire councillors found him to be remote, both physically and in his method of approach. I was therefore briefed to get on with him but to take a more active and assertive role on behalf of the eastern partner in the joint operation. One of the Nottinghamshire officers I worked closely with was the Public Relations Officer, Colin Slater, who was remarkably supportive, and we became great friends until his death in 2022. We shared a dedication to the job we were being asked to do, our Christian beliefs and a love of sport. Colin was a well-known local character, being a football commentator on local radio, amongst other involvements, and later head of public relations for Notts County Cricket Club. We worked together on publicity documents for the county's economic and inward investment strategy. Colin also worked with most of the county councillors and particularly with the leader and chairs of committees. I learned a lot from him. He had been present at my interview and went out of his way to make sure that I was able to settle down quickly, including finding us temporary accommodation. This was particularly helpful as Jill was more than eight months pregnant and gave birth to our daughter Claire on 10 April 1970.

One of my most challenging commissions was to compete for and attract to Nottinghamshire IBM's third major UK business development. From a long list of ten locations, we got down to the final two with Durham County. One of the two motorway junction sites we put forward had been memorably selected by the Deputy Director of Planning, in a brief meeting with Jack Lowe and me, when he stuck a pin in an ordnance survey plan mounted on his office wall. We were unsuccessful in our final presentation, as were Durham, with IBM opting for Hampshire. The deputy's 'instant' site choice at junction 27, M1, was subsequently chosen by Kodak for a major plant.

The district council chief officers I worked closely with were also very supportive of my role, which linked well with theirs in the regeneration of the former mining communities of the Notts/Derbyshire coalfield. Within the County Planning Department my role included involvement in drafting economic/industrial policies for local plans and commenting on planning applications for industrial development. During this time, I also became a member of the joint planning team preparing the Nottinghamshire/Derbyshire sub regional plan. This established a new economic growth area, known as the Mansfield-Alfreton Growth Zone, straddling the M1 and focussed on junction 28. This proved to be very successful in attracting new employment, though the developments now viewed from close to this motorway junction would not meet the more exacting environmental standards required of modern employment parks.

When I look back on my local government career, this was undoubtedly the most fulfilling and happy of times in terms of employment, with a very high degree of job satisfaction. Unfortunately, it all became soured by problems associated with the departmental structure, although I wasn't sufficiently aware of this at the time as I was enjoying the job so much. Prior to my arrival, the Assistant County Planning Officer had been given overall management responsibility for this area of work and beneath him was the Head of Development Plans. Neither of them were involved directly in the work or in briefing me on specific projects or aspects of my role. The Assistant County Planning Officer attended the joint industrial development committee meetings in Matlock, Derbyshire. My position in the departmental hierarchy therefore sat in the fifth rung as there was also the Deputy County Planning Officer, who was generally supportive but not too involved. I recall some difficult meetings with the Assistant County Planning Officer, who tended to drift from support to opposition dependent on the issue of the time. I did my best to keep these managers briefed and in the picture. On reflection this was a rather poor management structure, which lacked any formal and active line management for my role, but I had learned to cope with the situation.

Without realising it, matters came to an abrupt head when a new and somewhat revised post at a higher grading was advertised, and I was asked to apply. I didn't get the job and I can only deduce from the reaction of Jack Lowe, at my leaving do in a local pub, that he had been faced with a complaint from my two superior officers. By this point I had applied for and accepted a new role as Chief Assistant Planner at Swansea City Council, responsible for development plans and planning policy issues.

* * *

Swansea City Council and West Glamorgan County Council, 1972–76

When applying for the job in Swansea I had been motivated by two factors, the knowledge that local government re-organisation was pending in 1974 and Glamorgan County Council was to be split into three new counties of South, Mid and West Glamorgan, and secondly that Jill and I had both lived in Swansea for three years and had a great affinity for the area. We bought a house on the Gower Peninsula seven miles from Swansea, the first part of England and Wales to be declared an Area of Outstanding Natural Beauty in 1956. This designation was richly deserved as the coastal and inland scenery are stunning. Within a mile of our village house in Pennard we could walk to Three Cliffs Bay, subsequently voted one of the best views in the UK. Other smaller bays were also in easy reach, and all had to be accessed on foot, making them quite special. This was a mini paradise for our two children (Paul was born in Nottingham in November 1971) and we enjoyed weekends either playing beach games with other village families or walking the many footpaths on other parts of the Peninsula.

During my first week of acclimatisation, I became familiar with the large Department of Technical Services, of which the town planning division was a part. Bill Ward, the Director, was a multi-qualified professional in highways, civil engineering and town planning and I was soon to find that he was a man for detail with a limited propensity to trust the judgement of his middle managers. During that first week he introduced me to the current mayor in the Guildhall, with the passing comment that I didn't propose to go home for lunch, in common with the normal pattern for local Swansea employees, as I lived seven miles away on the Gower. This parochialism in the City Council organisation and systems was quite deeply ingrained. My role and responsibilities as Head of Development Plans included the local plan, conservation areas, the new Quadrant town centre development, and the Lower Swansea Valley project. The team of seven included two former colleagues from my geography course at the University.

The Lower Swansea Valley was, from the early eighteenth century to the early twentieth century, a centre for the smelting of copper and other metals. At the time of greatest industrial concentration in the 1880s, Swansea became known as 'Copperopolis'. The devastation these polluting industries caused left behind Europe's largest area of industrial dereliction, with heavy ground contamination. The air pollution experienced in the nineteenth century was devastating to public health. A prize-winning verse of the time sums up the dire situation:

> 'It came to pass in days of yore
> The Devil chanced upon Landore
> Quoth he, by all this fume and stink
> I can't be far from home, I think.'

The Prince of Wales visited the Lower Swansea Valley project in 1973, and I was asked to prepare a commemorative brochure setting out the progress achieved over the last 12 years. I had arrived at Swansea University in 1963, two years after the Project was commenced, and I will not forget the scene provided by the derelict valley of the River Tawe on a misty September afternoon as we approached Swansea's central station. The reclamation Project was a collaborative effort between the City Council, the University and the Ministry of Housing and Local Government, with additional funding from the Nuffield Foundation. Pioneering research was carried out to establish which plants would grow in the residual contaminated soils and to establish the extent to which the soil profile had to be improved by the removal of contaminants. The site is now home to the South Dock and Maritime Quarter of the expanded City Centre, the Parc Tawe shopping centre, the Morfa Stadium and sports complex and the enterprise zone industrial park. Green corridors and woodland/planted areas have transformed the biodiversity of what was formerly a barren landscape. New footpath networks and a Valley cycleway have improved public access and, together with careful interpretation of the industrial archaeology, have re-kindled public interest in the area. The project completion represents a success story of international scale and demonstrates how land reclamation projects can be truly transformative for the economy, the environment and health and wellbeing.

My role as Development Plan team leader covered the usual highs and lows of most local government departments. One memorable day was hosting Dame Sylvia Crow, former President of the Landscape Institute, who some 10 years previously had prepared a landscape enhancement plan for the expansive foreshore of Swansea Bay. I recall her comparing the expanse and visual qualities of the Bay to those of the Bay of Naples and pointing out where her proposals had not been implemented and how the project had only been partially successful.

Bill Ward's tight directorial control of his large and multi-faceted department was demonstrated in several ways. In his large office was a proportionately large Edwardian desk. As the week progressed his seating position moved around it, following piles of paperwork and small pink slips on skewers, or attached to a set of the papers. The eight Chief Assistants across the Technical Services Department received many of their instructions via the pink slip system (it's probably too gracious to describe this as a system). One of the Chief Assistants had instituted a competition for the highest skewered mountain of received pink slips. Papers and proposals on issues requiring the Director's attention were to be provided by a set deadline on Friday afternoon so that they could be taken home in the 'weekend bag'. I quickly learned that my ability to issue opinions and external correspondence was significantly limited when my out- going mail was intercepted and I received a remonstrance in one form or another from the Director. The 'eye for detail' was demonstrated one afternoon when I received a telephone call from the Director's home prior to his departure on holiday. He informed me that several paving slabs alongside a particular highway were displaced or broken and would I pass the message to the Chief Assistant Highways Engineer. A final reminder was an internal telephone call on a protracted planning issue when we were interrupted by the fire alarm going off in the building I was occupying. I said I would have to leave the building at once, but Bill insisted we continue our fraught conversation and said it was probably only a practice alarm drill. Not surprisingly, I received a rollocking from the Chief Fire Officer.

These problems were frequent irritants, and on occasion I did let the Director and my immediate superior, the Chief Planner, know my feelings and requesting what I thought were logical changes, but all to no avail. A somewhat bigger and problematic issue raised its head when, approaching Christmas, I observed a box of drinks being delivered at the Planning Division enquiries desk. I mentioned this to my two former University colleagues, and they said it was a fairly common practice each year and the odd bottle would be made available to the office party. I called my team together and insisted that they should not partake of any spirits at the party. I mentioned my earlier limited experience of 'gifts' arriving at Essex and Nottinghamshire Planning Departments. Adams Butter had gratefully sent a large tub of butter to the West Essex Planning Team to thank them for their help in granting a timely planning permission for their factory extension. The Area Planning Officer took a judicious way out of this dilemma when, realising the butter would go off if he tried to organise its return (as he meticulously did with other non-perishable gifts), he had it divided up into small packs to circulate around the wider office. At Nottingham, the head of Local Plans discovered an actual 'brown paper parcel' containing £500 and went to great lengths to discover the sender to return it.

The theme of bribery and corruption at Swansea continued when, in the Spring of 1973, I was approached in the staff car park by a gentleman who advised me that I was in a good position to make sure he was the fortunate future owner of a prime retail property in the new Quadrant Shopping Centre and that I should leave my car boot open to 'receive a large Easter egg'. I responded that I was not in the least interested in his offer and that in any event I had no influence over the sale or letting of the units. He implied that other officers had been accepting gifts. Soon afterwards I unwittingly became involved on the periphery of a much larger corruption investigation involving the Leader of the City Council, who was charged by the police with receiving various substantial gifts. I had attended a meeting of the Policy Committee of the City Council to take brief notes for the Director on certain agenda items. The police subsequently visited the Planning Division to go through files and took away various documents, including, as it turned out, my hand-written notes of this meeting. I was subsequently asked to be a witness at the initial Magistrates Court hearing and, in common with several others, including a former secretary flown over from Canada, I was not called to give evidence. The Leader was eventually convicted and received a lengthy jail sentence. This was one of several corruption cases in England and Wales around this time, including the well-publicised Poulson/ T Dan Smith Affair. I return to this topic in the final section of this chapter.

I had taken the post at Swansea knowing that Glamorgan was to be partitioned into three new counties as part of the 1974 re-organisation of local government in a two-tier system, with Swansea remaining as one of four new district councils. The attraction of working for an entirely new authority and remaining in our Gower home was attractive, and therefore I applied for one of the two Assistant County Planning Officer posts responsible for preparing the new County Structure Plan. Several members of my Swansea team were also keen for a change, and therefore hoped that I would secure this management post and subsequently be involved in interviewing for other planning posts.

The Local Government Act 1972 reformed the administrative authorities in England and Wales. The Act was one of the key measures of Ted Heath's incoming Conservative Government. The recommendations of the 1968 Royal Commission, the Redcliffe-Maud Commission, had not covered Wales and were not accepted in England. The partition of the former Glamorgan County Council was subject to considerable opposition, but it went ahead and elections to the new authorities were held in 1973. They acted as 'shadow authorities' until the handover date in April 1974.

The new Chief Executive, Mike Rush, and other chief officers were quickly appointed, with Graham King being selected as County Planning Officer. The interviews for the Deputy and two Assistant County Planning Officers were to follow shortly. On the afternoon of the interview, I left my office on the basement floor of Swansea Guildhall to attend one of the oak-panelled committee rooms close to the main ground floor entrance. The duty porter, who was a passing acquaintance in that he saw me arrive for work on most days and knew my name, pulled me aside. Clive, 'bach' he began, and advised that the Chairman, the Lord Heycock, would be seated at the far end of the room with nineteen Swansea councillors to his left hand and nineteen Glamorgan County councillors to his right. Facing him at the opposite end of the table layout was a chair. 'Whatever you do don't sit in the chair, stand behind it. There will be a piece of paper on the chair with two questions written on it, and the Lord Heycock will give you ten minutes to answer both questions.' A couple of minutes later I was standing behind the chair and being invited to read the questions by his lordship. 'Please commence by answering the first question, Mr Brook', which I think was asking me to set out how I would organise the new Planning Department. A little more than half-way through my answer I was interrupted and invited to move on to the second question, which I cannot recall.

I returned to my office somewhat dejected, to be quizzed by colleagues as to how things had gone. About half an hour later the new Assistant Chief Executive arrived in my office to say that I was wanted back in the Committee room to be interviewed for the second Assistant County Planning officer post. Somewhat surprised, I said I had not applied for this post. That's true he said, but you are on the shortlist, now come on. It transpired that the vote had been split 19-all on the previous interview, and Lord Heycock had used his chairman's casting vote in favour of the Glamorgan applicant. Three colleagues from the Swansea planning department were also on a shortlist of five, including the Head of Planning and the other Chief Assistant Planner in the Department, who oversaw development control. Towards the end of the afternoon, I was informed that I had been appointed to the second management post, which would be responsible for preparing local plans with the four District Councils, minerals and other county matter planning applications, landscape, and land reclamation. I had thought that my previous experience meant that I was better suited to the Structure Plan post, but I was delighted to have this new role.

For six months, until April 1974, I held two posts, my existing role for Swansea Council and my 'shadow' role for the new West Glamorgan County. While a very busy period of my life followed, I was temporarily rewarded with two salaries and a full involvement in the organisation of the new County Planning Department. Graham King proved to be a very engaging County Planning Officer with a forward-looking approach, and the new management team worked well together.

Most of my Swansea team accepted new posts in the County Planning Department. The relatively small department of around 35, combined with an open and relatively relaxed management style, produced a good working environment. This resulted in several major projects being completed over the following two years, including the production of the new County Structure Plan, joint local plans with two of the districts and the planning, design and commencement of land reclamation works in the Afan and Lliw valleys. The re-contouring and landscaping of former colliery tips at Abergwynfi and Blaengwynfi villages in the Upper Afan Valley were particularly transformational.

Richard Penn, Head of Policy in the Chief Executive's team, and I also worked on early community planning in the Upper Afan Valley. The two villages had been built to house miners and their families in the late nineteenth century, but the early demise of coal mining in this valley left a legacy of disadvantage. The area had been selected as the only Community Development Project (CDP) in Wales in 1969 and this ran until the mid-1970s. The socio-economic problems resulting from the mine closures, combined with the relatively remote location of these upper valley communities, meant that they were at a huge competitive disadvantage in attracting large industrial investment, particularly when compared with the South Wales coastal plain. Re-training, small business development and community enterprises were part of the planned way forward, but these initiatives had limited success. Between the 1981 and 2001 censuses the Upper Afan suffered a 24 per cent drop in its population.[9] The CDP was largely regarded as a failure.

The true residual assets of the area, following the reclamation schemes and work with the Forestry Commission, were its landscape and heritage qualities. A new Country Park, Afan Argoed, was established by the County Council (now the Afan Forest Park) and linked by footpaths with Margam Country Park. The mountain bike trails in the Upper Afan are regarded by Sustrans, the cycling charity, as 'world class'. Jobs have been created in leisure and tourism, including accommodation, bike sales, hire and maintenance, and at the South Wales Mining Museum.

After two and a half years the initial challenge of establishing the new County Planning Department, and subsequently completing a range of projects, left something of a vacuum and a tough decision for Jill and me when an attractive job vacancy occurred at Leeds City Council. Our children were six and five years old, happy at school and, as a family, we had an idyllic wider playground. It was not for nothing that Swansea was known as 'the deathbed of ambition', a combination of its wonderful scenery and a soporific climate. The Driver and Vehicle Licensing Centre had re-located to Swansea during our time there and several of the senior personnel who had established the new computer and management systems lived in our village. Most of them were subsequently offered promotions to move back to head up regional offices in England and the majority declined the move. While it was a wrench to leave, Jill was very supportive, and we could both see merit in moving back to the north of England where we had our roots.

* * *

Leeds City Council, 1976–1984

I was interviewed for the post of Assistant Director of Planning (Implementation) at Leeds along with five other candidates, four of these being group leaders within the existing department. The appointment letter arrived in Spring 1976 and, while pleased with the outcome, I quickly realised that I had a substantial challenge on my hands. I had to prove that I could carry out all that the job demanded as well as working successfully with my four group leaders, three of whom had been on the interview shortlist. Leeds was a Metropolitan District and had a large and reputable planning department with a staff of around 140 serving a geographically extensive and varied set of communities with a population of 730,000. The department was divided into three directorates covering policy and development plans, implementation and building control. My responsibilities covered three geographical divisions dealing with development control (planning applications, appeals and enforcement) and site-specific projects and an Environmental Design Group responsible for conservation, landscape, and design. In total there were 65 professional planners and technicians.

Prior to formally taking up my new position, I received several weighty postal deliveries. These were the statutory regulations, directions and circulars related to the new Community Land Scheme. This was introduced by the incoming Labour Government of 1974 and comprised the 1975 Community Land Act, which provided wide powers of compulsory land acquisition, and the 1976 Development Land Tax Act, which gave local authorities powers to tax the development values of land. This was a further attempt to address the key issue of compensation and betterment that had been raised by Ebenezer Howard in his landmark publication, subsequently by Winston Churchill in 1909, then in the Uthwatt Report of 1941 and in the 1947 Act (see Chapter 1). The 1947 Act had introduced the nationalisation of development rights and the associated land values, but the format intended in this Act did not last very long, with development charges being abolished in 1954. Subsequent dissatisfaction led to changes in 1959 and then, in 1967, the Land Commission Act was introduced, and abolished three years later.

The received parcels contained statutory regulations, government circulars and directions followed the passing of the two new statutes. I diligently read the documents and attended training sessions in Leeds, meeting colleagues in the Estates, Legal and Planning Departments who were to be involved in the implementation of this system. As with the predecessor Land Commission scheme, the Community Land Scheme had little chance to prove itself and was abolished three years later by the incoming Government of Margaret Thatcher. Apart from the build-up to its introduction and the intensive training required, I now have little recollection of the detailed work involved, but I do recall that many landowners deferred bringing their land forward for development in the hope that the scheme would be rescinded.

I started the job in the summer of 1976, and we moved as a family in the following year, initially to live with my stepmother in Huddersfield (both of my parents had died in previous years) and then to a new house in the market town of Otley in Wharfedale. Within a week of moving in our third child, Richard, was born in the local hospital.

The City Council was Conservative controlled in 1976, in contrast with the Spring 1976 arrival of Jim Callaghan's Labour Government. Callaghan had succeeded Harold Wilson as the next Labour prime minister on the former's retirement from office. Denis Healey, the Chancellor, and MP for Leeds East was a candidate for the leadership and a highly regarded politician on the national and international stage. Cllr David Hudson was the Chairman of the main Planning Committee, and we formed a good working relationship via our regular briefing meetings for committee agendas and individual projects of significance. Cllr Arthur Miller was the key Labour member on the main Planning Committee and he and David Hudson had a mutual respect, and we all had in common a love of Rugby, Arthur being a former League player and David and I rugby union. Looking back across local politics, Leeds City Council had been served by some stalwart and highly respected individuals, many of them became City aldermen in recognition of their service. They in turn had respect for the lead professionals who ran the Council departments and advised them. David Hudson was keen to get to know me as a professional and to understand my personality. Although there were differences in our political views, which emerged in good natured discussions, this never got in the way of our business relationship and an emerging friendship.

John Finney was professional to the core and became highly regarded at a national level as an advisor to the Association of Metropolitan Authorities (AMA) and within the Royal Town Planning Institute (RTPI). In 1983 he served as President of the Institute. He was in some respects quite a reserved and private person, but as I gradually got to know him, he became both a loyal friend and role model for my ambition to one day be a leading city planner in local government.

Mike Shields, the Deputy Director, had a strong role in the corporate affairs of the City Council and was a close advisor to the Council Leader on planning, land and development matters. Mike had an easy north-eastern style, was tremendously hard working and a good manager of people. Stan Kenyon was my opposite number as the Assistant Director Formulation (policy and development plans). He was much admired by his staff for his intellect, mixed with a modesty and slightly retiring personality. His knowledge of local and national politics was immense, and he might, in other circumstances, have been a professor of politics in some academic institution, brought out on election night to provide commentary and statistics. He was not party political but extremely politically aware and astute, an area in which I was rather deficient. He would succeed John as Director of Planning in the mid-1980s. The extra departmental roles of the Director and his Deputy meant that Stan and I had enhanced internal management roles.

This was my first management role where I had responsibility for the operation of the development control function on an authority-wide basis. We were typically dealing with up to 3,000 planning applications/annum across three divisions and six teams, organised geographically. A small specialist team of three officers dealt with enforcement, two having retired from the local constabulary. The retired Chief Inspector ran the team and reported directly to me and was so good at the job that I rarely became involved in the detailed issues. The Administration Division included a team, known as the Plans Processing Unit (PPU), who, as well as controlling the day-to-day systems of receiving, registering, and validating applications (planning and building regulations applications), were the first port of call for all enquiries from the public relating, for example, to whether planning permission was required for a particular form of development and how to go about making an application. In addition, they produced weekly and monthly systems reports covering performance against statutory and other internal targets for determining applications. Together with Les Parry, Head of Administration, I reported on performance at the weekly departmental management teams, affectionately known as 'Planman'. When managing performance on an individual, team, division, and departmental basis it was always essential, in my opinion, to try and balance this against the complexities of determining individual applications and arriving at the most appropriate 'planning balance' in each case.

For the system to work well the three Divisional Planning Officers (South and City Centre, East, and West) each had a high level of responsibility for reviewing the decisions within their teams within a system providing for a high level of delegation of decisions to officers. The level of delegation to development control officers increased across all planning authorities over succeeding decades as the number of applications and the material considerations to be considered on individual applications rose. Filtering systems have had to be applied to ensure that the planning committees are spending their time on the most significant and controversial applications. Typical filters include the chair of the planning committee counter-signing certain decisions taken by a lead manager and applications being referred up to the planning committee if either a ward councillor and/or the parish council had objected or requested referral for sound reasons.

There were repeated calls nationally from government, business, and the public to 'streamline' the planning system, which sometimes related to the development control system and the handling of applications and appeals and, more recently, to speeding up the production of development plans. We were very conscious, in the Leeds Planning Department in the late 1970s, of the need to enhance performance on all fronts while still trying to maintain the quality of decision-making. John Finney, in his national roles, represented the AMA and RTPI at meetings with government ministers and representatives of the private sector development industry. As a result, we received regular draft circulars, guidance documents, review reports and proposed legislative changes to comment on and discuss at 'Planman'. John would respond by combining our thoughts and, on occasion, consulting the Chair of Planning and even the full Committee.

John, being keen to maintain the national reputation of the Leeds Planning Department, asked me to meet him to discuss the design and introduction of a new householder planning applications system. This was to shorten the standard eight-week target then applied for determination of all planning applications (the eight-week threshold allowed applicants to appeal against the non-determination of their application if time was critical to them or had not received one which they considered to be satisfactory). We produced and trialled the Leeds Householder Application System, which involved a target determination fixed at broadly half that of other applications, simplified application forms, clear guidance notes, fewer consultations, and clear officer delegation responsibilities. Householder applications were defined as extensions to existing dwellings and all developments requiring planning permission that fell within the curtilage of a dwelling. The twin aims of deciding more applications quickly and freeing up more officer time to deal with larger/more complex applications were broadly achieved. The trial was largely successful, and this aided the introduction of a national system. However, one side effect I felt should be avoided, if possible, was to ensure that individual development control officers were not type-cast for years as only dealing with householder applications. This was clearly an important consideration in terms of the training and performance review of individuals, as well as their level of job satisfaction.

Most weeks I held a meeting with my four group leaders, where we were able to discuss organisation, systems, committee, personnel, and specific projects/site issues. These meetings were extremely valuable to me as a sounding board, for advising 'Planman' and the committees I attended. Throughout the week I had meetings with individuals and small groups on specific projects. I was fortunate to have an excellent secretary who did a brilliant job of managing my diary, filtering meeting requests while remaining a friendly contact to all my staff.

The scale and significance of the projects in which the Implementation Directorate was involved, either in a project coordination role or direct liaison with external developers, included:

- a plan for the re-generation of the Holbeck-Hunslet industrial zones, now the area accommodating the southern expansion of the City Centre via the South Bank master plan focussed on the planned national high speed rail line (HS2) and the integration of the new station into the expansion of the existing station, though the future of the project's arrival in Leeds remains uncertain.
- the development of a new village at Colton to the east of Leeds.
- the redevelopment of the former Quarry Hill flats area at the eastern end of the City Centre to provide a mixed-use leisure/conference/exhibition and business centre.
- the development of a new science park on the site of a former hospital in east Leeds and the accommodation of Systime Computers' new headquarters and incubator high tech units in south Leeds.
- working with a major housebuilder on the delivery of three large new residential developments on Council owned land.

Each of the three geographical divisions within the Implementation Directorate had a three-person implementation team who worked on a wide variety of site-specific projects usually involving Council owned land, which was either surplus to requirements, affected by an infrastructure programme, or subject to area change of some form. Typically, these teams would produce planning and development briefs to either guide future sales of Council owned sites or to improve or assist corporate programmes for development/renewal. The three team leaders reported to two corporate groups, the Land for Sale group, and Technical Board. Initially Mike Shields had co-ordinated this work and chaired the corporate groups involving representatives from all technical departments of the Council, Estates and, as necessary, Housing, Education and Social Services. Mike was shortly to leave to take up the chief executive's role at Trafford Development Corporation and he handed on the whole of this management role. With this expanded role, and following Mike's departure, I also inherited his role of informal advisor to the Leader of the Council, Peter Sparling, on planning and the corporate matters covered by the two groups described above. Cllr Sparling was a respected local solicitor and an effective and knowledgeable Chairman of key committees. We had a professional relationship of mutual respect with an appropriate degree of deference on my part. Once again, this formula was both effective and cultivated what we both considered to be an appropriate level of friendship. With hindsight this was to be part of my undoing.

Most appeals and public inquiries against refusals of planning permission were dealt with by my team leaders and divisional planners. There were a small number of major exceptions, including a large opencast mining application in the Aire Valley to the south, and the public inquiry into the extension of the main runway at Leeds-Bradford Airport. Leeds City Council, as planning authority for the area including the airport, had given a free vote on whether to grant planning permission to the 99 councillors constituting the full Council. The result was 55 against the grant and 44 in favour. The airport was owned and run by a consortium of West Yorkshire Councils including Leeds and Bradford metropolitan districts and West Yorkshire Metropolitan County Council. This was, to say the least, a somewhat bizarre situation. The majority vote had supported refusal predominantly because of the adverse impact of aircraft noise on communities in the north-western part of the District. West Yorkshire County Council were the main promoters of the application, arguing the business, investment and leisure/tourism benefits of the proposed extension designed to accommodate more and larger aircraft movements.

A large group of local objectors (LACAN) had employed a noise expert to support the community case against the development. The Deputy Chief Executive of the City Council called me in to brief me on his legal views on the content of the evidence I should be drafting. All we agreed on was the size of the formidable challenge the City Council and I faced in defending the position taken in the free vote. He identified the two key issues as being the noise impact and the surface highways access issues. I agreed the noise impact issue was key but stated that the umbrella issue was the need/demand and the ability to accommodate the increased number of air traffic movements and proposed destinations. After some argument he gave way and agreed, for no apparent reason. I advised him that I had no expertise in this highly specialist area and that the Council would need to employ an expert in air traffic forecasting. He responded that there was no budget to cover extra fees as these had all been spent on employing a QC to lead the City Council's case and I must find an expert who could provide me with adequate training over a very limited number of sessions. I was to be the only witness for the City Council. I tracked down an expert, Stan Maiden, who was the Research Officer at the British Airports Authority. While he agreed to help, he acknowledged the difficulties we faced.

The public inquiry was held in Yeadon Town Hall, close to the Airport. The significance of the technical and air traffic forecasting issues soon became apparent from the opening statement presented by Sir Frank Layfield QC, counsel for the applicants. He was the leading planning advocate of his time. The County Council had employed an ex-Air Chief Marshall as their expert on aviation matters, as well as several witnesses from the airport, the leisure industry and the County highways and planning departments. The Inspector, who was a remarkable likeness of the comedian Ronnie Barker, was assisted by a technical advisor, an Air Vice Marshall. The hall was packed largely by local objectors.

Sir Frank cross-examined me for a total of eight days, the first two of which were something of a character assassination attempt coupled with a forensic examination of the City Council's decision, the report to full Council and the supporting papers. As I left the hall at the end of each afternoon session the locals gave me various cheery supporting comments. The peak of the cross-examination came when Sir Frank moved on to the air traffic forecasts, new destinations and the types of aircraft anticipated. The question that has stuck in my memory ever since came up mid-afternoon, and Stan Maiden had come up to experience a session of the inquiry that afternoon. 'Mr Brook, a Brazilian Bandeirante Jet with a full passenger payload is making a flight to Frankfurt and returning to Leeds, what is the fuel requirement for this return flight.' A silence fell within the hall, and I rather hoped that the floor might open up beneath me. The Inspector broke the silence and asked me if I would please answer the question. Sir Frank then interjected sarcastically, 'Perhaps the parrot on your shoulder can assist you with the answer.' I turned to Stan, and we briefly conversed, and I then provided an answer, but I cannot recall whether it was correct. The might of the County Council's team was successful in their appeal and permission was granted with night-time operating restrictions. I vowed that I would never again be pressed into giving evidence where I did not have full professional command of the subject matter.

The May 1979 general election was won by Margaret Thatcher's Conservative Government, the first of four consecutive victories. The 'New Right' Thatcherite movement was a stark contrast with the Labour party's success in the local metropolitan elections held on the same day. Leeds City Council shifted from Conservative control to no overall control, and in the following year Labour made substantial gains and ended up with a working majority of 25. George Mudie, a NUPE trade union official from Dundee, became the new Council Leader. These proved to be seismic shifts at national and local levels.

In 1979/80, the Quarry Hill project had progressed to the stage of selecting a developer and approving a master plan scheme with all three political parties in agreement. The approved scheme would provide the City Centre with an international standard convention/exhibition and business centre. Following his local election victory in May 1980, George Mudie advised that he was not happy with these proposals and the decision was reversed. This was the first of a few signals that he was not, at this point, supportive of large planning projects and the workings of the Planning Department as currently organised.

A second and much stronger signal came at a meeting of the Industry and Estates Committee when I was to present two important reports on the plans for the regeneration of the Holbeck-Hunslet Industrial area to the south of the city centre, which were first on the afternoon's agenda. When I had briefly presented the first item, Cllr Mudie was the first to comment, including a statement comparing the Planning Department to a section of the crowd at a football match who were always complaining to the referee. I found this to be both strange and disconcerting. The chair of the Committee invited me to present the main report, which included a slide presentation of the plan proposals. I spoke for some 15 minutes and, as I concluded and sat down, the leader of the Conservative group on the Committee addressed the Chair. He complemented the Planning Department on the report and the proposals and suggested that Cllr Mudie owed the Department and me an apology.

For a while the role of the Department and its management progressed as normal. It became apparent over time that the new Labour leaders in several of the metropolitan boroughs were asserting their positions and authority in a variety of ways. In some cases, this produced tensions with the professional standing of directors and senior managers and the professionalism of departments. This was particularly the case in Leeds.

In 1982/83, departmental managers were advised that new staff structures would replace the existing ones. The posts of deputy director would be abolished. The local government union, NALGO and the departmental management would receive copies of the new structures for consideration and comment. If the proposed posts in the new departmental structure were very similar, in terms of responsibilities, to those in the old structure, then those personnel would simply be 'slotted-in' to the new post the following day. Ironically, I was the NALGO departmental representative for senior management and had therefore had an influence from two standpoints. This was of limited personal relevance as the new post of Assistant Director was 90 per cent the same as my old post and I was 'slotted-in' the following day. At the end of the week, I received a telephone call from the Director of Personnel to inform me that I and two other Assistant Directors had been 'unslotted'. Somewhat shocked I demanded to know why this had happened and who had led/influenced the decision. At first he was unwilling to say, but when pressed admitted that it was the Leader of the Council. This meant I would have to apply for my own post and be interviewed alongside any other applicants.

Subsequently, my group leaders informed me that they would not be applying as they regarded the post as rightfully mine. Prior to the interview I became aware that there was only one other candidate, one of my six team leaders, three levels below me on the departmental management hierarchy.

The interviews took place, and afterwards I was advised by the new Labour Chairman of Planning Committee that I had been unsuccessful and that I might think of applying for a post in the Industry and Estates Department. That evening, the former Leader of the Council, Peter Sparling, rang me to say how sorry he was, and that the decision had been taken on an organised card vote. I was later interviewed for the new Director of Industry and Estates post with an almost identical card vote outcome, despite having what the former leader described as a very good interview.

In the next few weeks, I had plenty of time to take stock of my unusual position. I resolved to seek a meeting with Cllr Mudie to try and obtain an explanation and enquire whether he was minded to offer me a new job. This was particularly relevant as he had astutely hand-picked three planners to join his inner Policy Team. All three would shortly become directors of departments. I did not fully appreciate at the time that my position was partly linked to removing the old management order. No offer of a meeting came and, after a few weeks had passed, I decided to present myself to his secretary on a regular basis to see if I could get an audience for a few minutes. Eventually this succeeded, and we had a short informal exchange in which he advised me that he considered me too entrepreneurial to work in the Planning Department. This impromptu discussion did not provide a positive route forward.

In total, several months passed during which I remained on full pay as an Assistant Director but without an active position. I had been moved to a 'new' office on the upper floor of our 1950s building on The Headrow. Each day, as I approached and left my office, I passed former departmental colleagues and was only able to engage about half a dozen of them in any form of conversation. A departmental 'mole' surfaced from time to time to provide position statements to the local government correspondent of the Leeds Evening Post, which resulted in three editorials in one week advising the wider metropolis that an unnamed Assistant Director in the Planning Department was on a full salary while not being productively employed. The national planning magazine made comment by way of the strip cartoon which appeared on the back page each month. This depicted a huddle of planners at a conference, with one member of the group questioning who the lone individual was walking away with a dagger in his back! Oh, that's a Leeds planner, remarked another!

The lack of any meaningful work and support from the Planning Department where I had been so actively and centrally engaged began to get to me. John Finney was unfortunately very ill, in hospital with viral encephalitis for some of this time. He was a fighter and made a remarkable recovery given the medical prognosis. He was able to take up his appointment as President of the RTPI despite some concern that he was not sufficiently recovered to fulfil the role and the annual programme. I did get some quiet moral support from other departmental directors. However, without the fantastic support from my wife, my morale would have ebbed into depression.

To conclude this formative episode in my career I did finally get a formal meeting with the Leader and was offered a new post with the exciting title of Head of UK and International Business Promotion. The apparent challenge was to attract new business into the City. Cllr Mudie seemed to be quite sincere in his wish to make this a successful operation. I was advised to meet with the Chief Executive and Director of Finance to sort out a budget and team to assist me. I researched the funding and operation of similar initiatives in other cities. I advised that Birmingham, Glasgow, and other cities had annual budgets for this work ranging from £250,000 to £500,000. I requested an initial budget of no more than £100,000, with the expressed aim of doubling this amount by injections of private sector support from professional firms within the City. My enthusiasm was short lived when I was advised that the budget provision would start at £1,500 for the first year, with an assistant and secretary as my new team. The message finally sank home that this was the end of the line with Leeds City Council.

I did take up the position to guarantee the continuation of my salary, determined to show what could be achieved, even within the limitations imposed by time and budget, while at the same time looking for a new local government management position. I was interviewed for a director's position in Gateshead and shortlisted for interview as Chief Executive of Stirling District in Scotland. After the first interview I asked the Director of Personnel whether the situation in Leeds had influenced the outcome, and he confirmed that without any doubt it had. The Stirling interview was preceded by a drinks-reception and dinner for the candidates. I was introduced to the Convenor (leader) of the Council who remarked 'Och, Mr Brook, the victim of the Stalinist purge in Leeds', to which I responded by saying that my fame travelled before me.

During the last few weeks with the City Council, I decided to bring a case for constructive dismissal. The permanent NALGO manager in the area was to handle my case and he advised that the percentage of cases won nationally was relatively low and that the compensation payments were limited to hundreds rather than thousands of pounds at this time. I stressed that I was not doing this for any financial reward, and it was a matter of principle. By the time of the second briefing meeting, he told me that our case had been leaked to the opposition, who were in the process of appointing a QC. I never found out how this had occurred. The witnesses to be called to defend the Council's position included the Leader, Chief Executive, Director of Finance, and the

Director of Personnel. At this stage I was in the process of leaving the Council and setting up in business as a planning consultant in the city. My NALGO representative had little appetite for the case. This and the negative impact of the leak, plus potential future press coverage, led me to conclude I should withdraw and concentrate on establishing my new business. During the week of leaving there was no big farewell drinks party, though I did have a couple of pints with a few colleagues. One of the committee clerks advised me what, in his opinion, had been my main problem. He said I tended to ride my 'white charger' down the committee room table. While I thought this was a little exaggerated, I recognised where he was coming from. I had been passionate about my job, and the roles of my colleague professionals. Other councillors, and indeed two of Leeds's greatest MPs, Dennis Healey, and Merlyn Rees, had not seen any problem with this approach. I had however failed to read and sufficiently understand the changing political scene. I now fully recognise that professionalism can become a blind to political awareness. Professional managers in public service need this understanding to successfully present their arguments and manage their departments. This is not an argument for unnecessary deference towards politicians. A balance of mutual understanding must be arrived at, which requires accommodation from both parties.

* * *

Planning, politics, and the public – 'an unresolved triangle', 1945–2000

In the concluding section of this chapter, I examine the evolving relationship between planners and the planning system, the general public and politicians. My short overview is an amalgam of my own experience, research mainly carried out by planners in academia, codes of practice and the Nolan Committee Report. There were few contributions available from practitioners covering this period, which is probably a reflection of their lack of time and inclination to put 'pen to paper' under the eye of local politicians.

In the introductory section of Chapter 1 I refer to the first two books written by David Kynaston under the joint title *Austerity Britain 1945–51*, the first volume of his projected sequence entitled *Tales of a New Jerusalem*, which will collectively cover the period 1945–1979. He refers, in his preface to Volume 1, to the period start and end dates as being 'justly iconic'. The year 1945 produced the Labour landslide election victory and the early introduction of a challenging socialist programme, which despite its 'big state' approach had strong public support. Some three decades later in May 1979, Margaret Thatcher became the new Conservative prime minister. Her aim was to reduce the size of the State and replace this previous level of control over our lives with her version of market economics based around individualism and entrepreneurship. Part of her approach was to cut regulations to achieve the desired individual and corporate freedoms of operation.

In considering the emerging three-way relationship between planning, politics, and the public, it is important to take full account of the socio-economic conditions and government policies of distinct periods of history. I have, for convenience, extended the period covered for this short review to include the Thatcher years and the following governments of John Major and Tony Blair up to the start of the new millennium.

At the time of the introduction and early operation of the 1947 Act planning system hardly anyone questioned the need for its content or proposed approach. This system was built, as part of the new welfare state, around particular political views of the role of the state and state planning, including its relationship with private sector interests. Modern town planning was, therefore, inherently political from its conception, and party politics at the national level would be a strong influence on its future content and implementation. The dominance of two-party politics, without sufficient consensus around the centre ground, has bedevilled much of the planning and policy making across various government departments. This has been a particular problem experienced by the town planning system, though parallels can also be drawn with transport and health planning. In this context it is important to recognise that the basic architecture of the 1947 Act system has survived substantially unchanged. The involvement of MPs and the Lords as legislators and scrutineers via parliamentary debate and select committees establishes and changes the statutes and regulations that govern the planning system. In our democracy, MPs are selected by the public and are there to serve both their constituents and the public interest in general. The level of public interest in town planning matters will influence MPs alongside their own interests and those of their party. These matters may be obvious considerations but are basic to the understanding of the interactions being examined.

By the 1960s there was a more open acknowledgement that town planning was a political activity and the formulation and implementation of plans and policies, and individual decisions on planning applications, involved value judgements that should include political debate and the participation of the public. The increased level of interest leads on, as we have seen, to the Skeffington Report. Its recommendations are not widely implemented, and many years are to pass before meaningful public participation occurs. Yvonne Rydin reviews the historical development of public participation in planning over this period.[10] She comments that some local authorities were keen to progress some of Skeffington's recommendations, but after drafting legislation, seeking advice and parliamentary discussion the result was a much-diluted system of basic publicity and consultation with no radical changes. The whole basis of the Government's approach to planning under Margaret Thatcher was, via deregulation, to limit the influence that planning placed on market activities. There was also an emphasis on speeding up the whole system. During this period there was a growth in local protest groups. One of the surprising features to emerge during this decade was the extent to which shire Tory voters were opposing the local impacts of the national government's policy approach. By contrast, in several of the new Labour councils there were attempts to enhance engagement with their public. As will be seen in later chapters, there are still considerable shortcomings to be addressed to make public engagement work and to better articulate, with the involvement of the public, what are the key components of public interest. The impact of the Thatcher Government was to overturn 30 years of welfare state politics and to greatly diminish planning for the public good. This had a negative impact for several years.

I now turn to examine the role of the professional planner in this triangular relationship. In subsequent chapters I introduce the role of private sector developers as their influence in the planning system grows. When the RTPI was established in 1914 its articles did not place any emphasis on public service, though as already identified in Chapter 1, one of the key drivers for an organised system of town planning, sought by the late nineteenth/early twentieth century 'movement for planning', was to improve the health and living conditions of large sections of the general public. The RTPI, in a recent Practice Advice paper entitled 'Probity and the Professional Planner',[11] state that: 'A defining feature of the planning profession is the duty to advance the science and art of planning (including town and country and spatial planning) for the benefit of the public under the RTPI's Royal Charter.' The royal charter was granted in 1959. The paper goes on to state that delivering this core duty of acting in the public interest 'has historically been defined in terms of protecting public health, public amenity, and the environment from "harm"'. This current paper expands the definition of acting in the public interest by stating that: 'Today, RTPI Members serve a range of interests. Acting in the public interest involves having regard to the expectations of the local community and politicians as well as future generations. Tensions can often arise when trying to reconcile these different interests and challenges.'

From the initial establishment and recognition of the profession in 1914, the articles of the Institute position town planning and its 'kindred subjects' as a mixture of the arts and sciences. This broad reach is maintained through to the Royal Charter of 1959 and then onwards to the current day with further expansion of the role. The lack of a clearer definition of the role of town and country planners has been questioned and attacked from various quarters. These negative reactions have come from other professions involved in design, the environment, urban and rural development, and conservation; from the general public; from politicians and from academics/planning theorists. The latter group have historically criticised both the poor definition of the town planner's role and the problems of the claim to serve the public interest when this is somewhat ambiguous and when we have increasingly been living in a world of difference. Heather Campbell and Robert Marshall in a review article, *Moral obligations, planning and the public interest: a commentary on current British practice*,[12] consider many of these issues and the triangular relationship between the town planner, politicians, and the public (public interest).

Campbell and Marshall, in the introductory section of their review, refer to the public dissatisfaction with the planning system identified in the Third Report of the Nolan Committee.[13] The Third Report, on page 69, includes the following telling statement:

'We have received more letters from members of the public about planning during our work on local government than on any other subject. Planning is clearly a subject that excites strong passions and for good reason. The planning system frequently creates winners and losers; it involves the rights of others over one's property; the financial consequences of a decision may be enormous.'

Campbell and Marshall state that the public dissatisfaction with the planning system 'has been reinforced by a number of allegations of impropriety, misconduct and out and out corruption in the operation of the planning machine at the local level. The current context of planning clearly raises fundamental value questions concerning its role and purpose.' The authors go on to balance these statements, which appear to tarnish the reputation of professional planners, by introducing the impacts of the environment in which they had been operating during the period covered by the paper. They conclude that planning theorists have tended 'to treat planners as though they operated in a vacuum, with little account taken of the socio-political and economic contexts in which such activities are embedded. Planning, as a form of state intervention administered at the local level, is inevitably subject to the pressures and vagaries of governmental and societal change. The recent past has been a particularly turbulent period for local governance, and this has inevitably impacted upon the role of planning practitioners and the expectations placed upon them.'

The authors of this paper review the obligations that at various times influence professional planners as individuals. They highlight a very important distinction between the range of sources of obligation that planners must consider in their daily activities compared with the traditional professions of law and medicine. They comment on an earlier paper by Bolan[14] who identified eleven moral communities of obligation for practicing planners. Campbell and Marshall reduce these to five main categories of obligations: individual values, the profession, the employing organisation, elected members and the public. They then use these obligations 'as an analytical device to examine the tensions which planners experience in their work.'[15] They then assess how these tensions have been influenced by the ways in which the relationships between the State, Society and the individual have been reconfigured in Britain during the 1980s and 1990s.

The role of the elected member and political affiliations in the triangular relationship has been the subject of several books and academic papers, as well as government instituted reviews and codes of conduct and guidance for the professionals and politicians. One of these books on the subject was written by my former planning lecturer, John Gyford,[16] and is quoted in the Campbell and Marshall paper when they move on to reviewing the role of elected members and their relationship with planning officers. Both sources refer to the long-standing contentious nature of the appropriate relationship between officers and elected members. Some observers have initially assumed that members should be responsible for policy formulation and officers for advice and the implementation of policy. As Campbell and Marshall state:

'In practice the relationship has always been messy and ambiguous and became a matter of some importance for the Maud and Baines inquiries into local government in the late 1960s and early 1970s.'[17]

Gyford,[18] in his review of the officer-member relationship, concludes that this remained generally unproblematic until the mid-1970s. There was widespread consensus both between national and local politicians, and between the latter and their advisors, regarding the merits of the welfare state and the mixed economy in serving the public interest. Campbell and Marshall[19] refer to an atmosphere of general consensus and, in these conditions, 'a formal relationship between professional staff and elected members, whereby the former advise the latter as to the appropriate course of action, tended not to be a source of tension. The combination of the respect for professional expertise and accord over the values and practices to be fostered led politicians to acquiesce to the advice of officials and to give individual planners considerable scope within which to act.'

The period following the mid-1970s is referred to by Gyford[20] as the 'period of re-appraisal in local politics.' The inherent tensions in the member-officer relationship had until this point remained dormant, but signs of strain began to show. He refers to many authorities, but particularly those with Labour administrations in the large metropolitan districts, witnessing an increased assertiveness by senior local politicians. Other factors are identified that added to the straining of the relationship, including the increasing involvement of councillors in the appointment of officers, greater stress on ideology as a determinant of policy and councillors being more assertive in day-to-day decision-making.

The conclusions from these two reviews and other commentaries align very closely with my personal experiences of the member–officer relationship during my local government career. This is particularly the case with the period 1970 to 1984. At Nottinghamshire and West Glamorgan County Councils and Leeds City Council, until 1979, I was provided with strong member support, understanding and appreciation of my particular roles. In the reverse relationship I had great respect for those councillors with whom I had regular contact as chairs and shadow leaders of those committees I reported to. In a few cases this relationship was so equitable and understanding that it was not surprising that a level of friendship developed.

I have briefly touched on my limited experience of impropriety and corruption within the operation of the planning system at local government level, where financial or other inducements in kind have occasionally come into play. From these experiences I am aware of gifts offered and being refused or being rapidly returned when sent, without any prior knowledge, to a planning department. I return to this subject again prior to summarising the considerable significance of the Third Report of the Nolan Committee on Local Government and its concentration in the final chapter on the planning system, the only topic-based chapter. One of the main reasons that the Nolan Committee was set up by Prime Minister John Major, was the level of public dissatisfaction with government at national and local levels, including deviations in standards of behaviour and the ethical approach required from elected politicians and employees. This was particularly the case in local government and the operation of the planning system. During my local government career, I became aware of two officers in influential management positions who had allegedly accepted bribes in exchange for favourably influencing development proposals. I use the word 'allegedly' as I am not sufficiently aware of the available evidence and whether successful legal proceedings were brought against the individuals concerned. During my eighteen-year period as a local government officer and my subsequent regular involvement with planning officers in my consultancy role I have experienced a high standard of professional ethics.

Standards of ethics involving leading politicians in national government are once again being brought into question as I write this chapter. The former Secretary of State for Housing and Planning in Boris Johnson's Government, Robert Jenrick, was accused of various forms of impropriety in relation to a major residential development in Tower Hamlets. I will return to this and related topics in my review of the operation of the current planning system.

I consider the Nolan Committee's Third Report on Standards in Public Life in Local Government (1997) to be an extremely thorough and erudite document and that Chapter 6 – The Planning System provides an excellent review of the officer–member relationship, the issues arising and the balanced recommendations that need to be implemented.

An important preface to the Third Report is the seven principles of public life set out in the First Report (1995). These can be summarised as follows:

1. Selflessness: Holders of public office should take decisions solely in terms of the public interest.
2. Integrity: Holders of public office should not place themselves under any financial or other obligation to outside individuals or organisations that might influence them in the performance of their official duties.
3. Objectivity: Holders of public office should make choices on merit.
4. Accountability: Holders of public office are accountable for their decisions and actions to the public.
5. Openness: Holders of public office should be as open as possible about all the decisions and actions they take.
6. Honesty: Holders of public office have a duty to declare any private interests relating to their public duties.
7. Leadership: Holders of public office should promote and support these principles by leadership and example.

All public bodies were recommended to draw up codes of conduct incorporating these seven principles.

In the introduction to Chapter 6 of the Third Report, Nolan remarks on the fact that issues relating to the planning system generated more correspondence from the public than on any other subject and the high incidence of planning cases dealt with by the Local Government Ombudsman (25 per cent of the total). With regard to misconduct and impropriety, reference is made to 'a substantial body of correspondence' (paragraph 271) and that 'public perception of impropriety is as important as the existence of genuine misconduct'(paragraph 273). There is a clear recognition that for corruption to arise this does not just relate to money changing hands. Decisions can be distorted in other ways such as those taken 'on the basis of personal friendship or shared interests, rather than for proper planning reasons' (paragraph 274), or 'by a desire for social engineering, by a prejudice in favour of a particular group in the community, or by a need for community facilities. Any of these aims, in proportion, can be praiseworthy: but if they trump all the other relevant planning criteria then public confidence in decision-making is likely to be casualty' (paragraph 275).

There is a recognition that the climate of public opinion has changed. Since its inception the 1947 planning system has carried a general presumption in favour of development. The limitation of an owner's rights to develop their property should only be restricted in the public interest. For some years post the introduction of the 1947 Act the need for post-war reconstruction was clear and development enjoyed broad public support. Nolan recognises that the climate of public opinion has changed. 'Development is now a term which has a pejorative ring, and the planning system is seen by many people as a way of preventing major changes to cherished townscapes and landscapes. If the system does not achieve this (and it is a role which the system was not originally designed to perform) then the result can be public disillusionment' (paragraph 277). Nolan states that there is little the Committee can do about this directly as it raises wide policy questions.

Councillors are recognised as quite properly exercising two roles in the planning system. As members of a planning committee, they make decisions on granting or refusing planning permission by using planning criteria. They also act as representatives of public opinion within their local constituency. It is further recognised that tension has always existed in this dual role and Nolan goes on to conclude that this tension has become much more acute as opposition to development projects has grown (paragraph 281). I would now add to this tension the expansion of the number of material planning considerations that potentially enter into the decision-making process. Leading decisions in the higher courts have been the main source of these additional factors to be considered.

The Report goes on to provide a very interesting and clear discourse on the roles of local councillors and planning officers in the decision-making process, making it clear that this is not a quasi-judicial process but an administrative one taken within a framework of law and practice, and this is a view that has been upheld by the courts. 'Councillors must bring to planning decisions a sense of the community's needs and interests. That is why they are there' (paragraph 290). The Nolan Report on local government takes the important decision to recognise the need for elected members to listen to their constituents and that they have to be able to marry 'their duty to represent the interests of the community with their obligation to remain within the constraints of planning law and only to take account of relevant matters'. This is recognised as often being a difficult task. For this reason, and because planning legislation and guidance are complex, Nolan recommends that it is essential that all members of an authority's planning committee should receive training (Recommendation 34). However excellent officer advice may be, the conclusion is that this is no substitute for training. I agree with many practitioners who argue that councillor training must be ongoing as the system is subject to frequent changes.

Nolan recommended the adoption of a Code of Best Practice in Planning Procedures (paragraph 295 and Recommendation 35). In England the Localism Act (2011) now requires all local authorities to maintain a code of conduct based on the seven principles of the Nolan Committee on Standards in Public Life, with a monitoring officer who has the specific duty to ensure that the Council, its officers, and elected councillors maintain the highest standards of conduct in all they do.[21]

The RTPI adopted a Code of Professional Conduct that sets out the standards, ethics and professional behaviour expected of its members who are to adhere to the following five core principles:

1. Competence, honesty and integrity
2. Independent professional judgement
3. Due care and diligence
4. Equality and respect
5. Professional behaviour.[22]

Several interim conclusions follow from this brief review:

1. Local government and the planning profession have responded positively to the recommendations of the Nolan Committee and public reaction by adopting or modifying codes of practice and guidance on professional ethics.
2. During most of the period 1945–2000, planning as a profession has been embedded in the state bureaucracy. However, over the last fifteen years the growth of private consultancy increased to the point where 18 per cent of all corporate members of the RTPI resident in the UK were employed as planning consultants.[23] It is also notable that for most of this period the planning profession has been predominantly made up of young white males

as revealed in a survey of the RTPI's membership in 1988.[24] Both of these characteristics of the make-up of the planning profession have changed considerably over the last twenty years.

3. In the long build up to the introduction of the first comprehensive town planning system centred around the 1947 Town and Country Planning Act, and in the years that followed, the concept that the planning system and the profession should be grounded in and focussed upon working in the public interest became established. This was relatively easy to comprehend when the movement for planning of the late nineteenth and early twentieth centuries was focussed around vastly improving the health and wellbeing of the urban and rural populations. The Attlee Government, in introducing the system as part of the welfare state, were also focussed on that central purpose, with the added challenge of post war reconstruction. The public were easily able to understand and support the drive for renewal as this was in their collective interest. As a result, two contrasting definitions of the public interest have emerged[25] (see the concluding section of the Campbell and Marshall Paper). In the first, liberal and utilitarian values are central. Individual interests are paramount, and the public interest is either the summation of all individual interests in a community, or the greatest good of the greatest number. This concept of the public interest based around individual choice dominated British society and saw a step change when Margaret Thatcher arrived as prime minister. The role of government in this concept is to maximise opportunities for individual choice. This requires market freedom and the delivery of local government services aligned to market conditions. Thatcherism promoted individual and market freedom at the expense of society, togetherness and the collective good.

The second definition of the public interest recognises the importance of shared values that transcend a simple summation of individual preferences and includes the conviction that there are circumstances where market processes are not the best means of determining desirable outcomes. While individual interests retain importance it is considered that an 'outside' overview is required as, firstly, individuals do not always realise what is in their best interests and, secondly, there are inequalities and different needs amongst groups of the population that need to be considered. This second definition has become increasingly significant over the last twenty years and is central to full public participation in the planning system.

4. The roles and relationships of elected members and planning professionals are better understood by each party, and Nolan and other reviews have assisted this process. The RTPI has had an increasingly significant training and communication role. However, tensions still arise, between the corporate

interests of local authorities and planning concerns. Budgetary constraints imposed by national government and reinforced by a culture of audit and efficiency of performance, came to dominate local government and other public sector agencies during and after the Thatcher era. These tensions resurfaced strongly in the austerity years following the 2007/8 financial crash and did so again as the impacts of the global Covid 19 pandemic became ever more apparent to the future of the economy and social and living conditions.

5. The tension between the speed and efficiency of decision making on planning projects and the quality of those decisions and development outcomes has been exacerbated by various government efficiency drives and a 'blame-game', promoted by various sources but mainly by leading politicians when in government, which identifies planners and the planning system as the prime culprit of development delays. I will return to this central theme, which has consistently been rebuffed by my profession with strong statistical and other evidence, as the search for quality of outcomes by national government has surfaced alongside continuing efficiency drives. As we shall see, these dual national government expectations are promoted without adequate recognition and attention to the resource implications within local authority planning departments.

6. Margaret Thatcher came to power in May 1979, declaring that she would cut 'red tape' and free up business with a particular emphasis on planning statutes and guidance, yet at the end of her final term of office in 1990 this had noticeably increased. This simplistic view of the controlling and negative influence of planning has resurfaced several times over the succeeding thirty years. Key factors influencing the expansion of town planning legislation and guidance have been the expanding reach of the issues that planning has been asked to deal with by national government and decisions in the higher courts.

7. The need to arrive at a clearer definition and understanding of the professional planner's role emerges from this review. As we shall see, this problem reoccurs and the RTPI have addressed this on more than one occasion. The phrase 'managing change in the built and natural environments' was one succinct attempt that did not adequately cover the role. Part of the problem relates to the changing nature of the role and the expansion of topics covered. Increasing collaboration between planners and other professions engaged in urban and rural development and environmental protection and enhancement has helped to dispel much of the competitive edge which previously existed. The real problem is persuading successive national governments, and to some extent the general public, of the explicit role and worth of professional planning when they attack the perceived barriers it creates.

Chapter 3

Health and Wellbeing – A Renewed Focus for Planning in the Twenty-first Century

The first chapter of this book traces the two main influences leading to the establishment of the modern town and country planning system as part of the post-Second World War welfare state. The first and most significant was the long gestation of the movement for planning, spanning the late Victorian and early Edwardian period, firmly based on the wish to improve the health and wellbeing of the UK population by improving living, working and environmental conditions. The socio-economic impacts of the Second World War were the more immediate catalyst, though several of the adverse impacts of the rapid urbanisation of the Victorian era remained to be resolved. The movement for planning was established and maintained over decades by a relatively limited group of philanthropists, writers, professionals, and a few politicians. The purpose and drive of the post-war Attlee Government and the strong support of the public in general for the creation of a better and healthier society for all, was the final catalyst that ensured that the town and country planning system was delivered in a series of statutes alongside the legislation establishing the NHS and social security system. The delivery of the comprehensive welfare state is regarded by most fair-minded commentators as a magnificent achievement. The extract from the 1944 White Paper, The Control of Land Use, and the content of the Beveridge Report (see Chapter 1) clearly set the health and wellbeing of the population as the central focus of both the wartime coalition Government and the subsequent Labour Government.

In this chapter I examine the re-emergence of health and wellbeing as a key topic for government and its increasing significance to planning policy and delivery at both the national and local levels. At the time of writing the impacts of the coronavirus pandemic (Covid-19), on this country and worldwide, have provided an acutely sharpened focus on a whole range of issues directly related to the health and wellbeing of the UK population and the existing and potential role of town and country planning in contributing to holistic and positive outcomes. I return to this important sub-topic and the lessons to be learned for future planning policy and delivery outcomes in the closing chapters.

In the pre-Covid months of 2019 there were a number of worrying socio-economic and health trends being highlighted, all of which were having, or were forecast to have, adverse impacts on the health and wellbeing of large sectors of the population. In addition to differential impacts on a demographic basis, geographical variations have also long been a matter of considerable concern, especially for those areas suffering the greater negative impacts. The first, and in some ways the most unexpected, trend is faltering life expectancy in Britain. Robin McKie, the Science Editor of the Observer produced a special report on this trend in June 2019.[1] He refers to the alarm caused amongst doctors and health professionals regarding this particular change 'because it reverses a trend that has continued, almost unbroken, for close to 100 years. Over this period lives have lengthened continuously blessing more and more British people with the gift of old age.' The trend identified by the Office for National Statistics prompted Danny Dorling, professor of social geography at Oxford University, to organise a meeting of researchers, statisticians, and geographers at University College London. The Observer Report quotes Professor Dorling's description of the situation as 'a perfect storm'. He goes on to explain that 'our faltering life expectancy rates show we have now got the worst trend in health anywhere in western Europe since the Second World War. To achieve that we must have made a lot of bad decisions.' Among many sections of the UK population there have been actual declines in life expectancy. While there is disagreement about the causes there is a growing consensus between leading medical practitioners and academics, geographers, and statisticians that this trend has a strong correlation with the period of impact of the austerity measures imposed by the Tory-led government in 2010. As well as this broad correlation, the experts identify specific causal outcomes of this period of austerity as leading to adverse impacts on health and wellbeing. These include reduced incomes, unemployment, removal of parts of the social care safety net and poor and deteriorating living conditions in many areas.

Obesity, and particularly childhood obesity, has been identified as a crisis in Britain, with British youngsters identified as the fattest in Europe. While diet and lack of exercise are usually identified as prime causes, there are other related causes, which include lifestyle opportunities and the differential geography of wellbeing including opportunities for outdoor exercise, household income and education.

Mental health problems have many causes, several of which are directly linked to negative ratings against wellbeing indices. There has been considerable research and comment, for example, on the benefits to our mental health and wellbeing of open spaces, access to and experience of nature, gardens and gardening and the uplifting effects of landscape and heritage quality. To this list should be added the important benefits of having secure and well-designed living accommodation that sits within a pleasing environment, and the ability to easily access healthy and economic transport modes. It soon becomes obvious to town planners, and indeed to many members of the public, that positive attention to all these aspects of land use and spatial planning has a huge role to play in enhancing our health and wellbeing.

Research by the World Health Organisation (WHO) and other bodies producing scientific evidence, establishes a much wider context in which the scale of the positive role of town and country planning to health and wellbeing outcomes needs to be considered. In 1998 the WHO, working with the International Centre for Health and Society at University College London, produced a report entitled 'The Social Determinants of Health – The Solid Facts'.[2] The essential message of this report is that policies and actions for people's health need to be more focussed on addressing the social determinants of health to 'attack the causes of ill health before they can lead to problems'. The report is linked to the WHO 'Healthy Cities Project – Health For All (HFA) 2000'. The HFA Policy on Europe: Target 14 entitled 'Settings for Health Promotion' states: 'By the year 2000, all settings of social life and activity, such as the city, school, workplace, neighbourhood and home, should provide greater opportunities for promoting health.'

In the foreword to the WHO report, Dr Agis Tsouros (Head, Centre for Urban Health WHO Regional Office for Europe) summarises the central theme of this key document:

'Recognising the health impact of economic and social policies and conditions could have far-reaching implications for the way society makes decisions about development, and it could challenge the values and principles on which institutions are built and progress is measured. The good news is that decision-makers at all levels increasingly recognize the need to invest in health and sustainable development. To do this, they need clear facts as much as they need strategic guidance and policy tools.'

While each individual is largely responsible for certain aspects of their own health, such as diet and exercise and avoiding smoking and excessive drinking, we now know that there are several social and economic circumstances that are important for a healthy lifestyle but often lie beyond the control of the individual. The aim of the authors of this report is to ensure that policy at all levels of government, in public and private institutions, workplaces and the community takes full account of the wider responsibility for creating opportunities for health. People's social and economic circumstances are known to strongly affect their health throughout life. People at the lower end of the 'social gradient' are known to run at least twice the risk of serious illness and premature death compared with those near the top. We have for several years been supplied with government statistics on deprivation and disadvantage at ward and enumeration district levels so that we can chart the detailed geographical distribution in terms of a number of indices including health, education, employment, and housing condition. Planners have regularly used deprivation indices, in an individual and cumulative form, to help frame regeneration and improvement policies in local plans and to formulate bids for national grant programmes. Poverty, unemployment, and homelessness have increased in many European countries, and these and other factors lead to social exclusion and adverse impacts on health. Section 10 of the Report considers the beneficial impacts of healthy transport and the policy and design implications for future city and neighbourhood planning and design. Reduction in car usage and planning and providing for more and safer walking and cycling, together with better public transport, promote health in four ways: they provide exercise, reduce fatal accidents, increase social contact, and reduce air pollution.

The Health Foundation, an independent charity dedicated to bringing about better health and health care for the UK population produced an infographic entitled 'What makes us healthy?',[3] which also emphasises the social determinants of health but incorporates an extended list including political, social, economic, environmental, and cultural factors. They also agree with the WHO report when they conclude that: 'Creating a healthy population requires greater action on these factors, not simply on treating ill health further downstream.' Their first infographic shows the extent to which health is primarily shaped by factors outside the direct influence of health care and identifies a healthy life expectancy gap of 18 years between the most and least deprived areas in England. Of the eight factors identified, four are directly relevant to the work of land use planners, including our environmental surroundings, housing, transport and family/friends and communities.

While there are considerable differences between the socio-economic and living conditions experienced in Britain in the second half of the nineteenth century and continuing into the first half of the twentieth century, and those which we have seen in the first two decades of the second millennium, there are several areas of great concern for the health and wellbeing of the population. These include increases in poverty, the incidence of poor living environments, air pollution, climate change, loss of biodiversity and increases in specific health problems including obesity, diabetes, mental health, and cardiovascular disease. The combined scale of these problems does merit comparison with the period 1850–1950 and the policy and investment programmes plus the implementation skills needed. These require action at and beyond the level achieved by the Attlee Government in the five years following the Second World War. Though the nature of the policies and programmes required are different, in the context of the national and global challenges we now face, the determined vision and coordination of strong government is vital. The cumulative position of mounting socio-economic, health and environmental problems was already apparent prior to the financial crash of 2007/8 and the challenges have been exacerbated by the period of austerity which followed, and which has still not ended. The known impacts of the Covid-19 pandemic and the impacts of Brexit are considered in the closing chapters of this book.

The positive and expanded role that town planning professionals can play in addressing the health and wellbeing issues identified is being actively promoted by the RTPI. Examples of this are the publication of the report 'Plan The World We Need',[4] June 2020, and a week of digital conference sessions: 'Planning for Post Pandemic Recovery', June/July 2020.[5] While these are directly related to the impacts of the Covid-19 pandemic, they also relate to the wider health and wellbeing issues being considered here. I joined a number of these on-line conference sessions, and, like many of my colleagues, I am more than ever convinced of the importance and significance of professional planners' skills and the contributions they can and should be making to enhancing the health and wellbeing opportunities of the UK population. We will see that the central definition of planning for sustainable development and climate change mitigation have very strong links with tackling adverse health and wellbeing impacts. While the national government has recognised these linkages in the NPPF, their current post-pandemic draft strategies for town planning seek to make fundamental changes to reduce planning policy and development management interventions. This worrying trend appeared to be increasing in significance under the leadership of Liz Truss, during the early weeks following her appointment as prime minister in September 2022. The governments of Johnson and Truss and their advisors saw a number of these planning interventions as a negative drag on economic and social progress, whereas the planning profession and those who support it are promoting the very positive and holistic outcomes that can be achieved by fully utilising the existing system and professional skill base, as well as positively examining how this can be enhanced to achieve even greater results in these most challenging of times. This is a central theme of this book. RTPI chief executive, Victoria

Hills, produced an impassioned open letter (2nd July 2020) responding to statements made by the government in relation to the profession. She opened the letter by stating:

> 'The planner-bashing rhetoric coming out of government this week has deeply concerned me – not only on behalf of the planning industry, but also on behalf of every single community in the UK.' She went on to refer to the attempt to tarnish the reputation of the profession in order to pave the way for the 'overhaul of the planning system'. She describes this as dangerous. In response to Prime Minister Boris Johnson's 'build, build, build' and 'project speed' strategy, she made the case that we must 'plan, plan, plan' to achieve joined up solutions for the communities we live in that address climate change, health and wellbeing and inequalities.

I totally agree with the contents of the letter and share the passionate message. We need to carefully monitor how the current Government responds.

In this chapter I consider what is meant by health and wellbeing given the varying definitions that exist and how national and local government have to date introduced some of the issues involved into planning policy and guidance documents. I examine the emerging recognition that government departments and professions, including health and planning professionals, need to work together to achieve positive outcomes. Finally, I present ways in which local planning authorities and the private sector are delivering and can, in future, continue to deliver plans and projects that have the health and wellbeing of communities and individuals embedded within them.

* * *

Health and Wellbeing – what does it mean and how has it been interpreted in planning policy

Wellbeing is defined in the Oxford English Dictionary as: 'the state of being comfortable, healthy or happy'. The UK Mental Health Foundation stress the importance of realising 'that wellbeing is a much broader concept than moment to moment happiness'. They go on to state that 'while it does include happiness, it also includes other things, such as how satisfied people are with their life as a whole, their sense of purpose, and how in control they feel'.

The New Economics Foundation consider that 'wellbeing can be understood as how people feel and how they function, both on a personal and a social level and how they evaluate their lives as a whole'.[6]

The UK Mental Health Foundation consider that 'most people would agree that wellbeing is something they strive towards, and the subjective measure of people's wellbeing is deemed so important that it is included alongside health and the economy in "measures of national wellbeing"'.

There are many definitions of wellbeing. The World Health Organisation (WHO) defines health as 'a state of complete physical, mental and social wellbeing and not merely the absence of disease or infirmity' (WHO-1948). This early definition of the relationship between health and wellbeing links the two explicitly and conceptualises health as a human right requiring physical and social resources to achieve and maintain.[7] It is clear from various sources that wellbeing includes a good state of mental and physical health, but wellbeing sits outside the medical model of health as it is not diagnosed. However, the2wq23re is general agreement that wellbeing can be measured via social attitude surveys.

A 2005 study examined how wellbeing was used in different disciplines and its potential for health research and health promotion, the latter being most significant in relation to the role of town planners. The research principally comprised a major literature review 'to examine the philosophical roots of wellbeing and the contributions of the main disciplines uncovered by the review; economics, psychology, health studies, sociology, anthropology and biomedicine'.[8] At this point in time these researchers concluded that, while wellbeing was a popular concept, it lacked a clear conceptual base and there was little agreement on how it can be identified, measured and achieved. Most disciplines were at this time biased towards limited aspects of wellbeing related to their own discipline. A particular lack of consensus and research was identified in the field of social wellbeing. The authors conclude that 'wellbeing' may offer considerable potential for unifying many sectors and interests around the central goal of improving health. The value of a multidisciplinary approach is identified, and they identify that Health Promotion Wales had, in 2003, moved to publicly incorporate wellbeing into the title of their website.[9] As we shall see, the Welsh devolved government were to make further groundbreaking progress. In this and successor Welsh documents, health is seen very much as a part of wellbeing rather than wellbeing being an add-on to health. The authors also conclude that wellbeing is a multi-faceted concept requiring clearer definition and agreement among both researchers and practitioners.

The appreciation of the significance and breadth of wellbeing was more readily apparent when Prime Minister David Cameron, in 2010, asked the UK Office for National Statistics (ONS) to measure national wellbeing, stating that we needed to 'start measuring our progress as a country, not just by how our economy is growing, but by how our lives are improving; not just by our standard of living, but by our quality of life'.[10] The ONS responded by setting up the Measuring National Well-being programme in November 2010 to monitor and report UK progress alongside the traditional measures of prosperity, which enabled policymakers 'to make better, well-informed decisions'.[11] A national consultation and debate took place in the early part of 2011 on 'what matters to you?' and the findings were developed into a wellbeing measurement framework.

In his introduction to 'Securing the Future – Delivering the UK Sustainable Development Strategy' (2005)[12] Prime Minister Tony Blair refers back to the 1999 UK strategic document, which had as its over-arching aim 'helping to deliver a better quality of life through sustainable development'. The Executive summary of the 2005 document re-states the definition of sustainable development arrived at during the 1987 World Commission on the Environment and Development, commonly known as the Bruntland definition, which states:

> 'Our strategy for sustainable development aims to enable all people throughout the world to satisfy their basic needs and enjoy a better quality of life without compromising the quality of life of future generations.'

Quality of life in these earlier strategy documents can be generally considered as equating to health and wellbeing. At the Rio Earth Summit in 1992, world governments committed to pursuing sustainable development and adopted Agenda 21, an unprecedented plan of action for sustainable development.[13] Ten years later, the Johannesburg Summit was seeking to establish key steps and indicators to make better world-wide progress. Following the Rio Summit, the UK government were the first to produce a national strategy for sustainable development in 1994. The 1999 strategy document, 'A Better Quality of Life', established the vision of simultaneously delivering economic, social, and environmental outcomes with specific indicators to assess progress. The UK Sustainable Development Commission reviewed the progress made on the 1999 strategy in a report published in April 2004 entitled 'Shows Promise but Must Try Harder', which identified 20 key areas in which we needed to take more decisive action. The 2005 review of the UK sustainable development strategy includes the following statement in pursuit of the main goal of the strategy:

> 'For the UK Government and the Devolved Administrations that goal will be pursued in an integrated way through a sustainable, innovative, and productive economy that delivers high levels of employment and a

just society that promotes social inclusion, sustainable communities, and personal wellbeing. This will be done in ways that protect and enhance the physical and natural environment and use resources and energy as efficiently as possible.'

The 2005 strategic review and the commitment to a UK Strategic Framework for sustainable development covered the period up to 2020, and this establishes five guiding principles including the following covering health and wellbeing and entitled 'Ensuring a strong, healthy and Just Society'. This includes 'Meeting the diverse needs of all people in existing and future communities, promoting personal wellbeing, social cohesion, and inclusion and creating equal opportunity for all.'

At last, there was formal government recognition of the focal need to promote the health and wellbeing of the UK population. These are fine words, and they identify a few of the dimensions of the expanded definition of health and wellbeing, but how and to what extent have these been translated into town planning policy where we would expect the national government to take a lead role? In the separate Planning Policy Statements (PPS), which preceded the first composite National Planning Policy Framework (NPPF) of 2012, there is an early reference to the positive role that planning plays in people's lives and to the promotion of personal wellbeing. PPS1, Delivering Sustainable Development, refers at paragraph 1 to the importance of good planning. It is regarded as making a positive difference to people's lives by helping in the delivery of homes and jobs and better opportunities for all, whilst protecting and enhancing the natural and historic environment. Good planning was regarded at this point as 'a positive and proactive process, operating in the public interest through a system of plan preparation and control over the development and use of land' (paragraph 2). Reference is made to the core principle of sustainable development, which is seen as underpinning planning with the better quality of life for everyone and for future generations recognised as being at its heart, as in the Bruntland Report of 1987. Paragraph 5 of PPS1 goes on to state that sustainable development should be facilitated and promoted by planning and this approach should include supporting existing communities, contributing to the creation of safe, sustainable, liveable, and mixed communities with good access to jobs and key services for all members of the community. This is further promoted at paragraphs 14–16 inclusive under the sub-heading 'Social Cohesion and Inclusion'. This approach means that the diverse needs of all people must be met in both existing and future communities, that personal wellbeing should be promoted with social cohesion and inclusion and the creation of equal opportunities for all citizens. This succinct policy approach puts sustainable development and the health and wellbeing of the population at the centre of a good planning system.

It is also interesting to look at an example of regional planning guidance and its relationship to the promotion of health and wellbeing through town planning. The Regional Planning Guidance for Yorkshire and the Humber to 2016 (RPG 12-2001), in common with the other RPGs covering England, incorporates a spatial planning strategy and this is recognised as having an important influence on the health of the region's people, which in turn is recognised as a key factor in the objective of seeking social equity and inclusion. Patterns of development, the quality of the environment and the use of transport are all considered to have health impacts. Cross reference is made to the Government's White Paper on health, 'Saving Lives: Our Healthier Nation', which was published in July 1999.[14] Paragraph 8.7 of RPG 12 quotes the White Paper's recognition of the following planning related factors that impact on people's health: provision of a safe, secure, and sustainable environment, reducing pollution, adequate housing provision, access to leisure and recreation, reducing social exclusion and increasing employment opportunities.

Current national policy is contained in the latest version of the NPPF. This was first introduced in March 2012, when hundreds of pages made up of former planning policy guidance (PPG's) and planning policy statements (PPS) were revoked and condensed into one national planning policy statement for England, the devolved nations having their own policy statements. Each PPS comprised a mixture of policy and guidance and interpretation was consequently not always straightforward. A separate system of National Planning Practice Guidance (NPPG) was established alongside the NPPF. The chief architect of this approach was Greg Clark MP, who was Minister for Cities from July 2011 to September 2012 and then subsequently the Secretary of State for Communities and Local Government from May 2015 until July 2016. In between these responsibilities for planning, he held two other ministerial posts, then in July 2016 was appointed Secretary of State for Business, Energy, and Industrial Strategy. A Cambridge economics graduate, he went on to gain his PhD in Economics at LSE and to work in consultancy and for the BBC. He clearly understood town planning, was directly involved in the drafting of the NPPF, and his affable approach meant he developed good relationships with the development industry and the RTPI. His government career and the very limited time he spent in each of his posts, prior to becoming Business Secretary, illustrates a key problem for the governance of planning, transport, and other national departments. Throughout my planning career there have been many ministers and secretaries of state holding responsibility for planning, but very few have stood out as being capable, having some passion for the job combined with a good level of understanding and empathy. The fact that, generally, they spent little time in these roles was clearly a contributing factor. Michael Heseltine was clearly a notable exception. This is one of the reasons why 'good planning' as espoused in national planning policy documents is difficult to deliver. Continuity on national policy formulation and delivery, together with associated decision making, is surely one of its essential pre-requisites.

The latest version of the NPPF was published in July 2021. As stated in the introduction to this book, the over-riding purpose of the planning system is to contribute to the achievement of sustainable development. The second of the three interdependent objectives, which are to be pursued in mutually supportive ways, is the social objective. This short statement essentially places the health and wellbeing of individuals and communities at the heart of the delivery of sustainable development and is therefore central to the realisation of good planning. Section 8 of the current NPPF, entitled 'Promoting healthy and safe communities', comprises, in three pages, the English government's policy on what planning should contribute to the health and wellbeing of the national population. This policy content can be summarised in the following way:

A) Planning policies and decisions should aim to achieve healthy, inclusive, and safe communities by (i) promoting social interaction, including opportunities for people to meet who might not otherwise come into contact with each other. Examples of how this can be achieved are given and include mixed use developments, strong neighbourhood centres, active street frontages and well-designed street layouts that allow for easy pedestrian and cycle connections within and between neighbourhoods. (ii) Creating safe and accessible places so that crime and fear of crime do not undermine the quality of life. Well-designed pedestrian routes and high quality public open spaces are seen as essential components. (iii) Planning is to enable and support healthy lifestyles and, importantly, local planning authorities are asked to consider local health and wellbeing needs. The provision of safe and accessible green infrastructure, sports facilities, local shops, access to healthier food, allotments and development layouts that will encourage walking and cycling.

To enhance the sustainability of local communities and residential environments it is necessary to make positive plans to incorporate all the local facilities already mentioned, plus meeting places and community facilities (including cultural buildings, public houses, places of worship and other local services). In preparing plans and policies and taking decisions on planning applications, local planning authorities should take account of and support the delivery of local strategies to improve health, social and cultural wellbeing for all sections of the community. The national policy also requires local authorities and other planning decision takers to guard against the unnecessary loss of valued local facilities and services and that the established shops, facilities, and services should be enabled to modernise. The aim is to ensure that a community can meet the day to day needs of all its residents. All this requires an integrated approach to planning the location of housing, economic uses and community facilities and services.

The creation of sustainable, healthy, and safe communities should also include estate regeneration and the social, economic and environmental benefits that this can deliver. Local planning authorities are to use their planning powers to help deliver a high standard of estate regeneration.

In the 2018 version of the NPPF, the need for planning to promote public safety and consider wider security and defence requirements was added and is retained in the current version. This and the other policy requirements for creating healthy and safe communities require close consultation and cooperation with local communities, health authorities, the police, and other agencies.

The final chapters of the book will demonstrate how the impact of the Covid-19 pandemic has brought these policy requirements and the need for skilful and positive town and country planning into sharp focus.

The current version of the National Planning Practice Guidance (NPPG) (16), which covers Healthy and Safe Communities, does so by posing a series of leading questions to assist planning practitioners to integrate these additional areas of work into the preparation of plans and policies and the evaluation of planning applications.[15] The first of these questions asks: 'How can positive planning contribute to healthier communities?' The following is the succinct and helpful answer provided:

'The design and use of the built and natural environments, including green infrastructure are major determinants of health and wellbeing. Planning and health need to be considered together in two ways: in terms of creating environments that support and encourage healthy lifestyles, and by identifying and securing the facilities needed for primary, secondary, and tertiary care, and the wider health and care system (taking into account the changing needs of the population). Public health organisations, health service organisations, commissioners, providers, and local communities can use this guidance to help them work effectively with local planning authorities to promote healthy and inclusive communities and support appropriate health infrastructure.'

The second key question asks: 'What are the main health organisations that need to be involved in considering planning for health?' Engagement between the plan-making bodies and health organisations is seen as essential to help ensure that local strategies to improve health and wellbeing are supported. The first point of contact on health and wellbeing issues, including health inequalities, is the Director of Public Health for the local authority. The NPPG proposes that plan-makers, working with the advice and support of the Director of Public Health, 'may need to involve the following key groups in the local health and wellbeing system'.

The first of these groups are the Health and Wellbeing Boards (HWBs), which were established under the 2012 Health and Social Care Act. These became operational on 1 April 2013 in all 152 local authorities that have adult social care and public health responsibilities. A short paper prepared by the King's Fund, 'Health and wellbeing boards explained',[16] describes them as a formal committee of the local authority 'charged with promoting greater integration and partnership between bodies from the NHS, public health and local government'. While they have limited formal powers, they do have a statutory duty, with clinical commissioning groups, which requires them to produce a joint strategic needs assessment and a joint health and wellbeing strategy for their local population. This strategic document is becoming a key evidence base document for preparing local plans and planning policies.

In 2016, NHS organisations and local councils joined forces in all English local authority areas to develop proposals for improving health and care known as Sustainability and Transformation Partnerships (STPs). Their purpose is to run services in a more co-ordinated way, to agree system-wide priorities and to plan collectively how to improve residents' day-to-day health. In some areas STPs have already become integrated care systems, which are a new form of even closer collaboration between the NHS and local councils. The NHS Long Term Plan aimed for every part of England to be covered by an integrated care system by 2021. The strong emphasis is on redesigning services around the needs of whole areas rather than individual organisations. This co-ordinated approach can agree system-wide priorities and plan collectively how to improve the health of local communities.

Consequently, a high level of integrated planning is emerging in which the town and country planning system must play its full part. With this in mind, the NPPG provides a broad but appropriate answer to the question 'What is a healthy place?' This is described as one which supports and promotes healthy behaviours and environments and a reduction in health inequalities for people of all ages. A healthy place 'will provide the community with opportunities to improve their physical and mental health, and support community engagement and wellbeing. It is a place which is inclusive and promotes social interaction.' In addition, a healthy place will meet the needs of children and young people to grow and develop as well as being adaptable to the needs of an increasingly elderly population, including those with dementia and other sensory or mobility impairments. This section of the NPPG cross-refers to the new National Design Guide (see Chapter 5), which is itself part of planning practice guidance 'for beautiful, enduring, and successful places'. This was first published by the Ministry of Housing, Communities and Local Government in October 2019 and its purpose is to demonstrate how well-designed places that are beautiful, enduring, and successful can be achieved in practice. It uses several examples and is the first time such a guide has been produced, though local planning authorities have for a long time produced their own design guides, particularly for housing development.

This new National Design Guide links to section 10 of the NPPF, 'Achieving well-designed places', and the NPPG separate guidance on the design process and tools. National planning policy requires the creation of high-quality buildings and places, and this is 'fundamental to what the planning and development process should achieve'. The National Design Guide emphasises that well-designed places influence the quality of our experience as we spend time in them and move around them. They can lift our spirits in a number of ways and contribute to our feelings of safety, security, inclusion and belonging and our sense of community cohesion. For all these reasons, well-designed buildings, places, and environments have been demonstrated as fundamental to our health and wellbeing.

In Chapter 5 I return to the significance of good design and planning's contribution. It has become increasingly clear that several other areas of town and country planning have a strong relationship with health and wellbeing. This is recognised by the NPPF in placing health and wellbeing at the centre of the delivery of sustainable development within the social objective, but interdependent with the economic and environmental objectives, so that all three are pursued in mutually supportive ways. As we have seen in tracing the evolution of modern town and country planning, the strong central link with the realisation of outcomes that are for the public good now expands into the wider understanding of health and wellbeing and all their facets. Several of the policy areas included in the NPPF have a distinct relationship with the policies and outcomes that need to be delivered to enhance the health and wellbeing of the population. One of the key responsibilities and skill sets required of town planners is to fully recognise these relationships and devise policies, plans and land use proposals that achieve integrated solutions and multiple benefits. Indeed, this is the focal approach that will deliver the three sustainability objectives in an interdependent and mutually supportive way. These related policy areas include delivering a sufficient supply of homes to a high standard, the promotion of sustainable transport, achieving well designed places, meeting the challenges of climate change and its associated risks such as flooding; conserving and enhancing the natural and historic environments and providing new urban open spaces as part of green infrastructure networks, including wherever possible connections to open countryside.

* * *

Bradford Metropolitan District Council- An approach to Health and Wellbeing

To assess progress at the local authority level I have selected my home authority of Bradford Metropolitan District Council (MDC) and examined relevant planning policy, guidance, and local initiatives, together with the interaction/joint working with the Director of Public Health, the NHS, local Clinical Commissioning Groups, and local communities. A partial review of the Core Strategy (the adopted version is dated July 2017) is underway and the Preferred Options version was published for consultation in July 2019. Together with the new Allocations Development Plan Document this will form the new Local Plan for the District. The draft Core Strategy Review includes a new Strategic Core Policy (SC10): 'Creating Healthy Places'. The introduction to this policy refers to the increasing evidence base which shows that 'the places where people live, learn, play and work are vitally important to health and wellbeing. The neighbourhoods, homes, schools, streets, and workplaces that we are born, work, live and socialise in have a significant influence on many factors that affect wellbeing.' The introduction goes on to outline the importance to wellbeing of several place-related factors including access to leisure facilities and greenspace, levels of traffic, infrastructure for public transport, walking and cycling, access to good quality housing, which is influenced by supply, location, quality, type, and design, as well as skills, jobs, and wage levels. As a result, Bradford planners, while supporting the NPPG, place greater emphasis on economic wellbeing by stating that healthy places are increasingly recognised as those where people want to live and work and as adding economic value and contributing to economic recovery. This is not surprising given the relative levels of deprivation, unemployment, wages, and poor health in many of the inner urban wards of the District.

The explanatory justification for this policy also recognises that the way places develop over time shapes people's exposure to both positive and negative impacts on health and wellbeing. Negative impacts include poor air quality, heavy traffic, noise nuisance and cold, damp housing. Positive impacts include local amenities, play facilities, parks and gardens, community, and neighbourhood assets. Also recognised are the differential vulnerabilities to negative impacts of children and the elderly and those living in urban areas.

The proposed policy approach recognises the need to create healthy places via good planning and design and seeking sustainable development. It takes on board the Council's responsibility under the 2012 Health and Social Care Act to meet local health and wellbeing needs and reduce health inequalities. The Planning Department have worked with the Council's own public health team, who have considered a wide range of evidence reviews to arrive at ten key principles for creating healthy places that are relevant to the District's needs. From this co-working and the development of the ten principles a comprehensive and ambitious strategy has been drafted, which should gain full public support. Developers will find it difficult to oppose the implementation of the policy via planning application decisions and for major applications (10 or more dwellings or where the site is 0.5 hectares or more and for non-residential development the creation of additional floorspace of 1,000 m² or more, or a site of 1.0 hectare or more). It will become necessary for applicants to produce a Health Impact Assessment (HIA) that demonstrates the positive and negative impacts of the development on the locality. It may be possible to mitigate negative impacts through planning conditions or obligations provided by the Community Infrastructure Levy (CIL) or section 106 planning agreements. Mitigation via financial contributions or conditions requiring expensive investment by the developer may prove to be areas of difficulty, primarily from a cumulative cost point of view, as there are usually requirements for education, transport, and other infrastructure contributions.

It is well worth summarising the content of draft policy SC10 – Creating Healthy Places as it demonstrates the careful thought that has gone into its preparation and the content of the ten principles, which adapt national policy and guidance to the needs of Bradford District. It is also appropriate to gain an understanding of the work that this will create for local authority planners when assessing applications alongside all other material planning considerations, which must be weighed in the 'planning balance'. Ministers and MPs and some civil servants fail, on too many occasions, to appreciate the individual and cumulative workload impact of new and altered planning policies.

The first paragraph of the policy sets out its purpose:

'The Council and its partners will seek to create healthy places by maximising health and wellbeing gains from development proposals and ensuring that negative impacts are designed out or mitigated. Development that provides opportunities for healthy lifestyles, contributes to creating healthier communities and helps to reduce health inequalities will be supported.'

The provision of supporting infrastructure in a major residential application development can, for example, provide new open space/greenspace, new or enhanced habitat and new footpath and cycleway connections. This provision can often be designed to link with existing networks and to provide health and wellbeing benefits for the new residents as well as adjacent communities. Good integral design can reduce the costs to the developer while still providing strong and positive health and wellbeing inputs as well as contributing to the end value of the development.

The ten principles are integrated into the policy in the following way:

'The Local Plan will promote health and wellbeing and the creation of healthy places in the District by ensuring that future development: i) Contributes to a healthy, sustainable, and well-connected District.' This is to be achieved by firstly directing new development towards well connected locations to enable active travel. Cross references to the relevant transport policies are provided. The second way is to ensure development contributes towards inclusive, welcoming, and safe neighbourhoods (cross reference to design policy DS5 – Safe and inclusive places). The third key relationship is with climate change with a strong linkage to strategic policy SC2 – Climate Change, Environmental Sustainability and Resource Use. The SC10 policy requires applicants to avoid factors within their development proposals that would negatively affect climate change while contributing to prevention measures that mitigate against the effects of climate change. The fourth contribution is to seek environmental improvements that minimise exposure to pollutants and the improvement of air quality (cross reference to environmental policy EN8: Environmental Protection). The final contributions to this first main policy criterion seek support for the delivery of housing, jobs, and essential community facilities.

Criteria ii), iii) and iv) collectively require developments to support pedestrian priority and active travel by designing places that promote walking and cycling, maximise opportunities for physical activity by supporting active-design principles and support for the Healthy Streets principle of promoting well-designed and safe places. Criteria v) and vi) respectively seek support for the provision, protection and improvement of multifunctional green infrastructure, open space and leisure and recreation facilities and an appropriate and varied mix of play for children of all ages. Criterion vii) requires housing developments to support the standards for new homes and neighbourhoods as set out in the Council's excellent Homes and Neighbourhoods Design Guide 2019. Criterion viii) seeks to give people access to the purchase and production of healthy food by the protection of allotments, the provision of new spaces for food growing, including new allotments and control over the distribution of hot food takeaway outlets. Finally, criterion ix) seeks the provision of decent jobs that support health and wellbeing and healthy working environments. The sixth principle relating to the creation of inclusive, welcoming, and safe neighbourhoods is included within the policy text under criterion i).

Providing a review of the full content of this local strategic policy, which seeks to comprehensively address health and wellbeing issues, demonstrates the centrality of the policy and the multiple linkages which exist with other strategic and topic-based policies in the local development plan. Health and wellbeing policy at the national and local levels is now at the very core of delivering sustainable development. The significance of this becomes much more apparent at the local level, where the specific health and wellbeing needs of local communities have been identified through consultation with the Director of Public Health, the local Health and Wellbeing Board and Sustainability and Transformation Partnership.

The Bradford and Airedale Health and Wellbeing Board is one of the partnerships established in response to the Health and Social Care Act 2012. Its members include senior officers and clinicians from local health organisations; senior officers and lead elected members from the Council and representatives from the Voluntary, Community and Faith Sector Assembly; Healthwatch and NHS England. The Board initially produced their Joint Strategic Needs Assessment, which provided a full understanding of the specific health and wellbeing challenges facing the city. This helped to shape the corporate Bradford District Plan, produced in 2016, with one of its five priorities being the realisation of improvements to health and wellbeing. Draft proposals for the Sustainability and Transformation Plan were produced in October 2016, though this covered the whole of West Yorkshire plus Harrogate, an area which equates closely to the Leeds City Region. This plan is more concerned with establishing integrated care systems.

Finally, it is important to mention 'Born in Bradford', one of the largest birth cohort studies in the world and based at Bradford Royal Infirmary. A total of 12,500 pregnant women were recruited to the study between March 2007 and December 2010 and the lives of their 13,500 children are being tracked through research studies and the use of regularly collected medical and educational data. This study was primarily driven by the fact that Bradford has some of the highest rates of childhood illness in the UK. 'Born in Bradford is helping to unravel the reasons for this ill health and bring new scientific discovery to the world.'[17] It is also providing a catalyst for communities to work with the NHS and the local authority to improve child health and wellbeing in the city. This is a hugely significant project, which will continue to follow the children until they are adults and help doctors to understand more about the big health challenges of the twenty-first century. The study has proven potential to promote change at the local level and has influenced the town planning policies and guidance referred to. It also has the potential to make major contributions to global knowledge.

* * *

Leading approaches to Health and Wellbeing- the examples of Wales and Amsterdam

To end this review of the movement of health and wellbeing issues to the centre ground of town and country planning and public sector policy and delivery, I examine two quite outstanding government initiatives, which are rightly attracting considerable interest around the world. These should give all UK planning professionals and their counterparts in other countries significant hope for the future, both in terms of their chosen profession and in the ability of governments to successfully address a number of the major challenges facing our planet including, enhancing health and wellbeing outcomes, achieving our climate change goals, and reversing the decline in biodiversity. As these two examples illustrate, a strong belief and sense of purpose is required from government and public bodies, together with a high level of engagement with the public throughout and great collaboration across a range of professions and expertise.

The Wellbeing of Future Generations (Wales) Act 2015 is a world first that places the delivery of sustainable development and improving social, economic, environmental and cultural wellbeing at the very heart of the duties of a wide range of public bodies, including all local authorities and local health boards, all Welsh Ministers and their departments, national park authorities, Natural Resources Wales, Arts and Sports Councils, the National Library and National Museum, NHS Trusts and Fire and Rescue Authorities. The Act puts in place a Sustainable Development Principle that is essentially based on the original Bruntland Principle. The addition of cultural wellbeing is at present unique to Wales. The Welsh Act came into force in April 2016.

This Act and the holistic approach being taken has had a long gestation. In many respects this originates from and then follows the United Nations progress from the original 1972 Stockholm Conference on the Human Environment, the creation in 1983 of the World Commission on Environment (the Bruntland Commission), the 1992 Rio Earth Summit and the development of the first agenda for Environment and Development and the Rio+20 Conference in 2012. The latter conference produced a resolution, known as 'The Future We Want' and the adoption of 17 Sustainable Development Goals (SDGs). These include Goal 11- make cities and human settlements safe, inclusive, resilient and sustainable. This hard fought for 'urban goal' ties the other 16 together by giving a focus on place -based action and integration. In 2006, Welsh Ministers were subject to a duty, under the Welsh Government Act of 2006, to formulate a sustainable development agenda. However, while Wales has been a pioneer in promoting sustainable development, reviews by the Wales Audit Office and the Sustainable Development Commission highlighted several weaknesses in delivery where sustainable development has often been one of a number of competing priorities. These reviews found that barriers to delivery needed to be removed by strong governance and effective mechanisms for delivery, reporting and learning. As a result of these findings the Welsh government, in 2009, introduced the national scheme 'One Wales: One Planet', which adopted a new vision for making progress on the country's commitment to sustainable development with the creation of an annual report reviewing progress.

In 2011, the UK's Sustainable Development Commission was closed following the UK government's decision to withdraw funding. From this point onwards the Welsh approach diverged significantly from that being taken by England. The Welsh government responded positively when the Welsh Minister for the Environment, Housing and Planning (Ms Jane Davidson) appointed a new Commissioner for Sustainable Futures in March 2011, stating that this demonstrated how seriously the devolved government took their duty towards achieving sustainable development. The appointed commissioner began a series of national consultations with the aim of drafting a new Sustainable Development Bill, which led to a White Paper in 2012. This set out proposals to introduce legislation that would make sustainable development the central organising principle of the government and public service organisations in Wales. Following on from the White Paper, a further nationwide consultation, the 'Wales We Want' conversation (2014), was initiated to develop the form and content of the Sustainable Development Bill, which was re-titled the Wellbeing of Future Generations Bill. These latest discussions focussed on a variety of intergenerational challenges including climate change, poverty, the ageing population, and health inequalities. This led to the examination of those opportunities, which required collaborative working and integrated solutions.

The national conversation resulted in some 7,000 people contributing through local communities and groups, and this included the testing of different ways in which the Welsh people could be engaged. One of the key features in the formulation of the Bill was the appointment of 'Future Champions' who have been given the role of advocates for future generations in their local communities.[18]

A brief examination of the contents of the Act demonstrates the holistic, determined, and pioneering approach of the Welsh government.[19/20] Seven wellbeing goals are established in the Act and all the listed public bodies must work to achieve all these goals. They cover a i) prosperous, ii) resilient and iii) healthier Wales; iv) a more equal Wales, v) with cohesive communities; vi) a Wales with a vibrant culture and thriving Welsh language; vii) a globally responsible Wales. The Act incorporates a 'sustainable development principle' under part 2, improving wellbeing. Where public bodies are required to act in accordance with the 'sustainable development principle', which means that they must act in a way that seeks to ensure that the needs of the present are met without compromising the ability of future generations to meet their own needs. All the public bodies are reminded of five considerations they need to consider to demonstrate that they have applied the 'sustainable development principle'. These include the need to safeguard the ability to meet long term needs and balance these with short term needs; act in a way that seeks to prevent problems; work with an integrated approach, which considers how their wellbeing objectives may impact upon each of the wellbeing goals, their other objectives and the objectives of other public bodies; act in collaboration with other bodies and individuals in order to help meet their wellbeing objectives and ensure the involvement of people with an interest in achieving the wellbeing goals making sure that those people reflect the diversity of the area which the body serves.

Those public bodies who are required to apply the sustainable development principle and deliver the wellbeing goals must demonstrate the progress they are making. To do this, they must produce a Wellbeing Statement setting out and explaining their wellbeing objectives, how they have been arrived at and what consultation has taken place. In addition, the Act requires each public body to prepare an annual report and to respond to the Future Generations Commissioner for Wales when that person has made recommendations to them. The Auditor General for Wales is given powers by the Act to carry out an examination of an individual public body to assess the extent to which that body has acted in accordance with the 'sustainable development principle' when setting their wellbeing objectives and taking steps to meet those objectives. Part 4 of the Act establishes Public Services Boards (PSBs) for each of the local authority areas in Wales, which must include members from the local council, the Public Health Board, the Welsh Fire and Rescue Authority and the Natural Resources body for Wales. In addition, each PSB is required to invite other people to participate, including Welsh Ministers, the Chief Constable, Police and Crime Commissioner and at least one body representing relevant voluntary organisations. Each PSB must improve the economic, social, environmental, and cultural wellbeing of its area by working to achieve the wellbeing goals via assessments, the setting of objectives and the production of a Local Wellbeing Plan.

The role of the Commissioner is to act as the guardian for the interests of future generations in Wales and to support public bodies in working towards the achievement of the wellbeing goals. Part 3 of the Act sets out this general duty of the Commissioner and the functions of the role, which include the preparation of a Future Generations Report. Sophie Howe, the current Commissioner, was appointed in 2016 and produced the first of these five yearly reports in 2020 with the assistance of her 23-strong, youthful team. This first report[21] extends to some 800 pages and has a wide coverage. It sets out the vision, how the culture of the public sector must change, the role of the Welsh Government and the progress made so far against the seven wellbeing goals. An assessment of seven 'areas of focus' follows, commencing with land use planning and placemaking, followed by Transport, Housing and Decarbonisation. The way in which communities and infrastructure for the future are planned, designed, and built is seen as critical to addressing long-term challenges and ensuring wellbeing at the national and local levels. 'Getting planning right can help us to meet a number of our wellbeing goals, by helping protect and enhance our ecosystems, strengthening our communities by ensuring they can get together and access the right services, facilitating healthy and active lifestyles, supporting a modal shift and identifying land for clean energy production and new ways of working and living.' The integrated and collaborative approach at national and local levels should ensure that town and country planning will rightly play its comprehensive and central role in this wholly realigned process.

The Future Generations Report 2020 covers only a three-year review period from May 2017, and it is not surprising that progress has been mixed. The Commissioner is candid in much of her overall assessment, pointing out where public bodies need to improve their performance and reminding them that they have 'a legal obligation to maximise their contribution to each of the wellbeing goals, not just to the one or two that are most relevant to their remit.' What emerges from her assessment is a need for an improvement in integrated thinking and working and the application of lateral thinking. No single action that a Public Body plans to implement should create only one impact. Too often the focus has been on one of the seven wellbeing goals rather than seeking to create multiple benefits. The Commissioner found it necessary to stress a fairly basic point that the goals should inform the objectives selected and the action steps to be taken rather than selecting objectives and steps, perhaps from existing plans, and then trying to retrospectively fit them to the goals.

The Report makes recommendations for each of the Areas of Focus. Those for land use planning and placemaking are progressive and recommend that the Welsh government and the other public bodies examine and implement the way in which the town planning system has been recast to follow the wellbeing goals. The need to better resource planning is stressed in the recommendation to Welsh Government. The more specific recommendations on land use planning stress that wellbeing objectives relating to planning should be based on placemaking and integrated with other objectives so that wider benefits are achieved.

Essentially this is the same approach as that contained in the NPPF for England, which places the achievement of sustainable development at the heart of the system and requires all three of the objectives to be satisfied in an integrated way. The problem in England is that the post 2010 Coalition and Conservative Governments have not put in place a comprehensive statute which embeds sustainable development and wellbeing goals into the national legislative system, nor have they revised/reformed the whole of the planning system to this end. While the 'new' Welsh planning system and the focal Wellbeing of Future Generations Act have imperfections, they represent a comprehensive and integrated approach with clear responsibilities and goals. They provide a means of implementation and monitoring which applies to the whole of the town planning and other related public planning systems (e.g., housing, health, and transport). The Welsh system is progressive and, in several respects, world leading.

It is illuminating to pursue this Anglo/Welsh comparison. The current town planning legislation in England is made up of the consolidating Town and Country Planning Act 1990, the Planning and Compensation Act 1991 and the Planning and Compulsory Purchase Act 2004. In 1990 there were two separate Acts covering listed buildings and conservation areas and hazardous substances and a third known as the Planning (Consequential Provisions) Act 1990. The four Acts are known collectively as 'the planning Acts'.[22]

The 2004 Planning and Compulsory Purchase Act contains the first attempt to provide a statutory definition of the purpose of planning in England. This is contained in section 39 of the 2004 Act under the heading of Sustainable Development. The General Note to section 39 of the Act contained in the Encyclopaedia of Planning Law and Practice[23] traces the early intent of this section of the Act and the very disappointing outcome. The original purpose had been to provide for the first time a definition in planning legislation of the purpose of planning. This was first proposed in a Green Paper published in December 2001 by the Department for Transport, Local Government, and the Regions (DETR) entitled 'Planning: Delivering a Fundamental Change'. The proposal was strongly endorsed by the Select Parliamentary Committee. The note goes on to state the legal view of the editors in the following way: 'However, it has arrived in the Act in a somewhat dilute form, buried half way through, rather than nailed to the masthead, limited in its extent and obscure both in its purpose and meaning'. The strong critique goes on to point out that, far from defining the purpose of planning, the section only applies to plan-making functions. It does not directly inform the development control process, including the determination of planning applications. The legislation is further diluted by not requiring planning authorities to pursue or even to observe the principles of sustainable development merely stating that they should act 'with a view to contributing' to its achievement. The previous Government's proposals for planning 'reform' (August 2020) did not present much hope for arriving at a definition of the purpose of planning or for embedding sustainable development and wellbeing goals into legislation. Instead, we continue to face certain destructive changes, which ministers, and their advisors, claim will speed up and improve the planning system.

I return to the issue of planning reform proposals in England in the final chapters. At this point I maintain my full support and praise for the Welsh initiative. Given that my wife had a Welsh father, and we have both spent very happy years in South Wales as university students and subsequently in our chosen professions, I might be accused of some bias. I would counter any such argument with my appreciation and recognition of why the Welsh people love their environment, culture, and collective character, this being gained via my participation in rugby, choral singing and life in Swansea and the Gower Peninsula.

Amsterdam is already recognised in Europe and worldwide as a city of culture with a tradition of tolerance. It has been described as the world's most liberal city by Russell Shorto in his affectionate portrait of the city and its people.[24] The liberalism he describes was born in the Protestant Reformation and the first age of scientific experimentation. It remains a liberal and free city of individual freedoms and individual rights, not just for its inhabitants but for everyone. More recently, Amsterdam has been recognised by the European Union as a leading centre for collaborative innovation connecting neighbourhood initiatives, start-ups and civil society with government, business, and knowledge institutions. The 'Amsterdam Approach' was the winner of the European Capital of Innovation award (iCapital) in 2016, which focusses on social issues and urban challenges. The city has also developed its Vision for Health and Wellbeing 2025 via a collaboration of the local council, the Public Health Service, care providers and the business community. The focal points in this strategy are helping people to stay healthy and active, becoming an age-friendly city and tackling obesity. The city has established a new Health and Technology Institute that combines innovative research with education and entrepreneurship to help deliver a higher quality of care, lower care costs and help to promote a healthy lifestyle. The Amsterdam Metropolitan Area was recently recognised by the European Innovation Platform for its innovations in the field of healthy ageing, a result of 37 organisations working on this topic.

The most far reaching and relevant approach to the promotion of health and wellbeing and sustainable development is the city's decision to employ a group of consultancies and work with them to pursue their transformative vision for a new way of planning the city and its economic, social, environmental, and cultural components. Amsterdam's stated vision is to be 'a thriving, regenerative and inclusive city for all citizens, while respecting the planetary boundaries. The result to date is the production of a public portrait of the city based on a wide range of existing data, plans and initiatives. The central question that the municipality and its collaborators wished to address was:

'How can our city be a home to thriving people in a thriving place, while respecting the wellbeing of all people and the health of the whole planet?'[25/26]

The group of consultancies employed by the Amsterdam municipality is led by Kate Raworth of the Doughnut Economics Action Lab (DEAL). Kate first published her concept of Doughnut Economics in a paper produced while she was a Senior Researcher at Oxfam.[27] Her book, *Doughnut Economics – Seven Ways to Think Like a 21st-Century Economist*[28] was published in paperback in 2018. Kate is an economist who has focussed on the thinking needed to explore the twenty-first century's social and ecological challenges. She teaches at Oxford University's Environmental Change Institute, where she is senior visiting research associate. She is also a senior associate of the Cambridge Institute for Sustainability Leadership. Her work has been recognised as ground-breaking by many institutions worldwide including the United Nations.

The basic concept of the doughnut in her 2012 paper was a simple world model that defined an inner ring around the central hole, which she termed the Social Foundation of wellbeing, that no one should fall below, and an outer perimeter, termed the Ecological Ceiling of planetary pressures, which we should not breach. Between the inner and the outer ring lies the safe and just space for all of humanity.[29] People exploring and embracing this concept have asked whether it can be scaled down to the country, region, and city level. Kate refers to the Amsterdam City event on 8 April 2020 as the 'launch of a new and holistic approach to downscaling the doughnut' and goes on to express her confidence that 'it has huge potential at multiple scales – from neighbourhood to nation – as a tool for transformative action.'[30] In *The Guardian*'s review of the Amsterdam launch, the environmental author and journalist, George Monbiot, describes Doughnut Economics as 'a breakthrough alternative to growth economics.'[31]

To address the central question for Amsterdam, four interdependent questions are posed, which become the four 'lenses' of the City Doughnut: what would it mean for the people of Amsterdam to thrive? (Local/Social); what would it mean for Amsterdam to respect the wellbeing of people worldwide? (Global/Social); what would it mean for Amsterdam to thrive within its natural habitat? (Local/Ecological); what would it mean for Amsterdam to respect the health of the whole planet? (Global/Ecological). For the City Doughnut model the inner ring sets out the minimum needed for the citizens to lead a good life, derived using the UN's sustainable development goals. The outer ring of the doughnut represents the ecological ceiling drawn up by earth system scientists, which indicates the boundaries that the citizens and organisations should not breach if they are to avoid damaging the earth's climate, resources, and biodiversity. The central premise of Raworth's thesis is that, rather than concern ourselves with the traditional economic growth model and its measures, centred around GDP, the goal of economic activity should be about meeting the core needs of all citizens in ways that mitigate climate change and reverse declines in biodiversity and the depletion of the earth's resources.

Having created the City-Portrait for Amsterdam by focussing through the four lenses, a useful tool is produced that is ideal for using in workshops to create new insights by involving change-makers, policymakers, and business. Raworth then demonstrates how the Amsterdam City-Portrait can be developed into a self-portrait by gathering resident's experiences, values, their hopes and fears for the future, their ideas and initiatives and their personal understanding of interconnections with the rest of the world. Raworth sees great potential for this self-portrait approach in Amsterdam, as a Doughnut Coalition of more than 30 organisations has been formed including community groups, SMEs, businesses, academia, and local government. This inspiring approach has grabbed international attention. The Covid-19 pandemic has increased the level of interest of other countries and cities in this pioneering approach.

The English Government has not, to date, demonstrated any outward interest in the 'Doughnut Model'. This is not altogether surprising as, apart from a broadly stated Sustainable Development Strategy and an equally broad National Planning Policy Framework (2019 version), there has been little sign of the emergence of a holistic delivery model for sustainable development and the health and wellbeing of all citizens, with Wales and to a significant extent Scotland being the notable UK exceptions. Oxfam, however, followed up Kate Raworth's 2012 discussion paper with another one in February 2015 entitled 'The UK Doughnut – A framework for environmental sustainability and social justice'.[32] The authors have selected social domains, indicators, and thresholds that they consider fit the UK context while recognising that these need to remain open to debate and revision. The authors rightly state that: 'The picture painted by the UK Doughnut Model is stark.' The UK significantly breaks through the proposed boundaries in nearly all the environmental domains identified, while at the same time inequalities in the distribution of the UK's wealth are a lead cause of deprivation across many of the social indicators. The authors go on to conclude that the Doughnut Model does provide a set of goals or objectives, which, if they could be delivered, would make for a much more sustainable society, organised in a way that would be capable of delivering a good quality of life for all without compromising the ability of others, now or in the future, to also gain an acceptable quality of life.

The questions and answers put forward in the NPPG for England on how positive planning can contribute to several health and wellbeing impacts is an encouraging sign of more holistic thinking from the UK government. The RTPI are placing health and wellbeing at the centre of several of their initiatives including a prize among their annual awards 'for excellence in planning for wellbeing', research and policy papers and, in 2017, the Young Planners Conference focussed on 'Healthy, Happy Places and People: Planning for Wellbeing'. The 2020 prize was awarded to Marmalade Lane, a 42-dwelling co-housing scheme near Cambridge, which as well as being awarded the wellbeing prize was the overall winner across all categories (the Silver Jubilee Cup). RTPI President, Sue Manns, in her presentation, referred to this development and other winners as shining 'a spotlight on the very best practice in delivering future places and spaces that work for everyone. They provide us with inspiration and confidence in the future of our profession.' This housing development combines state of the art housing design and layout with community facilities and beautiful shared gardens. All the residents are members of Cambridge Co-Housing Limited and have a stake in all the common parts of the development and contribute to the management of the community. Previous winners of the Wellbeing prize include the production of the Place Standard Tool by Scottish Government, NHS Health Scotland and Architecture and Design Scotland (2017); Seaham Harbour Marina, Watersports and Activity Centre – Durham County Council (2018) and Cuningar Loop Urban Woodland Park, a multi-activity environmental enhancement of 15 hectares of vacant and derelict land at Rutherglen, South Lanarkshire.

In 2019, Goldsmith Street, Norwich, a 105-dwelling social housing scheme for Norwich City Council, won the prestigious Stirling Prize for architecture. The judges described this as 'a modest masterpiece' representing 'a high-quality architecture in its purest most environmentally and socially conscious form'.[33/34] Designed by the London-based practice Mikhail Riches with Cathy Hawley, this project demonstrates how excellent housing design, incorporating Passivhaus environmental standards, can be combined with a very pleasing street layout for pedestrians with parking pushed to the perimeter of the development. The 'back street' has gardens and a fully landscaped pathway curving down the centre providing meeting and play spaces. This project is the first social housing scheme to win the Stirling Prize and demonstrates, as does Marmalade Lane, Cambridge, that it is possible to create beautiful, affordable, and liveable housing projects that deliver a range of the health and wellbeing objectives identified in this chapter. There are important lessons in these projects for the major housebuilders, for other Councils, the built environment professions and, above all, the UK government. These and other recent housing projects singled out for their qualities of good sustainable design confirm that very attractive neighbourhoods can be created without adding significantly to overall build costs while at the same time adding positive health benefits that reduce healthcare costs. When these benefits are combined with other on- and near-site environmental improvements, as part of a comprehensive neighbourhood design, then multiple objectives and benefits can be realised. The national policy requirements that developments should produce a net gain in biodiversity, mitigation of climate change impacts and enhanced greenspace networks are three of the key linkages producing health and wellbeing benefits while also meeting other sustainability objectives.

It is heartening to see that several local planning authorities have enthusiastically adopted detailed design guidance aimed at promoting quality places and developments. In 2019 I was involved in stakeholder review meetings on the draft 'Bradford Homes and Neighbourhoods – A Guide to designing in Bradford', which has been prepared by two leading design companies for the Council.[35] The introduction to this guide, adopted by the Council in February 2020, places strong emphasis on the significance of the health and wellbeing challenges facing the Metropolitan District, giving the examples of high rates of childhood illness and the serious concerns over air quality in certain areas. The overall aim of the guide is to ensure that new housing will create healthy communities. The resulting document is of considerable quality and justifiable ambition. It provides staged guidance, to be followed by developers and their professional advisors, which leaves little doubt as to how healthy neighbourhoods and the wellbeing of communities can and should be created. The Council state that 'these guidelines support the local plan policies and advance the government's policy agenda by placing high quality design, healthy and happy communities and inclusive design principles at the heart of future housing provision in the district'.

We can conclude that the sustainability imperatives, policies, and guidance are all in place at the national level and in most local councils to create higher quality and heathier communities with a full range of wellbeing benefits. The challenge is that these goals, policies, and national/local guidance are delivered on a more universal basis. We will see in several the following chapters further examples of land use change, development and environmental improvements contributing to health and wellbeing objectives. We will also encounter some of the challenges being faced and the impediments to delivery.

Chapter 4

Consultancy and the Private Sector 1984–2020

Spring 1984 – I have taken up occupancy of my new office on the top floor of Concourse House in Wellington Street, Leeds, at the heart of the city's business district. This location and address provide me with some street credibility for the launch of Clive Brook Associates, my new planning consultancy. There are at this stage no associates and no projects. The office has been provided rent free by Neil Pullan, the Chairman of a well-established Leeds construction company and an absolute gentleman. The agreement is that I will be responsible only for the annual service charge and will from time to time provide Neil with planning advice, which, as it turns out, he very rarely needs. Over my eight years as Assistant Director of Planning I had many meetings with developers and property professionals and consequently had a good range of contacts. Over the first couple of weeks, I devised a crude marketing plan, a very poor company logo and a letter setting out the services I could offer.

At this initial point there was no working capital in the company, fortunately my overheads were limited, and my wife had just secured a new permanent teaching post at Bradford Girls' Grammar School. I soon realised that a bank overdraft facility was needed. I selected the newly established city-centre branch of the Royal Bank of Scotland as my first port of call. An ebullient Scotsman, Boyd Brodie, had been posted from head office in Edinburgh to establish this important new branch and he readily granted me an 'interview'. I explained that I had not prepared a business plan but provided him with a copy of my CV and marketing plan. He agreed to provide me with a business account and an overdraft facility of £12,000, reasoning that should my venture fail over the next couple of years I was sufficiently well qualified to obtain a new job.

The first couple of weeks passed slowly and it is difficult to recall what further actions I took during this time. The phone didn't ring, I had no meetings arranged and no specific project plans to pursue. The office was freshly painted in white, and my new white telephone sat on an otherwise empty white desk. I was suitably shocked when one day in week three the telephone rang. It was Ken Morrison (later Sir Ken Morrison) the Chairman and Chief Executive of William Morrison Supermarkets, a company that had been founded by his parents selling dairy and other produce from a stall in Bradford Market. He explained that he had three projects to establish new superstores, two of these were to be in Leeds district and the largest, his flagship store, in Bradford. Given that I had only recently left Leeds City Council he invited me to join his team of professionals involved in the planning and design of the new Bradford hypermarket with linked retail warehouse units. This turned out to be the company's largest store and its flagship, as it was located in the northern part of Bradford urban area at Five Lane Ends, a few kilometres from the headquarters. Ken explained that he had checked out my credentials with two trusted professional advisors, Roger Suddards, a leading planning lawyer based in Bradford, and Peter Batty, one of the two founding directors of the appointed architectural practice Watson Batty. I eagerly accepted the offer. My first commission had turned out to be somewhat different to the small domestic or commercial project I had envisaged as the most likely starting point to my career in the private sector.

At the first project team meeting Ken chaired a discussion on the history of this industrial site and its location adjacent to two large residential estates, which had lost jobs following the closure of the International Harvesters manufacturing plant and exhibited other signs of socio-economic deprivation. The site had a long history of manufacturing, having previously been the home of the Jowett car factory. As a 'Bradford lad' Ken had a deep personal concern and commitment to provide replacement jobs and business opportunities beyond the retail jobs that would be created by the core development. Various views were given and towards the end of the discussion I suggested, with some reticence, the creation of an integrated business and training centre, linked into the core area of the retail development, with the aim of providing a full range of assistance to potential start-up companies as well as serviced space for established companies. Ken expressed enthusiasm for this idea and asked me to work with Watson Batty on the development and design of the concept. The result was the design and evolution of the Enterprise 5 Park, which contained a training and advice centre with a central exhibition space, through which all the hypermarket customers were to be routed. Several small start-up units linked into the exhibition area, and beyond the main retail complex a range of larger business units were proposed.

The concept was further developed by incorporating marketing opportunities for the new companies we hoped would be established, including use of the Enterprise 5 logo, the production of a regular newsletter, marketing of goods and services via the exhibition centre, organised displays of products within the Morrisons store and allowing public access to the 'street' accommodating the start-up enterprises. Ken provided an initial capital injection of £250,000 for the training and advice centre operation once construction was complete. I still think this was the first serious attempt in the UK to fully integrate the operation of a business development, advice, and training centre with a major superstore operation, thereby providing a unique set of marketing opportunities for start-up companies and established businesses. Unfortunately, within just a few months, the exhibition space close to the store entrance was taken over by more lucrative retail space lettings as Morrisons' Estates Department had not supported the fully integrated approach of the centre. The starter units and larger industrial spaces were successfully let in the first couple of years and the business services and training centre continued successfully for several years. In a recent site visit I was very disappointed to find that these focal enterprise services and the starter units had gone, leaving the still well-occupied industrial units to the rear as the only surviving element of the original concept.

In the early years of operation of Enterprise 5 I had tried to persuade Ken that this model could be replicated in other large urban areas throughout the UK where socio-economic deprivation needed positive attention and outcomes. My view that this combined investment would provide positive returns for the retail/commercial developer and the local authority was never put to the test, largely I suspect because of the opposition of the then Director of Estates. While this was in some ways a salutary experience I had gained much from this major early opportunity and remained very grateful for this. Ken and I subsequently remained friends over several years. The lesson learned was similar to that in local government at Nottinghamshire County Council. It is possible to be selected for a job or project by the chief officer of an organisation and to successfully deliver outcomes and then subsequently to be rejected by the intervention of other lead officers/directors.

In 1985/87 further commissions followed, particularly from the Yorkshire Regional Health Authority who appointed me as their planning consultant on the first of many hospital site redevelopment projects. These projects resulted from the rationalisation and modernisation of the NHS estate and the strategic decision to close most large psychiatric hospitals and replace them with smaller modernised units alongside what was known as the 'Care in the Community' programme. This well-intentioned modernisation of mental health services met with localised opposition when residents in neighbourhoods were confronted with planning applications for new build or residential conversions to accommodate those patients who could be re-housed in the community under some degree of warden supervision.

My company's involvement with health sector projects continued following the abolition of the Regional Health Authorities in 1996 and their replacement by eight regional offices of the NHS Executive (these also being later abolished by the Health and Social Care Act 2012). In 1990, secondary health care trusts were established to run district or area-wide hospitals. Our former key contacts in the Regional Health Authority gained leading estate/property management positions in the Leeds Teaching Hospitals Trust and the regional office of the NHS Executive, leading to new projects and eventually to work with the NHS Estates Department, which moved to Leeds. Subsequently I recruited one of the experienced estate officers from Yorkshire Regional Health as a permanent employee to enhance our contacts and project expertise.

In 1985 an office move proved essential following a serious fire at Concourse House, which started in one of the wholesale fashion houses located on the second floor. Fortunately, my files and other documents were not destroyed, but smoke and soot damage lingered on for months as a reminder of this event. I was again fortunate in being offered a sub-let of a small city centre office in Park Place by a friend and fellow professional involved in economic development projects. This occupation also came to an early end when my colleague experienced problems with the landlord. I decided at this point that I needed greater security of accommodation and that the only affordable way in which to achieve this was to move to a house that offered suitable accommodation for a family of five and separate space for an office. At this point I had not considered the potential expansion space that might be required. A search for suitable properties led to the selection of Burley Galleries, a Grade II listed building, dating back to 1745 and sitting at the historic core of the village of Burley-in-Wharfedale. The two previous owners had each carried-out alterations to the building, made up of four small former terraced houses, so that when it came on the market it comprised linked ground floor galleries with residential accommodation above. The first of the two former occupiers, previously a wool broker in Bradford, had converted the ground floor into an antiques shop and the second occupier, also a wool broker, established a retail art gallery centred around Victorian paintings.

We were able to secure the property at an affordable price given the requirements for various refurbishments and a modern damp proof course. The result was that I had abundant office space made up of two street-front offices and three linked galleries to the rear. Jill and I decided that we would use the rear galleries as a retail art gallery. At this point I also decided that I had sufficient work to justify the appointment of a secretary. The advert for this post was extended to include a love of art and the ability to manage customers alongside the main town planning business. Maureen Pollard, who joined the company in 1986, quickly proved to be an ideal fit for this dual role and Jill and I began our weekend tours of Yorkshire galleries to assess the competition, prices and gallery hanging fees charged to artists. This was a 'hobby business' due to our limited expertise and the time available, as Jill was still a full-time languages teacher. It was, however, a hugely enjoyable, if only marginally profitable, business experience. Maureen soon became a dependable bedrock of my expanding consultancy business and had a particular talent for aiding the development of client relationships and confidence in the business reflected in the numerous pleasing comments from clients.

Graham Connell was the first professional planner to join the company in 1987 and brought a range of additional skills particularly in the fields of planning policy and retail development. Sue Ansbro joined the following year. Sue and Graham had been planning students together in Birmingham and were subsequently appointed as graduate planners at Leeds City Council when I was on the interview panel. They both proved to be excellent appointments, becoming directors of the company. Further junior appointments followed, and this eventually meant that seven employees were based at Burley Galleries. As the office use expanded, planning permission was granted for a change of use from retail to office space and the art gallery business was closed.

In 1990 we decided that further expansion would necessitate a relocation of the office in or close to Leeds city centre and that renting an office was not necessarily the best solution, though this was very much the market norm. We explored the potential of acquisition, which would require us to obtain a commercial mortgage. Only two small, modern business developments had units that could be purchased. Having weighed the purchase price and the rental costs of comparable space, we decided to go ahead with the acquisition of a brand-new unit at North West Business Park, just to the north-east of the city centre, being developed by one of our housebuilder clients who were to occupy the unit next door. A commercial mortgage was finally arranged with the Riggs Bank of Washington. At one point the interest rate charged rose to 14.5%. Despite this, the balance of advantage against rental costs remained over the period of occupation.

Reverting to the 'start-up' period in 1984/6, it is important to mention other key factors that were extremely helpful to the development of the business. In 1984 there were only three other planning consultancies in Leeds district, two being sole trader companies run by retired city council planners and the other a significant competitor, Weatheralls, a leading firm of chartered surveyors, who had for some time operated a town planning team. My experience as Assistant Director of planning at the City Council provided me with a range of contacts in locally based development companies and professional practices involved in property and development.

During this period, I was contacted by Andrew Williamson of Walker Morris solicitors, who had recently been promoted to the position of planning partner in this well-established legal practice. For some time, the full development of planning consultancies was, in part, dependent on receiving project briefs from planning lawyers. Andrew's predecessor, as the lead planning partner, had a strong relationship with my main local competitor at Weatheralls. We met and he decided to introduce me to a couple of new projects, commencing a strong working relationship and friendship. As well as being an experienced planning lawyer, Andrew became a fully chartered town planner in the early period of our working relationship (rather than simply a legal associate of the RTPI – the route taken by most of his contemporaries). His knowledge and approach were certainly better informed as a result and it was a factor contributing to our mutual level of understanding, immensely valuable when we subsequently worked together as advocate and witness in many public inquiries. He was often rated in the top half dozen planning lawyers by his peers. His northern background in Burnley was like my own on the other side of the Pennines. Although I subsequently worked with several the top planning barristers in later years, I frequently recommended him as my advocate of choice, where I had the confidence of my client. Sharing common values in life, together with a strong professional understanding, is certainly a very good recipe for a successful working relationship.

These factors supported the expansion of the company via introductions to different components of the private sector marketplace, including several of the major housebuilders and two or three leading commercial developers. Mountleigh Developments, based in Leeds, was formed following the successful redevelopment of the Leigh Mills site in Pudsey to the west of the City. Tony Clegg and Ernest Hall, the founder directors of Mountleigh, had been involved in the textiles industry. Following this first notable development, the two directors went their separate ways, Ernest Hall purchasing Dean Clough Mills in Halifax, previously the largest carpet manufacturing site in Europe, and turning this into a multi-use business and leisure development, providing space for established companies and start-ups, an art gallery and concert hall, bars and restaurants and a range of other uses. Ernest was a successful concert pianist and supporter of the arts and was very keen to support new business enterprises. I was asked to interview him for a local government magazine and his passion for business and the arts was readily apparent. It was rumoured that he often helped start-up companies at Dean Clough by supporting them financially and providing business advice and he did not deny this. He regarded all school leavers as individuals with innate talent and was concerned that many were leaving school with 'failure stamped on their forehead'.

Tony Clegg was a very different character, with considerable drive and business acumen. He took control of Mountleigh and transformed the company into the largest and one of the most successful commercial operators in western Europe. One of his early deals was to acquire some 800 houses occupied by the American forces based at RAF sites in East Anglia. I was engaged as planning consultant on projects at Huntingdon and Woodbridge in Suffolk, seeking to expand this housing provision. One meeting involved flying in the Mountleigh private jet (shared with Asda Stores) to the USAF headquarters at Ramstein in Germany to meet with three colonels responsible for the property side of the operation. Two were dressed in daytime uniform, while the other wore a flying jacket, boots, and his pistol, and maintained a Lee Marvin pose throughout. Our flight approach to the base was interrupted by advice over the intercom that a NATO exercise was in progress and we would have to circle well above the base for some 45 minutes before we could land. On departure, Tony informed his co-director that he would have to leave us to find our own way home as he was taking the plane to Madrid for a hastily arranged meeting with the directors of Galerias Preciados, Spain's largest department store chain, for which he had recently made a bid. The property press in the UK could not understand the reasoning behind this move, which was driven by the extensive freehold property owned in Spain's main cities.

Mountleigh were to become involved as the lead developer in my company's largest project to date. In the late 1980s Andrew invited me to act as planning consultant, promoting part of the Earl of Harewood's landholdings, on the edge of the north Leeds urban area, as a residential allocation through the North Leeds Local Plan. The site was in the Green Belt, but we were able to make a powerful case based around the lack of alternative sites to meet the five-year land supply requirement and the limited harm caused to the main Green Belt functions, largely due to the physical and environmental containment of the land parcel and its relationship with the urban area. During the inquiry Andrew cross-examined the head of local plans at the City Council. The previous evening, we had discussed the content of the questions to be pursued including where development land could be found in the longer term beyond the standard five-year review period. The response given, when the question was posed, included the potential alternative of a new settlement somewhere in the northern half of the district. After the close of the inquiry, we discussed the likely need for and the potential advantages of this development option. We decided that the merits of this strategic option should be quietly investigated by examining the policy issues, looking at other national examples and carrying out a preliminary site search. By this point my company had sufficient resources to risk carrying out this work on an unpaid basis. We embarked on a large-scale mapping of the relevant constraints and opportunities, and from the overlay three potential sites emerged, two being within the adopted Leeds Green Belt and one to the north of the River Wharfe, near the boundary with Harrogate District, which was beyond the Leeds Green Belt and relatively unconstrained. This third site to the east of Wetherby and its racecourse, and west of a large industrial site at Thorp Arch, was selected for further detailed investigation. The other two sites were at Micklefield, to the east of Leeds near the A1 (subsequently upgraded to the A1M) and at the junction of the A1 and the A64 Leeds to York Road.

We progressed work on the project by inviting Richard Fletcher, the leading partner in a large, Leeds-based architectural practice, to produce some initial plans for a 5,000-dwelling new country town, which would be a private-sector-led development. Richard knew the area well and armed with initial plans, we approached the main local landowners and followed up the initial approaches with a private meeting in a local hotel. All the four main farming families were supportive. The next stage was to seek a lead investor/developer. Our initial approaches were to two large pension funds, but neither of these expressed an interest. I suggested to Richard and Andrew that we approach Tony Clegg at Mountleigh with a proposal that they take on the lead developer role, with a 51 per cent share, potentially followed by a subsequent approach to three of the six major national housebuilders. The three of us met Tony one evening in his London flat and we unveiled the outline master plan and associated ownership and analysis plans, including our constraints mapping, which by now covered the whole of Leeds Metropolitan District. When we came to the key proposal for the lead developer role and share there was hardly any discussion. Tony quickly proposed that Mountleigh would take on 100 per cent of the costs through the planning stages and then lead negotiations with the housebuilders. Mountleigh agreed to appoint our three companies to develop the project and pursue the case for a major residential allocation through the Leeds Unitary Development Plan (UDP). We were also authorised to appoint environmental and transport consultants to work with us in preparing an environmental impact assessment and topic-based reports on landscape, ecology, and transport. We named the proposed town 'Springswood'. A well-established urban design practice, Price and Cullen, were later introduced by Richard, and we agreed to ask them to review and, where possible, enhance the master plan proposals. Gordon Cullen, the founder of this practice, was by this time in the sixth decade of his working life and was an eminent urban designer, artist, and writer. In Chapter 5 I return to various aspects of Cullen's life and works that have influenced my career, as well as producing some interesting linkages that I was not aware of until I was introduced to and purchased David Gosling's anthology of his work.[1] Price and Cullen produced some new perspectives on the location of the proposed development in a parkland setting, as a counterbalance to the old established market town of Wetherby, together with a number of examples of Gordon Cullen's quality urban design drawings, which enhanced the visual potential of the design proposals and their settings.

Tony Clegg became ill during 1989, and though he made a recovery from a benign brain tumour this did limit the previous dynamic scope of his work. He was, however, strongly supported by his two lead directors, Geoff Goodwill and Rob Stansfield. The property-led recession of 1989/91 was another major blow to Mountleigh and other leading commercial property companies, including another client, Rosehaugh. Both major property development companies went into receivership in 1992. In 1990/91 my company felt the effects of this recession and we were particularly fortunate that both Mountleigh and Rosehaugh paid two of our largest outstanding invoices. At this time, I held a meeting with our employees to advise them of the seriousness of the market position and that competitors were making members of their planning teams redundant. I promised that there would be no redundancies if we worked together and accepted that pay rises were not practical at this difficult time.

Following Mountleigh's demise, the options they possessed on Springswood and other land they owned in Leeds passed to an individual property entrepreneur, who then sought an agreement with a developer who could fund the planning promotion work required on these potential development sites. The first draft of the Leeds UDP had favoured the allocation of land at Micklefield as a new settlement, based around its railway station and the regeneration arguments arising from this former mining community.

The Springswood proposal, together with those at Bramham and Micklefield, were eventually considered at the Leeds Unitary Development Plan Inquiry (1995/96), the second longest-running development plan inquiry after the Greater London Development Plan. Two Principal and one Senior Planning Inspectors were appointed to run the Inquiry. Principal Inspector Peter Shepherd presided over the assessment of the three new settlement proposals using a set of criteria agreed with the lead participants. The Springswood proposal was outside the Green Belt and there was limited public opposition and significant potential benefits, including the linkages between the proposed 5,000 new homes and the existing 5,000 jobs on the Thorp Arch Trading Estate, less than two kilometres distance from the centre of the new town, and the proposed by-pass for the town of Boston Spa just to the south. The Springswood allocation proposal was rejected, as were the two competing schemes in the Green Belt. Inspector Shepherd concluded that the Springswood proposal had merits but was somewhat ahead of its time. There were two main reasons for his rejection. The first was the amount of good agricultural land that would be lost in the grade 2 and 3a categories, and the second was the poor quality of the public transport links despite the introduction and proposed early funding of a new 'figure of eight', ten-minutes frequency, local bus service linking the new town with Wetherby, the Thorp Arch Trading Estate and Boston Spa. The Inspector's concern regarding transport links was more related to those with Leeds and its centre and main employment areas. The Micklefield option found some favour on sustainable transport grounds as it had a railway station. While this result was a great disappointment after several years of dedicated teamwork there were elements of hope for the Springswood proposal in that the Inspector had not criticised the scale and design and potential need in the medium to longer term.

In 1994 the Department for Transport reviewed the national road transport programme and removed several schemes including the Seacroft–Crossgates by-pass to the east of Leeds, which had been proposed to relieve congestion and improve the accident rate on the eastern stretch of the A6120, Leeds Outer Ring Road. The City Council had made it clear in the early stages of preparing the UDP that their preferred area of growth was to the east of the city, and I was aware that a large proportion of the land that lay inside the by-pass line was in the draft, rather than the adopted, Green Belt. An outer orbital relief road to a dual carriageway standard would bring several transport benefits and the potential release of large tracts of land between the urban area and a new relief road. The transport benefits went some way beyond those envisaged by the Department for Transport and included the strategic link to the extended M1 motorway, the reduction of air pollution and severance for residents along the eastern section of the Outer Ring Road, the ability to improve public transport routes and the links between the large and expanding Thorpe Park business park and potential residential development in the form of an urban extension. Initial work determined that there was capacity for some 5,000 dwellings, primary and secondary schools and two local shopping centres. Government planning policy at this time had started to look favourably on urban extensions as a potentially sustainable way of accommodating development.

I was introduced to the landowner of the central section of this potential urban extension by Paul Winter of Eversheds, also a leading planning lawyer, whose reputation was to be further enhanced when he moved to London and worked on the London Olympics, the new Arsenal football stadium, and other notable projects. The landowners were a local Leeds construction, property and development company, J J Harrison, established in the 1930s with extensive landholdings in the north of England. One of their main directors had boasted that you could walk from coast to coast across northern England without stepping outside their land. Gwen Fuller, then company secretary and lead manager, and later chief trustee for the family of owner directors, was very supportive of these outline proposals. We discussed the limited number of the majority landowners including the City Council, Persimmon Homes, GMI, then owners of the Thorpe Park business development, and White Rose Developments, a joint venture between Yorkshire Water and Evans of Leeds, who had also become involved in the Springswood options. I had previously worked with the lead directors of these property companies.

Early meetings were held in our board room, and I chaired these, often experiencing some difficulty in bringing the parties together given the strength of the personalities of the leading players. The male directors were somewhat taken aback by Gwen's depth of knowledge of property, development, and construction in these early meetings. J J Harrison agreed to support Clive Brook Associates and Eversheds to promote the comprehensive transport and urban extension proposals, covering all the landholdings, through the UDP. The other developers were not prepared to establish a formal consortium and initially did not provide funding at this key stage.

There was agreement that our involvement did not lead to a conflict of interest with the Springswood proposal given the scale of development required in the Metropolitan District as a whole and the anticipated longer-term nature of the new settlement scheme. The Second of the Principal Planning Inspectors on the panel for the 1995/6 UDP Inquiry, John Bagshaw, was a planner and transport specialist. The outcome was support for the development and the formal identification of the East Leeds Orbital Road (ELOR) in the UDP Proposals Map. Essentially the land and the route were safeguarded subject to certain policy requirements and subsequently allocated as a phase 3 development in the UDP Review of 2006 following a further development plan inquiry.

The northern quadrant of the development, covering land holdings between the A58 Wetherby Road and the A64 York Road, was subject to an outline planning application by a local consortium of landowners and other interests led by Persimmon Homes. This application proposed 2,000 homes, the initial section of ELOR, and physical and social infrastructure. The lack of an overall consortium of developers and main landowners for the totality of the urban extension, the absence of a formally adopted composite master plan and a clear negotiated basis for the delivery of ELOR and the other necessary support infrastructure, in the form of legal heads of terms and a project-wide development brief inhibited progress on the development. The City Plans

Panel approved this outline application in principle in March 2015, deferring and delegating the final decision to the Chief Planning Officer subject to a number of caveats. By 6 January 2020, the complex section 106 legal agreement[2] was at last nearing final agreement with all the northern quadrant participants apparently ready to sign. Planning applications for the Central and Southern Quadrants are now being progressed and ELOR opened to traffic in 2022.

During the 1990s, Clive Brook Associates gained further clients working throughout the UK but with a concentration in the north of England. Our Manchester office was opened in the mid-1990s via the sub-letting of a small city centre unit and the appointment of Kate Bailey, a planner and landscape architect, who had been chief planning assistant at Trafford Borough Council. Initially via Kate we gained some good contacts and were appointed to act on the second stage of the Manchester Supertram to Salford Quays. This was, however, a difficult time to establish a strong market presence as 15 planning consultancies opened an office in Manchester City Centre during the 12-month period following the opening of our new office. Leeds was experiencing a similar strong expansion and in a relatively short period there were more than 20 planning consultancies in the city. If this expansion is extended to include the city regions the number of companies involved in offering planning consultancy services were more numerous. We had a strong background and client base in Leeds and the wider Yorkshire and Humberside Region, and a number of these clients had land and property interests in other regions.

Expansion of the number and range of projects occurred in several ways, with many of these, especially those from individuals and small companies, coming from previous clients. Via a Leeds development and property management company we became involved in a review of Jack Walker's property portfolio, which mainly comprised a UK-wide spread of steel stockholding sites, which he had acquired from GKN. An initial review of some 30 sites led to them being graded in terms of future redevelopment potential, followed by specific projects to achieve development plan allocations and/or planning permissions, where the opportunities were good and the timetable for action was short term. This led to a number of successful outcomes, and we jokingly calculated that the returns had enabled Jack, by now the owner of Blackburn Rovers, to purchase Alan Shearer.

Terry Glazebrook, who ran a design company based in Scarborough, was a lead consultant advising Bourne Leisure, the largest privately owned leisure company in the country. Via acquisitions of holiday caravan sites from companies including Butlins, Rank and Haven they had gained control of the majority of the large coastal caravan parks in the UK. They devised a strategic programme of site and facilities improvement. Terry introduced me to the company and, following interview, we were appointed on our first three projects in Scarborough District. We became part of a team preparing and negotiating planning permissions that, via master plans and a suite of planning and environmental documents, would upgrade the sites in terms of design and layout quality, landscape, and visual appearance, and enhanced modern facilities. In several cases this involved acquiring additional land to make the improvements, which included the introduction of golf courses, nature areas and green space. These were all used in the design to change the developments from row upon row of static caravans into flowing freeform groups, which had a much better fit with the local landscape and character of the countryside.

As a result of several meetings with company directors I was asked whether we could prepare the business and tourism case reports that were necessary to promote the many socio-economic advantages of these schemes. I readily took on this challenge and we were able to make strong arguments in support of several site proposals in relation to the projected expenditure in the area by holiday makers, direct and indirect job creation, purchasing of local services etc.

During the expansion of the planning consultancy market in Leeds, Sue Ansbro and Graham Connell were offered director posts by two of the incoming national practices. Their departure within a period of less than six months lead to a rapid review, a new marketing initiative and the promotion of Andy Rollinson and Richard Baxter as new directors. The departure of two experienced directors, who were long-standing friends, was an initial blow to my confidence, but I knew that we had to quickly regroup and ensure we remained fully competitive in the new market conditions. We had spent some time devising a draft company succession policy, which involved Sue and Graham taking on the company when I reached retirement.

Thankfully, we did not lose any of our existing clients and the market expansion created new opportunities for us. We also expanded our services by employing Stephen Heward, a landscape architect, and John Hemsworth, a chartered surveyor with extensive property and development experience in the healthcare sector. Our work in the residential sector continued to expand and included several the leading housebuilders as well as regional and local builders.

In 1999 I was contacted by Paul Winter, who advised me that a national planning practice, Robert Turley Associates, who had eight UK offices, had expressed an interest in acquiring the company to fill a large remaining gap in their national market coverage. The work to be covered by this new regional office would extend across the four city regions of Newcastle, Leeds,

Sheffield, and Nottingham. My first reaction was to reject this approach as we were making reasonable progress. My daughter, a planning lawyer in Leeds, reminded me that our original succession policy had failed, and I should give this approach serious consideration. At this stage in the development of planning consultancies in the UK, I was not aware of any previous acquisitions that might constitute part of a guide to the way forward. Following a meeting with Rob Turley and discussions with our accountant, I started to give this proposal serious consideration. Rob and I quickly established a good rapport and identified that our companies had similar ethical values. Further discussions followed and I requested a position on the main board, no staff redundancies, and the inclusion of the proportion of value in our office building, which had already been paid for via the commercial mortgage, in the purchase price. The latter requirement was initially a bit of a sticking point but following valuations a deal was agreed. A new rented office was selected in the city centre and a new chapter in our consultancy experience commenced.

The newly expanded company now had more than 100 staff and, employing around 75 chartered town planners, was one of the three largest planning consultancies in the UK. The hoped-for synergy between our client base, contacts, and regional expertise, and those of the national company's, was realised within two to three years. The nine offices all worked to a centralised business plan requiring monthly reporting of financial and other outcomes. This introduced a strange mixture of collaboration and competitiveness, and it took a few months for the new Leeds office to fully adapt. The various offices exhibited different strengths and talents as well as general planning and project management skills. Just before joining the main board, I was invited to undertake a half-day set of psychometric tests, which were targeted at assessing how my personality and general abilities would fit and interact with the seven existing board members. This was carried out by an Oxford-based consultant and observed by Maggie Godfrey, the Head of HR, planning director and existing board member. Prior to this session I was asked to complete responses to various questions that examined my life and career from primary school onwards. Generally, this was an enjoyable and relaxing experience, during which I was to discover certain things about myself that had, till then, been partially or totally hidden.

Maggie helped to further prepare me for my first board meeting by providing a synopsis of the key characteristics of my fellow directors. I had to temper and refocus my thoughts and inputs, in part because I was no longer a managing director, and also to reflect that, while I had more years of experience and had recently gained a post graduate management qualification in running private companies, I did not have certain skill sets which they individually possessed.

My four and a half years at Turley Associates (the prefix 'Robert' was removed in a remarketing of the company) combined a series of learning curves that illustrate the complex and often demanding roles that town planners face when pursuing leading roles in both the public and private sectors. Rob Turley was a delight to work with and the best marketeer I have met. He combined vision, determination, and charm, with a positive outgoing approach to planning that we shared. The freedom to take initiatives was welcomed but the degree of control that others sought to exercise through the financial and business planning cycle too often led to a culture clash. This remains a common issue in many professional service companies. Rob's approach to running the company was comprehensively demonstrated when the company structure was changed to establish an employee-owned trust into which he transferred a large part of his shareholding.

Rob had established his company in Manchester in 1983, a year prior to Clive Brook Associates. A couple of times in our one-on-one discussions he warmly suggested that, had we met up in those early years, we would have created a very successful partnership. This was particularly gracious when I reflect on his achievements in setting up a company that grew to a 200-employee business operating from ten offices. He retired from the business in 2008 and very sadly died from cancer in 2016. His obituary refers to his truly pioneering work in establishing and growing the company, the transition to employee ownership and the philanthropic culture he developed. John Rose, who wrote this, refers to Rob as 'a consummate planning professional who was able to communicate his enthusiasm for planning to all he met' and goes on to describe him as 'one of life's great encouragers who recognised that everyone had a contribution to make. Rob was a committed and very active Christian, believing that work and faith go hand in hand and that there should be no sacred/secular divide. He passionately believed that planning should be a Godly process as it involved human creativity, using the earth's resources wisely, creating beauty and space in an environment where humans and the natural world can thrive.'[3] Town planning, in its search for a rounded and defined purpose, would do well to express this as its primary goal, even though it includes the roles of other environmental and design professions. I feel proud to have worked alongside him for a small part of his career and shared his vision.

When my four-year period on the main board came to an end, we agreed that it was time for me to move on at the age of sixty. I was not ready for retirement but could see the attractions of combining a new challenge with a slower pace of life in terms of travel and working hours. We had just moved house in Ilkley to live on the back of the Moor, a five-minute walk downhill to the town centre. Dacre Son & Hartley were Yorkshire's largest residential estate agent, as well as running commercial and rural practices, and their headquarters office was based in Ilkley town centre. Jonathan Isles, the managing director, was an old friend from my early jobs with the Yorkshire Regional Health Authority and subsequently various hospital trusts and housebuilders. Dacres previously had a planning and architectural team but, due to various changes, this no longer existed. I arranged a meeting to ask Jonathan whether he saw scope for re-establishing a planning team, and he was enthusiastic and very supportive. This would be a new challenge and we could both see considerable synergy with the land and development team that Jonathan led. This had proved to be the source of several earlier jobs for Clive Brook Associates.

The new planning team was established when Jennie Hanbidge joined the company as my secretary. She had a business degree and quickly demonstrated the personality and aptitude required to assist in establishing this new team. Jennie's interest in town planning grew, and she would later qualify as a planning graduate at Leeds Metropolitan University and subsequently become a member of the RTPI.

Jonathan recognised that I had an interest in corporate strategy and business planning and asked if I could work with the company directors to produce a corporate strategy and potentially subsequent business plans. For a couple of years, I worked on developing the planning team's role alongside that of Business Development Manager, a non-executive role, reporting to the MD and the Board of Directors. The corporate strategy was drafted, revised, and agreed, though establishing business plans for each main team proved more difficult.

The company had a total of 21 agency offices throughout West and North Yorkshire, and there was considerable potential for spin-off projects in addition to the steady flow of jobs from the Land and Development Team and the Commercial Team. I became involved, as Jonathan's planning and development advisor, in several residential land deals, including the selection of the housebuilder/developer to partner the landowner in either an option or planning promotion agreement. The latter usually involved the appointment of a strategic land developer who would take full responsibility for securing an appropriate development allocation/planning consent via all the necessary stages required to achieve the best scheme, subsequently marketing the site by seeking a combination of best value and most deliverable solution. Housebuilders would traditionally follow the option agreement form of contract with the landowner. Several of these residential projects involved firstly advising the landowner and then, with the agreement of both parties, transferring to work for the residential developer. This approach rarely produced a conflict of interests.

At the age of 65 I had no wish to retire, but we decided that there was merit in appointing my successor as head of the planning team. Mark Johnson, a planner, and chartered surveyor, joined from the House Builders Federation northern regional office where he had been the lead planning advisor and had built up a very strong group of contacts within the residential market. He was an excellent fit, quickly adapted to his new role and further expansion followed.

Mark subsequently became a director of Dacres based on his skill set, experience, and performance. In 2013, on our return from a meeting in Bradford, he announced that he had decided to leave Dacres and set up his own planning practice in Leeds and asked if I would like to join him in forming Johnson Brook. I had reached the tender age of 69 and, confessing that I was attracted by this proposition, replied that I would discuss this with Jill. Mark and I had worked harmoniously and successfully at Dacres and, given our combined experience and contacts, we were convinced that the venture would be successful despite the competition in the Leeds City Region. Jill was, as always, supportive, and my fourth consultancy role began with a mixture of surprise and enthusiasm for what was happening.

Three years on I resigned as a Director of Johnson Brook but have retained a small consultancy role with the rebranded company as well as a few longstanding projects on a freelance basis. My passion for town planning is relatively undimmed and now includes some voluntary work with planning, property, and environmental links. I also feel a compulsion to respond to government consultations seeking radical changes to the planning system. Planning, design, and the environment seem to be somehow embedded in my genetic makeup. My daughter, Claire, is now a leading planning and environmental lawyer and main board member, having spent her first 17 years of training and practice with Andrew Williamson at Walker Morris. Paul, my eldest son, is an interior architect who works across a range of projects renovating and modernising buildings. The youngest, Richard, is a chartered landscape architect with skills in landscape design, visual and character assessments, computer software and photography. Following their careers and discussing some of the projects they are involved in helps to keep me up to date, but there is a need for a careful balance with many other aspects of family and community life, particularly the individual development and wellbeing of our four grandchildren.

* * *

The expansion of planning consultancy

In recent decades planning consultancy has expanded in terms of the number of companies, the planners that they employ, the breadth of the topics covered, and commissions pursued.

The role of planners in the immediate post-war period was predominantly embedded in the public sector, and town planning was largely, though not exclusively, seen as a public sector role required to deliver the 1947 Act system through local government and to work for the public good. The work of the few planning consultants who were involved in the major post-war reconstruction projects of large towns and cities collectively gave the planning profession a bad name, which still lingers on today.[4] Many of these lead practitioners were dually qualified as architects and planners and produced grand designs, while a few were engineers leading on highways and building projects. This general denigration of the post-war planning profession is by no means universally justified. It is, however, true that many of the city plan schemes had destructive impacts on heritage, the compactness and character of centres, and the displacement of pedestrians by intervening dual carriageway routes through and around the centre.

My own experiences lead me to reflect on the loss of character of Huddersfield's Victorian heritage, in particular the wrought iron work of the market and the narrow, cobbled streets, referred to by the town's consultant architect, Sir George Grenfell-Baines, as the 'Montmartre of the north'. He was the founder member of Building Design Partnership, now one of the largest architectural practices in Europe. Bradford and Swansea city centres suffered major loss of connectivity within their centres due to a dominant system of highways planned and built in the 1950s/60s. At Swansea I was involved in early schemes that sought to reconnect areas and improve pedestrian experience, but the city had to wait for later plans before real improvements were realised. In Bradford I am involved in the Property Forum's work with the city planners to reconnect sub zones of the centre, which are still divided by dual carriageways. Will Alsop, the visionary but maverick architect, launched his master plan to reconnect Bradford city centre in October 2003. Part of his vision has been realised in the form of the City Park around the focal town hall building, but the cost of implementing the rest of his grand ideas remains a deterrent.

Professor Otto Saumarez Smith has recently followed up his contribution to the lives of leading post-war urban planners for the Oxford Dictionary of National Biography by writing a book on the subject,[5] which, while it does not seek to exonerate these leading urban planners, does approach the subject in a more sympathetic way and considers the importance of understanding how the ideas of these radical planners led to the realisation of these major projects. In his summary for the Dictionary of National Biography he concludes:

'The subject is hugely significant because of the profound effect it had, and continues to have, on everyday lives and places across the country.'[6]

The current membership of the RTPI now exceeds 27,000 spread across 88 countries, with the great majority being based in the UK. Based on the most recent survey approximately 60 per cent are male and 40 percent female. Women, in my experience, make particularly good planners and I tend to think this is at least in part due to their practical and common-sense approach combined with their abilities to multitask and communicate effectively with a wide range of people. The growth of the number of women employed in planning is only paralleled by the legal profession, where women now make up 49 per cent of practicing lawyers. By comparison, only 17 per cent of registered architects are women.

In terms of ethnic diversity, only 6.24 per cent of the RTPI's members are from BAME communities, which is significantly below the proportion in the UK population (at the time of the 2011 census, 13 per cent of the UK population identified themselves as Black, Asian or Minority Ethnic).

The geographic spread of the membership in the UK is 19,610 in England, 2,089 in Scotland, 1,162 in Wales and 735 in Northern Ireland. There are more than 1,000 international members spread across a total of 88 countries. While this number is not particularly large it does reflect the world-wide influence of the UK planning system, both in terms of one that has been copied in other national administrations and one that remains as world-leading and worthy of being directly associated with.

A particular focus of this section is to examine the rise of the private sector as a lead employer of planners. In 1993 there were 17,763 members, of which 13,153 were corporate members of the Institute (74 per cent). A total of 75 per cent of these corporate members worked in the public sector and some 20 per cent as consultants in private practice or in the development industry. In the mid-1980s the proportion of consultants in private practice was significantly below 20 per cent of the membership, with a major concentration in London.

An illuminating review of the 50-year career of Geoff Smith, who started his town planning work at Nathaniel Litchfield and Associates in a two-room office in Portland Place, London, in 1969, provides a useful insight into the development of planning consultancy in the 1970s and 1980s.[7] The founder of the firm, Nathaniel Litchfield, was one of the leading planners of his time, establishing his reputation with the creation of 'the planning balance sheet' and the development of land use economics, which were on my curriculum as a planning student in the 1960s. Lichfields has grown from that small office of 6 staff in 1969 to its leading position in today's UK market with more than 215 staff spread across eight offices. This review describes how, back in the 1970s, planning consultancy 'essentially combined two main offers: master-planning and land use economics.' The firm developed the application of surveying and economic techniques to land use planning analysis, which led to retail impact assessments, housing need studies and the use of cost-benefit analysis, which was in vogue in the 1970s. Planning applications are described as relatively simple affairs with architects making most of the submissions. London is confirmed as the dominant location of private practices, though many of these had a national reach. The economic problems of the late 1970s led to the firm working in Europe, the Middle East and Africa. The 1980s is characterised as a time dominated by a freer market approach to planning, which led to the further growth of private sector planning.

An RTPI study of 2019 provides data on the employment of planners in the public and private sectors over the period 2006 to 2018.[8] This is based on two data sources, the Annual Population Survey (APS) and the RTPI's records and its specific survey of members in 2017. This research Paper shows that 44 per cent of planners worked primarily in the private sector compared with 56 per cent in the public sector in 2018 (these figures for the private sector combine private practice and the third sector to achieve full comparability with the APS data). Figure 4 in the paper illustrates the decline of employment in the public sector from around 70 per cent in 2006 and the increase in the private sector from around 29 per cent in 2006. The RTPI and other researchers agree that this shift towards the private sector is partly a direct result of the cuts in public services associated with the long period of austerity following the financial crash of 2007/8. This shift in the balance of planners employed in the two main sectors is not, however, a simple cause and effect relationship. The growth of the private sector has also been caused by an expansion of town planning's functions and the creation of new market opportunities, while some moves from the public sector are likely to have resulted from retirement of individual planners or their transfer to other types of employment. For example, a friend of mine left local government to run a garden centre.

The loss of planning staff from local government, particularly those at a senior/management level, is part of a fluctuating longer-term trend. Reorganisations of local government in 1963 (in London), 1974 following the Redcliffe Maud Commission, when two tier authorities were created, and 1985 when the metropolitan counties were abolished, led to losses of planning staff amongst the most skilled and experienced in the profession. During this 22-year period, several senior planners took early retirement, usually with a redundancy package/pension enhancement, and then moved into private practice. London experienced most of these transfers in the first ten years.

The growth in planning consultancy continued steadily from the mid-1980s to the early 1990s, followed by a more rapid expansion up to the start of the new millennium. The expansion in the 1990s was driven by a combination of the expanding scope of planning work and a trend amongst established companies to set up new regional offices. In the following decade ('the noughties') many local planning authorities had to outsource parts of their planning services to the private sector due to reductions in public sector budgets creating weaknesses in the resourcing of planning departments with a consequent loss of staff, knowledge, and skills base. The 2008–2013 recession saw the most dramatic cuts in public services, with up to a 40 per cent loss of resourcing and over one third of staff leaving some planning departments between 2011 and 2015.[9] The effects of the recession on the numbers employed in planning consultancy were, by comparison, slight, while local government still experiences some of the impacts. While some recovery was apparent over the last couple of years, the full effects of the Covid-19 pandemic and Brexit are still to be felt, and already the UK has again experienced periods of recession due to the economic impacts of the ongoing pandemic. The Government, as part of their socio-economic response package, have published a White Paper: 'Planning for the Future'.[10] This proposes a far-reaching and radical overhaul of the planning system and improvements to the resourcing of public sector planning. These and related proposals are considered in the end chapters of this book.

The outsourcing of planning services has recently decreased in popularity, with many services being brought back in-house as local authorities have increasingly realised that they could often provide particular services at a lower cost. This has not, however, restrained the continued growth of private consultancies. A major factor driving this continued growth has been the many government 'reforms' and changes to planning and environmental legislation, policies and procedures, all of which has added to the operational complexity and fragmentation of the planning system.[11] Examples of new roles for private practitioners arising from these changes include the production of Neighbourhood Plans (the Localism Act 2011), the need for development viability appraisals arising from the NPPF 2019, and sustainability appraisals and environmental impact assessment processes resulting from European legislation being translated into UK planning law, regulations, policy and guidance.

As a result of the growth in private practice it is now possible to identify at least five distinct types of company:

1. The sole trader planning consultant, where the practitioner takes a conscious decision not to employ associates and to retain certain personal freedoms as well as reducing risk. These sole practitioners are often very skilled in general planning activities due to solid experience in local government, or

they work in niche specialisms such as heritage, minerals, or countryside planning. I have found that a number of these are prepared to pass on work from new clients or projects arising outside their normal geographic area of operation to maintain their independence and avoid the responsibilities and costs of becoming an employer.

2. The majority of planning consultancies are limited companies employing graduate and chartered town planners with a strong regional base and a range of project expertise. Many of these operate from a single office but in some cases can offer wider geographical coverage beyond their home region or even a UK-wide service where they have a particular expertise or a client base that includes developers with nationwide interests. A few may operate from a couple of regional offices, but they remain distinct from the third category identified. It is not unusual for these companies to seek to expand their services by employing one or more specialist planners and/or individuals in professions allied to planning, such as landscape architecture, ecology, archaeology, or transport. Typically, this category of company will have 5 to 20 employees.

3. Large UK-wide planning practices operating from multiple regional offices. These have normally started out as smaller practices based initially in one region. My former company, Turleys, started in Manchester and initially served the North West of England. Litchfields started in London and the South East region. A few of the companies in this category started life in other professional disciplines and expanded into the provision of planning services. Turleys and Litchfields are examples of companies that are now operated as employee-owned trusts. Geographical expansion has been key to the growth of these companies. In researching the setting up of a new regional office, these companies will typically seek to appoint a well-established planner in local government or private practice to lead the establishment and growth of their new office. Occasionally this will be achieved by the acquisition of an established consultancy with strengths in the core region and a good existing client base. These independent companies tend to retain a strong professional and ethical approach, which means that they remain in tune with the core values of planning and, in particular, the goal of serving the public interest.

4. This group are very large internationally based firms that are publicly traded companies on one or more of the world's stock exchanges. 'After extensive consolidation through mergers and acquisitions, these companies are now considered "mega-firms" with offices on every continent, dozens of practice areas, and thousands of employees."[12] Examples of these firms quoted in Linovski's short article are WSP and Aecom, other examples are

Arup, RPS Group, Atkins, and Savills. There are about eight to ten of these global consultancies, providing services in a multi-disciplinary company operation. These companies can take on very large and complex projects requiring a well-coordinated and mixed team of professionals. The firms are often involved in major transportation, infrastructure, and environmental projects. These mega-firms are in a strong position to take on some of the world's most pressing planning issues. Climate change is a leading example of a complex problem leading to national and international scale projects. While Linovski's research has concentrated on North American experience, there is strong evidence that these firms tend to prioritise shareholder value over staff and client care and conflicts can arise between shareholder's interests and planning values. There is some concern that further growth of this type of firm has the potential to undermine the ethical foundation of the planning profession.

5. The final category that I identify are planning teams established within another professional organisation that remains a relatively small but profitable unit for the host company. My experience of re-establishing a planning team at Dacres, Yorkshire's largest estate agency, is an example of this category. A number of national and now international firms of chartered surveyors/agents, having established a planning team, have been able to grow this into a company falling into category 3 and, in a couple of cases, category 4. A few large law firms, as well as having a legal team dedicated to planning, have established a separate small team of planning practitioners.

Many of the tasks resulting from the extension of the planning system, and to some degree its fragmentation, at a time of local government retrenchment, are being performed by the private sector. Over the last three decades justifications for these private sector inputs have included: 'increased capacity and efficiency, cost-effectiveness, specialised skills, knowledge and expertise, creative and visionary inputs, increased legitimacy and increased independence and critical distance'.[13]

Academic papers on the growth in activity of the private sector have argued that many of the consultants involved in performing these functions are not trained, qualified or chartered planners. My experience, based on working in various types of planning practice and several sub-markets, leads me to the conclusion that this is an over-statement that needs qualification and review. On larger projects, experienced and qualified planners are invariably required to project manage one or more stages of the job and to lead a team of specialist consultants. Other planning projects are led by consultants with dual qualifications, the architect planner or urban designer being a common example. Traditionally, architects were responsible for many projects involving the preparation and submission of applications, particularly where this was centred around the design and construction of buildings. This trend has markedly declined as the recognition of planner's skills and the complexity of the application requirements have increased. In the smaller number of cases where the architect project manages the application process (preparation of plans and reports, submission, validation, and negotiation), they usually make sure that an experienced planning consultant is working alongside them.

In the relatively new area of producing Neighbourhood Plans, introduced by the 2011 Localism Act, the Qualifying Body (Parish Council, Neighbourhood Forum, or community organisation) have often started out with great enthusiasm and assumed that the use of volunteers with amateur specialist interests in, for example, transport, heritage, or environmental matters, will be able to produce the finished approved plan. Many have, part way through the process, engaged planning consultants who are specialising in this field.

Planning consultancy did not have a particularly auspicious start and some of its early post-war contributions still present problems in reordering city and town centres. The subsequent growth of the consultancy sector, within the context of an overall growth in the various membership categories of the RTPI, has strengthened the skills and knowledge base of the profession and increasingly placed it in a position where planners are able to take on new and often challenging roles. Over the last twenty years this has primarily been achieved via the private sector, as public sector planning has suffered from successive cuts in resourcing. National and international research on planning issues, knowledge and skills development has made a recognisable contribution. However, in the UK there is a recognisable degree of disconnection between academic research and planning practitioners and a wish to fix this.

A useful contribution from academia is to highlight the potential problems associated with the increasing fragmentation of the planning system. This has been caused to a large extent by many government changes to the system, which have failed to recognise that a more holistic approach to system change has been required. The expansion of the topics covered by planning and their significance, and the failure to fully integrate them within the system, has also made a significant contribution to this fragmentation.

There have been, in the last two or three years, some encouraging signs in the form of the reviews carried out by the TCPA (the Raynsford Review), the RTPI in a series of papers and, most recently, by the government in the form of the new White Paper on the reform of the planning system. This has opened up debate and consultation and there is new emphasis on strengthening the roles and responsibilities of town planners.

Chapter 5

Planning, Design and Beauty

Town and country planning has an integral role in the design of places, buildings, and the environment, and consequently has a shared responsibility for both the realisation and preservation of beauty in order to enhance the lives of existing and future generations. The impact of design, the achievement of beauty and the avoidance/replacement of ugliness in the places where we live, work and play are an essential component of our health and wellbeing and the achievement of sustainable development.

In the first chapter of this book, we have traced the origins of the modern town and country planning system of the UK and how its emergence was a caring response to the adverse living and working conditions, including lack of access to sanitary facilities, public open space, and the wider open countryside. The daily experiences for most of the men, women, and children in the urban areas of Britain during the last half of the nineteenth century and the early decades of the twentieth century were ones of wearing routine, exposure to various types of pollution, poverty, disease, and lack of education. There was little, if any, beauty in their lives to lift their spirits and few amenities available to lighten their burden. A small number of Victorian industrial philanthropists were determined to improve the lives and living conditions of their workers and established exemplar new settlements such as Saltaire, Bournville, Port Sunlight and New Earswick.

In Chapter 1 I make significant reference to William Ashworth's book, *The Genesis of Modern British Town Planning,* because I consider this most eloquently presents the staged reasoning and the public arguments made for a system of town planning that includes but goes well beyond aesthetics to cover health, social and economic wellbeing. There are some remarkably strong parallels between the arguments made in the late nineteenth and early twentieth centuries for a publicly funded system of town planning, and those that have emerged in the last two years for the need to re-resource and repurpose town planning with a focus on the creation of beauty and the health and wellbeing of citizens. The essential contributions of good design and quality planning across the built and natural environments are once again being given the high level of recognition they deserve.[1,2]

Dame Fiona Reynold's book, *The Fight for Beauty*,[3] is an excellent and passionately presented case for making the protection, enhancement, and creation of beauty a central requirement of planning and environmental policy. She uses her experience as the former Director General of the National Trust and in earlier posts, firstly at the Council for National Parks and then at the Campaign to Protect Rural England (CPRE), to chart 150 years of the history of campaigning and the leading personalities and events that have given life and purpose to this key issue. While her book majors on landscape and the countryside there is recognition of the need for beauty in the urban scene and in new built development. She recognises the important linkages between the Victorian campaigners for beauty and the movement for town planning chronicled by Ashworth, and she is a strong advocate of the current need for good land use planning to achieve a range of enhancements to the natural and the built environments.[4]

Dame Fiona's appointment as a lead advisor to the Building Better Building Beautiful Commission has clearly had a beneficial impact on the Living with Beauty report.

Much has been written about beauty and design as both separate and related subjects. There is widespread agreement and understanding of the aesthetic pleasure that can be derived from experiencing beauty and good design in all its many forms. Aesthetic pleasure arises from activities involving looking at or listening to things. This is how we appreciate/enjoy contemplating landscapes, paintings, a well-designed house, good architecture, urban design or listening to music. Aesthetic pleasure can elicit physical pleasure. Beauty has been shown to be a very influential factor in our lives and is well summarised by Eric Carlson editor of the online magazine *Modern City* in a recent article:[5]

> 'Our brain function, mood and general livelihood are usually improved when we live and work in places we find aesthetically pleasing.'
>
> 'Beauty isn't just superficial and is actually a key part of how we understand and interact with the world around us.'
>
> 'People's wellbeing and sense of home is often deeply attached to landscapes and environments.'

Sir Roger Scruton, the Co-Chair of the Building Better Building Beautiful Commission, was a well-known author and philosopher (Sir Roger died on 12 January 2020, some two weeks after the completion of the final text of the Commission's Report). Through his writings he aimed to raise public awareness of the relationship between the built environment and the happiness (or wellbeing) of human communities. He spoke of beauty in the following way:

> 'Like the pleasure of friendship, the pleasure in beauty is curious: it aims to understand its object, and to value what it finds.'[6]

William Wordsworth, the leading English romantic poet, expresses his love of beauty in much of his poetry, as well as the impact that his appreciation had on his physical and mental wellbeing, and this appreciation is passed on to successive generations via his writing. While best known for his strong attachment to the English Lake District and its landscapes, he also appreciated urban landscapes and urban design. The latter is well expressed in the poem *Upon Westminster Bridge* written on 3 September 1802, where he describes the calmness and beauty of the panorama.

The opening two verses of his well-known lyric poem *The Daffodils* (1804) describe an idyllic lakeside landscape. The final two verses express the lasting personal value of this visual experience. Wordsworth in common with other poets and artists in the Romantic movement, was in part praising the qualities of nature and the landscape and in part rebelling against the ugliness of much of the extensive urban and infrastructure development fuelled by the Industrial Revolution. He was a great promoter of beauty and the need to protect our cherished landscapes, particularly his beloved Lake District.

The new search for beauty in the built environment in 2020 and the added strength of the public's wish to protect and enhance the natural environment, have many parallels with the movements for planning, environmental protection, and improved access for all to the countryside in late Victorian and Edwardian England. The appreciation of beauty in the countryside, in parks and gardens and the built environment, and the need to achieve enhancements and the creation of more beautiful places, have all been given added value during the months of the Covid-19 pandemic. There is growing appreciation and public understanding of the impacts of climate change and loss of biodiversity, which pose on-going threats to the quality and beauty of the natural environment. The impacts of varying degrees of Covid 'lockdown' have deepened our awareness of the many values of our natural environment, our need for access to greenspace of all types and sizes as well as beautiful landscapes and townscapes. When good or outstanding design of the urban fabric is combined with new and enhanced access to a network of interconnected greenspaces then new, distinctive, and beautiful places to live can be created.

To date, the positive health and wellbeing benefits of living in places of character and beauty have been denied to large proportions of the population. The report of the Building Better Building Beautiful Commission ('Living with Beauty') once again exposes the substantial negative impacts which many modern housing estates have had on the lives of their occupants and their more immediate neighbours due to poor, ill-considered designs, a quantitative and qualitative lack of public and private open space and poor connectivity via sustainable transport modes to essential services. The same problems exist for those living in older urban neighbourhoods where monotonous and functional buildings, lacking form and good design, at best do not provide any daily uplift in the spirit of those living there and passing through and, at worst, become depressing and affect an individual's mental health.

There have recently been many studies of beauty and its relationship to our health and wellbeing. This question is now often being asked: is the experience of beauty a basic human need and a right? Helicon is an American consultancy working with artists, cultural leaders, and other sectors to mobilise culture and creativity for positive change. In December 2019, as part of the Art of Change Initiative, working for and with the Ford Foundation exploring the interplay between the arts and social justice in the world, Helicon brought together a small group of experts to discuss the topic of beauty. The group invited were from a variety of fields including psychology, economics, art, philosophy, and public policy. Their stated goal was 'to question the dominant paradigm of our society, which overvalues economic metrics as an indicator of wellbeing, and ask how we might more effectively articulate, value and nurture beauty as a basic need and right'.[7] The conversation in this workshop had two parts: a) 'making the case for why beauty is essential to the health of human beings and society, and how it can be a catalyst for justice' and b) 'expanding the space for beauty in our contemporary discourse and policy-making and understanding the role of art and artists in that shift.'

The workshop concluded that: 'Contemporary society has overvalued economic growth and technological innovation, equating these with progress in human development and prioritising those factors at the expense of the things that most people agree make life worth living, among them human connection, beauty, nature, love, and art. We have always known intuitively – and now we have research data to back this up – that material wealth alone does not lead to happiness. Yet our hyper-capitalist society has made it increasingly difficult to talk about and champion the more humanistic elements of life as worthy of investment and development.' This conclusion aligns well with the Doughnut Economics approach outlined in Chapter 3.

American artist Theaster Gates, in a TED Talk entitled 'How to revive a neighbourhood: with imagination, beauty and art',[8] makes the case for why we should treat beauty as a basic service and happiness as a metric of success. In this online talk he provides the example of a Chicago neighbourhood where housing development has failed to 'take account of our human thirst for beauty, for the sublime, the emotionally enriching, the spiritual'. As reflected in studies in this country and abroad, building homes and neighbourhoods without beauty, culture and a social agenda creates its own kind of problems and the residents of such areas do not value or look after where they live.

These views from across the Atlantic are shared in the Living with Beauty Report. In the introduction to the Executive Summary, under the sub-heading 'Ask for Beauty', the Commission state that:

'We do not see beauty as a cost, to be negotiated away once planning permission has been obtained. It is the benchmark that all new developments should meet. It includes everything that promotes a healthy and happy life, everything that makes a collection of buildings into a place, everything that turns anywhere into somewhere, and nowhere into home. So understood beauty should be an essential condition for the grant of planning permission.'[9]

* * *

Design in Planning – Personal Reflections and Involvement

All aspects and fields of good design contribute to beauty and happiness in our lives, from products used in our domestic and working lives and the buildings and places we occupy, to the appreciation of art and architecture, the access to beautiful landscapes and nature reserves and quality urban and countryside renewal. The recent renewed emphasis on the role that town and country planning should play in the achievement of well-designed places, buildings and landscapes is most welcome. It is also somewhat surprising, following many bleak years when opportunities for good design in our urban and rural environments have been missed and in too many cases avoided for financial expediency, that there is a reluctance to engage skilled designers and a lack of focus in government, policy, and guidance on the importance of good design.

I express my surprise at this overdue new emphasis because it has taken so long and it comes from the Johnson Government who, along with their predecessor, and now successor governments, have been critical of planners on several fronts and have concentrated on speed of delivery and the economic benefits of development at the expense of quality, beauty and health and wellbeing returns. The emphasis on making the achievement of beauty a mandatory requirement to be delivered via a strengthened town and country planning system, contained in the August 2020 White Paper, and the strong recommendations and evidence base of the Living with Beauty Report, were rather exciting. The realisation of this requirement at national policy level in the latest revision of the NPPF is most welcome and the emphasis must now be on the delivery of these qualities.

Reflecting on my own career in planning it is, to say the least, disappointing that it has taken so long for politicians and policymakers to realise the many benefits of good design, allied with good planning. Design has played an important role in my training and my various roles in public service and consultancy. I have been fortunate to gather a wealth of understanding of design and its integration within town and country planning via inspiring lecturers, professional colleagues in architecture, landscape architecture and urban design and in my project management and negotiating roles.

The school of planning I attended was based in an institute of higher education rather than a university and, as a result, had a relatively lowly ranking alongside the well-established planning courses in several universities where the bachelor and master's degree courses had for some time been fully recognised by the RTPI as the first requirement for gaining chartership. Looking back, I think that this disadvantage was outweighed by several factors, including the small group of six graduates, the sandwich format of the two-year concentrated syllabus, visiting lecturers, including practitioners with Essex County Council, and the outstanding qualities of our two permanent lecturers. Dale Robinson, the course leader, was an architect planner with an active professional practice and, as well as his keenness to pass on his design expertise and techniques, was also able to provide live project examples. His passion for his subject and his determination to get us through the challenging master plan project, forming a main component of the RTPI external exams, was infectious. We were introduced to master planning and its design components from the first week onwards.

During my early years of local government experience at Essex and Nottinghamshire County Councils I had limited project experience involving design, with the exceptions being town centre plan preparation and involvement in the macro design of new industrial estates. The Essex County Planning Officer during my three years at County Hall was an architect planner, and it is notable that during this period a small team was established to produce the ground-breaking Essex Residential Design Guide, which was used as a template for other design guides.

At Swansea City Council I was responsible for conservation areas and town centre planning alongside a small drawing office. My eight years at Leeds City Council proved to be a major boost to my involvement and understanding of design within the context of assessing planning applications, the preparation of design policies and guidance and specific projects. My Director, John Finney, was a qualified architect planner. It was quite common in the 1960s and 70s for chief planning officers to be dually qualified. One of the four groups within my Implementation Directorate was the Environmental Design Group (EDG) comprising architects, landscape architects and heritage specialists. They were internal consultees on key planning applications where design advice was regarded as essential to achieving well-balanced decisions and participated in negotiation meetings with the private sector developers and their design advisors. They also produced design policies for local plans and design guidance and briefs for development projects. My responsibilities included arbitrating on cases where there were differences of professional opinion between the planning case officer and design specialists and negotiating with developers. During this period, I chaired the corporate Technical Board where specialists from various departments came together to advise on City Council developments from design, engineering, and property perspectives. The aim was to achieve integrated solutions to a wide range of interlocking issues. This was a period when local authorities and the planning service were relatively well funded and were able to employ specialist design officers.

Throughout my career as a planning consultant, I have worked closely with architects, landscape architects and urban designers on a wide array of projects, frequently in the role of project manager, including the briefing and appointment of the specialist team on behalf of my client. The role of many town planners as the project manager of developments through the planning application stages was, in my experience, acknowledged and respected by the great majority of design professionals, house builders and commercial developers. There were also schemes where there was a strong case for design leadership and the architect acted as project manager. In particular this applied where the project was based around a single landmark building or building group in an urban setting.

My interest and involvement with design continues via discussions with my two designer sons. Visits to European city townscapes, rural landscapes and reading books by or about great designers including Antoni Gaudi,[10] Charles Rennie McKintosh[11] and Cesar Manrique[12] continue to inspire me. Over the last 13 years I have been involved in a voluntary capacity as the client-side project manager of a complex refurbishment and extension project for All Saints Parish Church in Ilkley.

Planning, Design and Beauty 121

My two designer sons frequently remind me that 'design is design is design' and when we think about this statement it is borne out by many individuals and examples. Paul and Richard both have wider design capabilities than their core profession, Paul via painting and sculpture and Richard through photography, graphic design, and a degree and first career stage in product design.

Charles Rennie Mackintosh is today recognised as a truly great figure of early twentieth century architecture and design. Professor McKean, in his one of many books on Mackintosh, describes his range of work:

'From jewellery to graphics, from wall decoration to exhibited paintings, from pottery vases to wood engraving. He designed all kinds of objects of domestic use: tables, chairs, cutlery, and napkins; carpets, mirrors, curtain fabric and light fittings; beds, hat stands, wardrobes and clocks. He designed complete buildings – their foundations and structural steel, their sophisticated ventilation systems, and their plumbing. He painted landscapes and flowers. But the heart of his achievement was the design of places to be inhabited, rooms and sequences of rooms, their form and light and material.'[13]

The triumph of his architectural period is the Glasgow School of Art, considered to be a masterpiece by many. Sadly, this has twice been damaged by large fires in recent years. Mackintosh's genius was not recognised in this country during his lifetime, as it was abroad, and he died in relative poverty and obscurity.

Gaudi, the Catalan architect, who lived from 1852 to 1926, was recognised as a design genius in his own lifetime. Indeed, the Chair of the architectural faculty at Barcelona University, on hearing the views of his student and inspecting his early work, declared that he was either a madman or a genius. While his earlier architectural work reflected the eclectic styles of the period: Romantic, Gothic and Art Nouveau, he was later able to develop his own unique style, which is present in his many works in his home city of Barcelona. He was commissioned to design houses, their interiors and, in several cases, their furniture. Gaudi devoted 43 years of his life to his internationally renowned and still unfinished masterpiece, the great cathedral of Barcelona, the Sagrada Familia. Trewin Copplestone, in his book *Twentieth Century World Architecture*[14] describes this church as 'one of the most extraordinary structures of 20th-century architecture'. When I visited the unfinished building some years ago, I was immediately struck by its scale and great originality, and by his original three-dimensional models in the exhibition hall beneath the site, which he produced in the place of plans and drawn elevations to guide the detailed construction. Gaudi was fatally injured when he was knocked down by a tram in Barcelona and, for his funeral, thousands of citizens lined the streets, for the half-mile-long procession, to pay their respects to this great architect of the people.

Cesar Manrique was born in Lanzarote, part of the volcanic Canary Islands group in the Atlantic Ocean. He attended the local university to study architecture but found the course too restrictive and subsequently graduated from the University of Madrid as a teacher of art. Two years spent in New York City, from 1964 to 1966, confirmed an international reputation as an artist, with exhibits at the Guggenheim and other metropolitan galleries. He returned to his island home, where he developed his free art through paintings, mobile and static sculptures, architecture, and garden designs by fusing them with the natural and geological environment of Lanzarote. He did not rely on working drawings for his architecture and designed his structures and details on site via collaboration with his favourite builder. Through his passion for the island's environment and friendship with a leader of the provincial government he had considerable influence on planning and design coding of the built environment of Lanzarote, influenced by the destructive environmental impacts of high-rise tourism development along the coastlines of mainland Spain. Building height restrictions were introduced, together with a limited palette of colours for windows and shutters, marine blue for those facing the sea and forest green for inland focussed facades. He was in close communion with nature and a campaigner for the protection of the island's ecology. This central theme to his life and work is communicated in the intimate thoughts that he sporadically committed to the written word:

> 'The public has begun to realise that the things that matter lie in the essential beauty which nature reveals to us constantly, communicated through the principal of biological sensitivity natural to mankind.'
>
> 'I believe that we must promote quickly the characteristic differences of every place on the planet. Otherwise, in the near future, we will have a boring, standard culture, lacking in all creative imagination.'[15]

Manrique was also to die tragically in a car accident, close to his stunning villa carved into a volcanic landscape and now the home of his Foundation.

In 2006 I was involved in a welcome lunch for new members of All Saints Parish Church in Ilkley and, while gazing out of one of the windows in Church House, a hand tapped me on the shoulder and a voice enquired what I was looking at. My reply was along the lines that I was contemplating the relationship between the buildings in the historic courtyard. The enquirer, John

Tinkler, an acquaintance from my years at Leeds City Council, was the project manager on a building plan to expand and improve the Church property. The project gestation had commenced in the mid-1990s, but a start was made on drawing up plans in 2000. Within a couple of weeks, I was a member of the project steering group and quickly became engaged in drafting a development brief as part of the justification for this project and to steer it through the many permissions it would need. The Church was a Grade II* listed building and the separate Church House Grade II. These buildings and the land around them sat on the site of the Roman Fort of Olicana, a scheduled ancient monument. Across the courtyard to the rear of Church House is the Grade I listed Ilkley Manor House. The whole site is located within the Ilkley Conservation Area. Inside the Church are three Grade I listed Saxon crosses and various other items of heritage significance. This was certainly the most complex heritage design project I had encountered in my professional career, and at times the weight of responsibility was not without its problems.

In 2011, I replaced John as Chairman of the Building Project Steering Group, a formal sub-group reporting to the client body, the Parochial Church Council. Following strategic, design and financial reviews of the project, the macro design for the three interlinked components was agreed: a reordered Church, a new link building and a refurbished Church House, all to be accessed at a new common level. Earlier options for extending the Church had been opposed by English Heritage (now Heritage England). I worked closely with the original architects (Overton Architects, a local Ilkley practice) on the design concept for the link building and the refurbishments of the existing buildings that were to be joined. We agreed that the link should reflect the character of the buildings that it was joining but do so in a form that was, in itself, a modern architectural statement that would stand the test of time. Planning and listed building consents were required, plus scheduled monument consents for the siting of the new building and investigative archaeology associated with the foundations for the link building and the installation of a new underfloor heating system in the Church. In addition, faculty approvals had to be negotiated with the Diocesan Registrar and approved by the Chancellor of the Diocese. By the end of 2014, all the key permissions required were in place and it was possible to arrive at a firmer cost appraisal of the total project.

I was grateful, in 2015, when Derek Twine (former Chief Executive of the Scout Association) arrived and took on the role of Development Co-ordinator responsible for fund raising, negotiations with the Heritage Lottery Fund (HLF) and public relations, and subsequently the appointment and direction of heritage interpretation staff (following approval of a substantial grant from the HLF). My responsibilities as Chair of the PCC sub- group and member of the PCC included the selection and PCC approval of a new professional team, liaison with the project manager, quantity surveyor, architect, and other technical advisors on the preparation of detailed construction plans and tender documents. Following the receipt of tenders, bid proposals were evaluated and a well-established, locally based contractor (Dobson Construction), with the required experience on heritage projects, were awarded the contract. Over 100 plan, elevation, infrastructure, and contract drawings had been prepared by the team prior to the commencement of the contract and were reviewed by my team on behalf of the PCC. The gross value of the contract was circa £1.8m at this point and it became apparent during the construction that funds raised would not cover all the works. Scheme revisions were devised and proposed by the contractor and verified by the architect in detailed design amendments. The changes were agreed by the client, which enabled the full refurbishment of the Church, the construction of the new link building and the adaptation of the eastern end of the Church House proposals to proceed. The design and contract amendments provided a completed access and service hub for the total project. The 'new' buildings were opened and consecrated on 23 June 2019.

While the project is not yet complete and further finance has to be raised to finish the 'west wing' of Church House, there is great collective pride that this landmark stage has been reached after more than 25 years of planning, design optioneering, negotiation and hard work. This has undoubtedly been a huge part of my recent design learning curve and I was delighted to work with the project architect on the selection of the millstone grit floor paving slabs (the hardest and one of the most decorative forms of sandstone in terms of its patina and millions of years in the making), sourced from a quarry in my hometown. A consensus has arisen from visitors and the Ilkley community that the finished product combines beauty with added flexibility of use.

A final design influence on my planning career has been the work of Gordon Cullen, including surprising place and project linkages with this supremely talented draughtsman and urban designer. I was unaware of the linkages with parts of my life until my landscape architect friend, Stephen Heward, introduced me to David Gosling's anthology of his work.[16] Stephen brought this book into the office one day and pointed out these linkages. I immediately determined to track down a copy of the book, aided by another friend who was a collector. A telephone call to Janette Ray, a rare and out of print bookseller specialising in architecture and related subjects and based in York, close to the Minster, confirmed that she had a copy that was in very good condition. She enquired whether I was sitting down and then revealed a selling price of £120. She had researched availability elsewhere and the only other copy she could trace was on sale at a similar price from a bookshop in Manhattan. I considered this a reasonable investment and was delighted when I unwrapped the parcel a few days later to reveal a book in quality condition.

In his foreword to the book, Lord Norman Foster declares that Gordon Cullen (GC) is one of his heroes when he thinks back over the factors that influenced him to study architecture. He goes on to say that, nearly forty years later (October 1995), he is still entranced by the magic of Cullen's sketches:

'They influenced the way in which architects expressed themselves graphically and the way in which they thought, especially their sense of social values. In fact, the messages that Gordon Cullen promoted in these works about the importance of the urban landscape are more relevant now than ever before.'

'His critical analysis of the visual squalor that we are always in danger of taking for granted, was the conscience behind his pencil.'

Gosling, who worked with Cullen, ranks him alongside Charles Rennie Mackintosh in the quality of his draughtsmanship and pioneering design work. Cullen was born in Bradford in 1914 and was educated at Prince Henry's Grammar School in Otley. We lived in this Yorkshire market town from 1977 to 1987 and all three of our children attended this school, a mere 800 metres walk from our house. Cullen moved to London with his family in 1930, showing an aptitude for art at an early age. He first studied at the London School of Arts and Crafts, transferring to the Regent Street Polytechnic architecture to RIBA intermediate level course. His skills as an architectural designer and illustrator were soon recognised by, amongst others, Walter Gropius.

Cullen then worked for a small number of London-based architectural practices. He joined the MARS (Modern Architectural Research) group, an informal society comprising several leading figures of the Modern Movement. Though he was a shy person he seems, through his work and enthusiasm, to have made friends and influential contacts. In 1946 he joined the staff of the Architectural Review, described by Gosling as an elegant and influential monthly magazine. Between 1947 and 1953 he produced a catalogue of studies under the heading townscape, ranging from topics such as the London squares and 'Westminster Regained' (described by Gosling as a revolutionary study in pedestrianisation) to his final study published in November 1958, which was the full development of his theory of townscape. The theory and its practical application have a primary focus on the pedestrian experience and design priority over other traffic and the appreciation of serial vision as one moves through urban spaces in an uninterrupted sequence of views. His full development of the theory, and its practical application potential in his numerous articles in the Architectural Review, led to the publication of his book *Townscape* in 1961. This seminal work was well reviewed in the UK and the USA. Cullen was recognised as both a great artist/draughtsman and a persuasive wordsmith, and subsequent editions were published in several languages.

In April 1965, Cullen worked with Shankland Cox, Liverpool City Council's consultant planners. The editors of the *Architectural Review* considered that this was the first time that a 'townscapist' had been incorporated into the planning team and Graham Shankland is recorded by Gosling as saying that 'Like any artist and all good urban designers, Cullen had two sides to his nature: the objective and descriptive analyst of the power and magic of a place and the personal vision of its future.'

In the 1960s and into the early 70s, the Cullen family spent their summer holidays in the French village of Biot, a historic hill settlement, situated in Southern Provence on the Cote D'Azur, in the hill country that rises from the coast between Nice, Antibes and Cannes. It was here that Cullen produced some of his best artwork, five paintings of the village, which have pride of place in David Gosling's book.

In April and June 2000, Jill and I visited Antibes with a view to buying an apartment or small house. During our first visit, while we appreciated the character of the town and its marine setting, the crowds and in particular the traffic were significant detractors, as were the property prices. Antibes desperately needs a traffic management plan to remove visitor traffic from its narrow central streets. We decided to explore inland and caught the bus to Biot. Soon after alighting the bus at the foot of the village ramparts, we came to the display windows of a local estate agent. The price of properties was significantly cheaper than Antibes and the coastal littoral. Details of two almost identical Provencal terraced houses caught our eye, as did the asking prices at around 110,000 to 120,000 francs (at this time 1 French franc was roughly equivalent to £1). In we went to book an appointment to view later that afternoon. The properties were in a small domaine and sited around a large triangular-shaped park with an outdoor pool and beautiful specimen trees, including a large Kashmiri pine. It was love at first sight and we returned in June to complete the purchase of the second house we had viewed.

After 20 years of holidays, with family and friends, based in this very special village I can totally appreciate why Cullen and his family loved this place, its people, the dramatic landscape setting, and the village townscape. Cullen ran two summer schools in the village and, subsequently, his Biot series of paintings were printed in the Architectural Review in December 1974.

David Price and Gordon Cullen established their urban design practice in 1985. David had studied at Sheffield University where Cullen had been a visiting lecturer. The partnership of the younger, talented, and outgoing Price, with the great skills and experience of Cullen, resulted in a very successful combination. I met David Price and Gordon Cullen when they were introduced to the Springswood new country town project (see previous chapter) by Richard Fletcher, the architect director for the project. Richard had suggested to me that, having worked with Price and Cullen in London, he felt they could add to the master plan and the concept presentation of the proposals, via an enhanced vision for the landscape setting of the new town. I agreed, and our client added them to the project team for a discrete period. Our main meeting was memorable for the enthusiasm displayed by Price and the much quieter, almost subdued, input of Cullen, greatly enhanced by his early sketches. Gosling's brief reference to this project and the drawings reproduced in his book imply that the study was in some way centred on Wetherby and does not mention the counter-balancing new settlement proposal. Price and Cullen's full set of drawings for the project extended the proposed parkland from the new settlement to the River Wharfe at Boston Spa, in the form of the proposed arboretum, and to the west by encapsulating the existing racecourse and a proposed new golf course, thereby creating the counterbalance with the existing market town of Wetherby.

* * *

Planning and Design: A new Relationship

Considerable evidence has been gathered from a very wide range of sources and expert opinion to demonstrate that the planning of the built and natural environment has a major impact on the health and wellbeing of people's lives. There is a growing recognition that town and country planning is a key force for achieving positive outcomes via development and environmental projects within urban and rural settings. Such outcomes can only be achieved by working closely with other professionals who are active and skilled in the design and management of the built and natural environments. There is also great scope via such cooperative working to reverse some, or all, of the negative impacts of past development via regeneration and renewal projects. In the previous chapter I have only been able to bring together a small proportion of this evidence. Further strong evidence exists to support the creation and protection of beauty and its many components as a key contribution to our everyday health and wellbeing. Those components over which good planning has a strong influence include the protection, enhancement, and creation of landscapes; the improvement of biodiversity, natural habitats, and ecological networks; the protection and enhancement of our built heritage and developments, which create new buildings and places in urban and rural settings.

It has taken a long time for government to arrive at a full recognition of the important contributions that planning can make to our health and wellbeing, and the central significance of the roles that good design and beauty play in this fundamental relationship. As we have seen, this aligns closely with the experiences of our late Victorian and Edwardian counterparts.

The Concise Oxford English Dictionary definitions of the word 'design', both as a noun and a verb, demonstrate from the outset there are strong links between design and planning (used in its widest sense). When used as a noun design is defined as 'a plan or drawing produced to show the look and function of something before it is built or made'. It is also 'the art or action of planning the look and function of something' and 'the purpose or planning that exists behind an action or object'. Used as a verb, design is to 'decide upon the look and functioning of (something), especially by making a detailed drawing of it' or to 'do or plan (something) with a specific purpose in mind'. If we substitute the word 'somewhere' for 'something', the relationship between planning and design, in the field of town and country planning, becomes clearer. The clarity increases when the relationship is considered in the specific context of the current statutory functions of town planning in the built and natural environment.

Town and country planning's contributions to the achievement of good design, quality places and environments and to beauty will be achieved by the following mechanisms all of which are recognised and promoted in 'Living with Beauty':

1. Strengthened national planning policy and guidance requiring beauty and placemaking to be strategic policy requirements and integrated into the central aim of achieving sustainable development.
2. The government should introduce mandatory requirements to deliver good design and beauty into planning and environmental legislation.
3. Ensuring that local and neighbourhood planning authorities include specific policies in their plans which require well designed places, buildings and environments which make a positive contribution to beauty.
4. Each local planning authority should be required to produce a series of design guides covering placemaking, housing and environmental improvement, demonstrating how to prepare development proposals that will achieve quality designs and places and contribute to beauty.

5. The incorporation of area and project specific master plans within local plans. Those for specific projects will relate only to large scale or complex projects, recognising that the great majority of site and project-specific master plans will be commissioned by developers working to local authority policies, design guides and design codes.
6. Ensuring that suitably skilled staff are employed by local planning authorities including urban designers, landscape architects, architects, and ecologists to work on the preparation of design policies, design guides and the negotiation of quality proposals via the development management process. There is a lot of lost ground to be made up in the reskilling of local authority planning departments. This applies to the appointment of design specialists as well as the general training of planning officers through graduate courses and via continuing professional development.
7. Carrying out much more considered consultation with the public at several stages of the planning process to obtain their specific contributions to the design of local areas and development proposals that arise within them. This will require more intensive and earlier consultation with greater continuity to bridge the gap between the local plan process and development management.

Certain of the above components have been around for a few decades, including generalist design policies in local plans, the preparation of design guides and master plans. The Essex Design Guide, initially covering residential development only, was first published in 1973 by the County Council with subsequent main revisions in 1997 and 2005. Throughout the 45 years of its development and use it has received wide praise and been the template for several other design guides. These guides supported best practice and the achievement of quality in designs that gave full account to the local vernacular architecture whilst supporting modern interpretations. Master planning of residential, commercial, and mixed-use sites is also long-established, and I have personally been involved in or project managed several scheme designs. Historically the problems have been: firstly that the use of these techniques have been sporadic in their application; secondly that local planning authorities have been unable to employ or retain specialist design staff; thirdly developers, particularly most of the large housebuilders, have chosen to standardise their product with designs that are marketable, meet the financial returns demanded by their banks and avoid perceived additional costs on master planning and urban design and the environmental components they would advocate.

From the introduction of the 1947 Act planning system, the issue of promoting or controlling the design of development has been a fraught one.

The extent to which 'good' design can be achieved through the planning system has proved problematic. John Punter, while Professor of Urban Design in the Department of City and Regional Planning at Cardiff University, contributed a chapter that examines the issue of design within the practice of town and country planning in a book which reviews 50 years of urban and regional planning policy.[17] In this review he considers the limited and changing conception of design within the planning system and why design has been such a controversial area of planning practice. He explores the evolution of design control in the planning system and changes in its status over these five decades. He concludes his chapter with what he considers to be the achievements of control and the problems and barriers that still confronted the achievement of good design at the turn of the century.

There are some important lessons from Punter's review as we move towards the most positive and wide-reaching changes for planning's future role in the achievement of good design and contributions to beauty. A review of pre-2010 government planning circulars and advice relating to design control demonstrates a widening of the areas of interest and involvement from elevational design of individual buildings in 1966, to wider aesthetic involvement in 1980 and on to the much wider concepts of urban design and master planning. All this was still some distance from a holistic approach, which would ensure sustainable development, attractive outcomes for individual's health and wellbeing and contributions to beauty in their lives. Gordon Cullen, in his work, understood and promoted the holistic approach to urban design and its wider sphere of influence as did several architects and other designers.

* * *

Essential actions to ensure the planning system contributes to good design and beauty

The final section of this chapter commences with a review of some of the key findings and proposals of the 'Living with Beauty' Report and continues with my personal commentary on the actions required to ensure the desired outcomes based on my experiences of the role of planning in the design process for urban and rural environments. Earlier in the chapter I have summarised seven main components of town planning's contribution as recognised and promoted in the Report. I strongly endorse the Commission's main findings and policy propositions in relation to these key components. I also agree with the need for a holistic approach involving not just the selected team of professionals, the public-private sector cooperation required, extensive public consultation and the new addition of ongoing stewardship from existing and new landowners.

All my colleague members of the RTPI will, I am sure, endorse the Commission's conclusion with regard to the recognition of planning officer's central role 'with beauty as their primary responsibility' and that 'their powers, their rewards and their education must reflect this role' (page 24 of the Report).

The Report, in common with others, criticises the bland and sub-standard housing layouts and designs of most of the major housebuilders, and responds to the arguments which they often advance in support of a standardised approach and the avoidance of what are seen as design 'add-ons'. They refer to customer choice, market preferences, pricing, affordability, and the costs of the design inputs. The need for a thorough review of the major housebuilders business model, largely driven by the profit margins insisted on by their banks, is long overdue. The Report highlights the need to take into consideration the social costs of poor housing environments and build quality, together with the lack of on-site facilities and poor connectivity. The long-term health and other social and environmental costs need to be considered (at page 51 of their report the Commission state two of their strong evidence sources for these conclusions).

The first two of the Commission's policy proposals seek to embed a requirement for beauty and good placemaking in national policy (Proposition 1) and to require a net gain rather than the current approach of minimising harm (Proposition 2). Policy Proposition 4 seeks to embed the national policy requirement for beauty and placemaking into local and neighbourhood plans. This Proposition concludes with the proposal that the basis of using the call for sites process as the pool for selecting development allocations should be replaced by 'a more coherent sustainable approach based on an analysis of constraints and opportunities'. While I agree with the constraints and opportunities mapping approach as a starting point, as it is one of the two principal ways of initially identifying the areas of greatest development potential and least harm, I consider that there are benefits in retaining a modified version of the call for sites in the early stages of site selection. Landowners and developers should be encouraged to put sites forward in the knowledge of the constraints/opportunities mapping results. This will enable valuable evidence to come forward on the actual opportunities that are available and, most importantly, deliverable. These submissions should include information on the ability to mitigate constraints and present environmental enhancement opportunities.

Proposition 5 of the Report seeks the inclusion of policies and proposals in local plans that encourage clarity on the scale and design features of new development proposals and the production of area-based master plans, which would examine optimisation of opportunity and the ability to create beauty, thereby avoiding a piecemeal site by site approach.

In 2014 I advised one of our major housebuilder clients of the benefits of producing an area master plan encompassing the whole of the Principal Town of Ilkley including its rural surroundings. The client had at this point a small number of development options on sites being promoted as residential allocations through the local plan. There were many environmental and development opportunity arguments for such an approach. A previous growth study had been produced by Bradford Council's consultants, Broadway Malyan, which covered all settlements in the District hierarchy.[18] In starting this project we had knowledge of the main sites being promoted in and around the town via landowners and developers. However, we went back to basics in producing detailed GIS mapping of all primary and secondary constraints.

Our initial team included architects, landscape architects, transport, and flood risk consultants. At the outset we were aware of the potential for the Parish Council to produce a neighbourhood plan, but they had not yet initiated the formal process. We advised our client that we should prepare our Area Plan on an open basis by making it available in draft form to the Parish and District Councils, as well as the public. Following the subsequent declaration by the Parish Council that they were to prepare a Neighbourhood Plan, we engaged with them in a presentation of our work to date and participation in a public exhibition, which they organised. This was a long way from the level of public engagement required in these plan-making exercises.

Emerging from our area plan team working were strategic environmental proposals for two 'bookend' country parks to the east and west of the town. Primary environmental constraints to development included the Nidderdale AONB to the north, international habitat designations to the south and north of the town and designated river floodplain areas along the valley of the River Wharfe. These, together with secondary environmental and heritage constraints and the lack of brownfield development potential within the urban area, determined that residential and employment development would be required within the Green Belt designated areas to the east and west.

Detailed discussions with our landowner and developer clients enabled us to develop area master plan proposals demonstrating how a number of interlocking environmental improvements could be achieved together with new recreation areas for both residents and the large number of day visitors from the Leeds City Region. A key policy driver was a requirement to protect the internationally designated areas of the South and North Pennine SPA/SACs. The adopted policy in the Core Strategy (2017) creates offset zones, covering the whole of the urban and surrounding rural area of the town. Within these zones development can only proceed if habitat and recreation pressure mitigation can be achieved within or near to the development site. The detailed guidance setting out how the mitigation areas can be provided, and the subsequent management requirements, was published in January 2022 in the form of a Supplementary Planning Document. Our clients are prepared to designate very large areas for habitat enhancement and recreation provision, which are considerably in excess of the net residential development areas proposed. The area master plan demonstrates how this can generally be achieved together with linkages to greenspace and wildlife corridors along the river valley. Site-specific master plans have been produced to further demonstrate the advantages of the proposals and how they can be delivered.

Area and locality master plans for a geographical part of a town or city will provide more comprehensive opportunities to develop environmental and transport connections and improvements. Mitigation and renewal opportunities, to which new developments may be able to contribute, can also be identified in this way.

Getting landowners and developers to work together and agree to commission an area or project-based master plan can, in my experience, be a difficult and lengthy process. The four main landowners in the Springswood new town proposal were very cooperative and enabled the master plan to be brought together relatively quickly. By contrast, the four main landowners of the East Leeds urban extension did not enter into a formal collaboration agreement and, while they initially funded a project master plan, the lack of a formal agreement, particularly related to the programming and funding of the orbital road, which defined the 5,000 dwelling project, led to a fragmentation of the programme with the developer of the Northern Quadrant submitting a planning application and master plan in advance of the rest of the land to the south. This has necessitated repeat work on a master plan for the Central and Southern Quadrants. A more detailed and specific policy in the Leeds development plan, either including or requiring a single project master plan, would have saved considerable expense and delay. The City Council's ownership of two large tracts of the total development area did not help the early programming.

In the location and site-specific allocation policies of local plans, the responsible planning authorities need to include a requirement to produce a project or site-specific master plan for all major single use developments and in mixed use proposals. The threshold size of such a requirement should be determined by the local planning authority. The need for and benefits of site-specific master plans will, in many residential development cases, include sites of 10 or more dwellings, the current major development threshold as defined in the NPPF. For other uses the threshold site size of one hectare could be used. One of the key drivers for this approach is to ensure that proposals for development take full account of all the local context features and issues that include and surround the site. The allocation proposals in the local plan should include a series of design considerations/requirements appropriate to the locality of the site.

The matters to be considered in producing a site-specific master plan will, if the 'Living with Beauty' Report's proposals are adopted, include the contents of relevant local development plan policies and supplementary guidance documents including, for example, residential design guides. Neighbourhood plan design guidance is also relevant.

The constraints and opportunities presented by the locality of the site in terms of land use, community characteristics, transport routes, environmental conditions and qualities is the next group to be surveyed and appropriately mapped. The full mix of these considerations will differ according to the general location context in terms of urban, rural, or urban/rural fringe. Connectivity opportunities are important considerations in relation to the potential to create new footpath/cycleway links and the extent to which these will benefit the new development and any adjacent neighbourhoods. Access to the open countryside and any ability to extend the countryside into the urban area via the development are also important connectivity issues. Similar considerations apply to habitat improvements and ecological network connections, open space provision and the associated potential for green corridor links. The progression of the site-specific master plan should consider any opportunities to contribute to the renewal requirements of nearby neighbourhoods.

The additional groups following on from: 1) planning policies and guidance and 2) mapping of constraints and opportunities of survey and analysis, to be undertaken in producing most site-specific master plans, can be summarised as follows:

3) Baseline surveys of the site, including topography, habitat quality, ground conditions/land contamination/land stability, drainage and flood risk, and existing utility services (siting and capacities).
4) Access – highways access and network connections, public transport facilities, local services and employment centres including schools, nurseries, convenience stores, cafes, small business hubs, churches, and community centres.
5) Landscape – including character assessment; landscape and visual impact assessment of the proposed development (NB this together

with other assessments may need to be iterative, starting with a more general assessment to define the general siting of cells of development and subsequent qualitative refinements); on and near-site landscape features including trees and hedges, water bodies and open spaces.
6) Planning history of the site and locality – this can provide some important insights into problems encountered in previous attempts to realise development on the site.
7) An assessment of the built development character of the locality and any elements of heritage significance that requires the potential impact of the development on their settings to be taken into consideration. This assessment will identify considerations influencing design in terms of the vernacular characteristics of the wider area; scale, massing, and height of nearby development; materials used; any landmark buildings or spaces to be considered; townscape and the urban grain of the area.

Based on all or most of the above considerations the positive design progression of the proposals can commence with the aimed for end result being the realisation of a deliverable development that provides the necessary net gains to biodiversity, connectivity, greenspace, visual and built characteristics. To put this another way, the overall aim is to create beauty and a place that will enhance the health and wellbeing of existing and future residents. Many of these considerations must also apply to mixed use, commercial, leisure and retail developments where the challenges presented are often greater than those for new residential developments.

I am delighted that many of the proposals in 'Living with Beauty' have been adopted by government, firstly by the incorporation of a requirement for Beauty into the July 2021 version of the NPPF as part of a revised definition of sustainable development, and secondly by the production of a National Design Guide in January 2021. This delight will be shared across the professions involved in delivering quality developments. However, this will require a significant investment of design, planning and project coordination skills and resources by local planning authorities, developers, and their professional teams. These decisions by government will require local planning authorities to incorporate them as policies and guidance in their development plans and supplementary planning documents. Investment in this area will have lasting positive impacts on all three objectives of sustainable development. This much needed investment in staffing levels, skills and training is predominantly a re-resourcing of planning departments making up for losses over several decades.

Integrated professional teams will need to be established/re-established within individual planning authorities, or a single team shared between smaller authorities. Private sector developers and major landowners already employ such teams, some coming from multi-disciplinary consultancies and others assembled by the project coordinator on behalf of their client. Usually for the great majority of master plans these will include an architect or urban designer, a landscape architect, a town planner, an ecologist, and a transport/highway design specialist. All these professionals have a direct contribution to make to the design of a development. Other professionals will need to be added who have more of an indirect contribution to make to the design process. These include civil and drainage engineers, archaeologists, land surveyors etc.

The core design team must work on a well-coordinated and interactive basis from the outset to a project brief prepared by or on behalf of the client. All too often in the past, the appointment of one or other design professional has been delayed with adverse consequences for outcomes in terms of quality and cost. This has been particularly the case in relation to the appointment of a landscape architect towards the end of the master plan process, or beyond, who is then asked to retrofit a landscape design. Such an approach almost always results in a compromised scheme that misses genuine enhancement opportunities. The new aim to achieve beauty in developments means that integrated, well-led teamwork is essential and individual professional egos will need to be subordinated to a team ethic dedicated to this aim.

Having identified the team skills required, the client needs to appoint a project coordinator. Deciding which professional is most suited to this role is dependent on a combination of the nature and relative complexity of the project, the need for early and effective public engagement, and the skills and experience required, including technical, professional, design and interpersonal skills. The planning consultant can often provide the best combination of skills at this pre-application, or even pre-allocation stage (bearing in mind that site-specific master plans are often prepared to promote sites for allocation in the development plan). In urban contexts the architect may have the best combination of skills to take on this role.

Each local planning authority will need to decide whether they have the skilled staff and supporting resources to produce specific types of master plan as part of, or to support, their local development plan. The considerations they will need to consider include whether they can produce the end quality results to a specific programme. They will be faced with the alternative of appointing consultants to act on their behalf and early comparison of the most cost and product effective way of moving forward will be essential.

The revised NPPF now makes it clear that the creation of high-quality buildings and places is fundamental to the outcomes of the planning and development process. The National Design Guide identifies the ten components that make up good design. These now give full recognition to all the key factors which contribute to individual and community health and wellbeing. These design components include active travel provision and access to sustainable transport; provision of high-quality open spaces providing access to nature and recreation; safe and attractive spaces for social interaction, cohesion, and inclusivity; healthy, comfortable, and safe internal and external living environments; the provision of energy efficient and climate resilient developments and identity-creating attractive and distinctive places. The Guide emphasises the ways in which good design comes together to influence the quality of life (wellbeing). A well-designed living environment is now fully recognised as a key contributor to individual happiness and a feeling of contentment, which, when combined with ease of access to nature, recreation, and local services, contribute to positive physical and mental health outcomes.

The integration of good design and the achievement of beauty alongside other improvements and adequate resourcing, will all contribute to the revitalisation of the current planning system. The benefits to be derived for both national and local economies, for individual and community health and wellbeing, for the environment, biodiversity and our cultural heritage are huge and of long duration. While the social and environmental benefits are clear, the government needs to fully appreciate the economic benefits and avoid the contrary approach of planning deregulation.

Chapter 6

Development Management and the Scope of Planning Control

The 1947 Town and Country Planning Act introduced the control of the development of land, thereby removing the previous rights of landowners to pursue developments without permission. From this point onwards express planning permission was required for most types of development alongside the creation of permitted development rights in specific sectors such as agriculture and within residential curtilages. Rights of appeal were also introduced for applicants against the refusal or non-determination of their planning application. Given the general rights previously available and the choice of words, this part of the 1947 Act had a negative tone. The positive aspects of the 1947 Act system were seen as the production of development plans and the encouragement of comprehensive renewal following the impacts of the war. While the private sector was encouraged to invest, the great majority of the comprehensive development schemes for renewal were public sector led and funded. The control of development did have elements of a protective role in relation to heritage and the environment, which has subsequently been considerably expanded by new planning and environmental legislation.

The 1990 Town and Country Planning Act is the third and most extensive of three consolidations of planning legislation since the introduction of planning control in the 1947 Act. The two previous consolidations were the 1962 and 1971 Town and Country Planning Acts. The 1990 legislation repealed the few remaining sections of the 1947 and 1962 Acts as well as a mass of related legislation. Three other Acts were introduced in 1990 dealing with listed buildings and conservation areas, hazardous substances, and Planning (Consequential Provisions). These four Acts taken together are known as 'the planning Acts' with the Town and Country Planning Act being known as 'the principal Act'. The 1990 planning Acts comprise very detailed and at times complex controls over all types of development that are subject to the requirement for planning permission.

In 2004, the Planning and Compulsory Purchase Act was the next major planning legislation, which as well as introducing a reorganised development plan system made several amendments to development control in part 4 of the Act. Schedule 6 of the 2004 Act identifies the changes which this Act makes to the 1990 planning Acts. This summary of the main legislation, which currently comprises the operational planning acts, illustrates the growing complexity of the development control system. The complexity increases when numerous extant Statutory Instruments are added to the legal context in which the development control system must operate. These statutory instruments deal with a wide array of matters. Some of the main ones include: the Use Classes Order, the General Permitted Development Order, the Development Management Procedure Order, fees for applications and appeal and inquiry procedures.

Against this legislative backdrop of the emergence and growth of the development control sub system and its operation some key questions arise and are addressed in the following pages:

- What should be the extent of local and national government's control over the use of land?
- What should be the prime purposes of this part of the planning system?
- What have been the benefits of the development control system?
- Has the development control system and its implementation resulted in adverse impacts on the economy and have the various attempts to streamline the system been justified?
- Is the move away from development control to development management a total repurposing of this part of the planning system and an attempt to change the culture of this part of the service that local planning authorities provide?
- How significant is the control and/or management of development for the health and wellbeing of communities and individuals?

Development is defined in the 1990 Act as: 'the carrying out of building, engineering, mining or other operations in, on, over or under land, or the making of any material change in the use of buildings or other land.'

The control of development is set out in 51 sections of the 1990 Town and Country Planning Act, as subsequently amended by the 2004 Act. Local planning authorities consider and determine planning applications, participate in planning appeals, have powers to take enforcement action where development has been carried out without the requisite grant of permission or conditions of the permission have been breached. In addition, they have several miscellaneous powers or responsibilities conferred by this legislation, though it is not necessary to consider them in this context.

Several textbooks and legal guides have been written on the development control system. The many changes in the planning law governing the control of development have led to leading planning lawyers and their publishers producing revised editions of their legal texts. Cullingworth and Nadin's central text on the whole planning system, including development control, was first published in 1964 and there have been regular reviews. My own copy is the eleventh edition issued in 1994, extending to 343 pages. The extensively revised fifteenth edition was published in January 2015 with an expanded author group and covers 654 pages.[1] This demonstrates the greater coverage and complexity of the British planning system.

Throughout my career in local government and private practice I have read or referred to several legal textbooks on the planning law relating to the control of development. None has compared with *Urban Planning Law*, written by Malcolm Grant, for its clarity and coverage.[2] Morgan and Nott, in their text book entitled *Development Control – Law, Policy and Practice*, attempt a wider presentation of the subject to students of planning and related disciplines.[3] In the preface to the second edition, they set out their attempt to analyse the development control process and its context by including planning theory, philosophy, policy and practice, as well as planning law. This textbook recognises the background framework provided by planning law but gives emphasis to the wider dynamics and perspective necessary to better understand the operation of the development control process.

My aim is to provide, via my experience as a planning practitioner and policy analyst, an assessment of the extent to which development control, and its partial transition to development management, has operated, and will in future operate, for the public good and the health and wellbeing of individuals and communities.

The link between the adopted development plan for local planning authorities is established in section 70(2) of the 1990 Act. In arriving at planning decisions on applications and appeals, the local planning authority or planning inspector are to 'have regard to the provisions of the development plan, so far as material to the application and to any other material considerations'. This link was strengthened by the first composite statement of national planning policy in the original version of the National Planning Policy Framework (NPPF March 2012) and then further clarified and updated in the NPPF (February 2019). Section 4 of the updated NPPF (Decision-making) at paragraph 47 states that 'planning law requires that applications for planning permission be determined in accordance with the development plan, unless material considerations indicate otherwise'.

The consideration of the up-to-date development plan when examining a planning application sits within the general presumption in favour of sustainable development applied by paragraph 11 of the NPPF. This means that development proposals that accord with an up-to-date development plan should be approved, 'or where there are no relevant development plan policies, or the policies which are most important for determining the application are out of date, granting permission unless:

i. the application of policies in this Framework that protect areas or assets of particular importance provides a clear reason for refusing the development proposed; or
ii. any adverse impacts of doing so would significantly and demonstrably outweigh the benefits, when assessed against the policies in this Framework taken as a whole.'

These steps represent the current main framework for deciding planning applications and appeals. It is therefore necessary to: a) test the proposals and their context against the three interlocking sustainability objectives and whether gains are achieved in relation to all of these; b) assess whether there are up to date and adopted planning policies which are relevant to the proposals and whether the proposals are in accordance with these policies; and c) decide whether there are other material considerations which need to be assessed. Within this framework there are often many other considerations to be assessed in the 'planning-balance' before arriving at a decision.

The 2012 NPPF, at paragraph 186 under the heading 'Decision-taking', required local planning authorities to take a positive approach to foster the delivery of sustainable development. The 2019 strengthens the required approach to a 'positive and creative' one. It is important to emphasise at this point that all the main sections of the NPPF comprise statements of government policy. Prior to March 2012 there existed a mixture of Planning Policy Statements (PPS) and Circulars produced by the English government (the devolved nations have their own policy and guidance documents). The PPS and circulars were often a mixture of policy and guidance, and it was often difficult to discern what constituted a specific policy approach and which was simply guidance. Now guidance has been separated out into a compendium of regularly updated advice and procedure – the National Planning Practice Guidance (NPPG). The combination of the NPPF and NPPG documents make it somewhat easier for development control officers and planning consultants to discern the government's approach to different types of development, applicable national and local policies, and other material considerations. This is particularly helpful in the context of an increasingly complex array of issues, material considerations and policy areas to be considered in arriving at decisions. Over time the number of decisions delegated to officers has increased substantially, freeing up planning committees to spend more time on major and complex proposals. Some decisions are delegated to the chief development control officer in consultation with the chair of the planning committee.

* * *

What are the arguments for controlling land use and what are the benefits?

In the opening chapter I refer to the Scott Report on Land Utilisation in Rural Areas, and more significantly the 1944 White Paper – The Control of Land Use. The summary case for a comprehensive land use planning system (quoted at page 14) succinctly answers this question in a context of the impacts of the Second World War and the need for reconstruction. The importance of retaining as much agricultural land as possible is recognised. Significantly, the summation of the case concludes with the statement that the competing uses for land should be harmonised in order 'to ensure for the people of this country the greatest possible measure of individual wellbeing' alongside national prosperity. Harmonising competing land uses in a well-designed plan in a way which promotes individual and community wellbeing is a good way of summarising the vision for town and country planning as the modern system of planning legislation was introduced.

Prior to the 1944 White Paper, the Restriction of Ribbon Development Act 1935 was a recognition of the pressures arising from the scale of land use change in the inter-war period (1919–1939) leading to the need to reconcile the pressures for house and road construction with the preservation of rural amenity. Part of the 1947 system of planning legislation included the National Parks and Access to the Countryside Act 1949, which provided for protective designations of the countryside including the establishment of national parks, Areas of Outstanding Natural Beauty (AONBs) and nature reserves. In addition, this Act helped to secure enhanced public access to the countryside following pressure for such rights during the inter-war years, which at its peak led to the Kinder Scout Mass Trespass of 1932. When introducing the National Parks Bill to parliament, the minister, Lewis Silkin, referred to the countryside recreation and access, which the legislation was to provide, as being important to the health and wellbeing of the people. The next chapter considers the emergence and use of Green Belts as a development control mechanism to counter the spread of urban development via a ministerial circular 42/55. These pre- and post-war measures demonstrate national government's wish to control the pace, location, and nature of development for a variety of reasons. The establishment of Green Belts was initially seen almost exclusively as a measure to control the spread of urban development, while the other designations had a number of purposes including the primacy of their environmental qualities as beautiful landscapes or important habitats.

Current national planning policies for conserving and enhancing the natural environment are set out in section 15 of the NPPF.[4] When determining applications and appeals in national parks and AONBs, great weight is to be given to conserving and enhancing landscape and scenic beauty and additionally to wildlife and cultural heritage in national parks. The scale and extent of development within these designated areas should be limited, and applications for major development should be refused unless there are exceptional circumstances which justify granting permission.

National Park Authorities are the planning authorities for the designated areas of the English countryside, and they must prepare a local plan that includes policies for the control of development. In addition, they must produce Management Plans, which are required by the 1995 Environment Act. Working in cooperation with a number of partners, they will produce a forward-looking strategy covering the next 5 to 10 years, including proposals for environmental management and actions covering landscape and habitat protection and enhancements, plus recreation and public access alongside the socio-economic needs of the people living within the park boundaries.

Stringent land use and development control policies have been applied in national parks for many years. In the late 1980s/early 1990s I was involved in planning applications in the Yorkshire Dales, Lake District and Snowdonia National Parks. As planning consultant for a housebuilder, I applied for a new residential development near Grassington in the Wharfedale area of the Yorkshire Dales, which was to replace a disused former education establishment comprising several wooden lodge style buildings sited in a fold in the surrounding landscape. The scale of the first application proposal was significantly less than the footprint of the existing buildings but was refused and dismissed on appeal as being out of character with the landscape and settlement pattern of the locality. A further application was made for seven dwellings in a farmyard cluster design, approximately one quarter of the number of replacement dwellings originally proposed and paying great attention to the landscape character and visual impact. This was also refused, and the subsequent appeal dismissed, with the Inspector concluding that the landscape should be allowed to revert to its original undeveloped condition.

Around this period, the local Grassington Hospital was closed by the NHS and various plans were proposed for its redevelopment for a mixed residential and employment use, generating much needed homes and jobs. All the schemes considered were successive reductions in the amount and disposition of development, taking account of landscape quality and enhancements. The end result, following a public inquiry, was the dismissal of the smallest economically viable scheme on the same grounds that the landscape should revert to its pre-development state. The decision was upheld following a high court challenge. Eventually a development of three very separate detached houses was approved and every time we take the walk through the restored grounds, incorporating a nature reserve, I reflect on the discordant fit with the settlement pattern of this part of Wharfedale and the missed opportunity to provide jobs and a small mixed housing scheme including affordable housing.

In the 1990s, similar results followed planning applications to Harrogate District Council for the conversion of a small number of disused and relatively isolated barns within the large estate of Yorkshire Water, which is situated in the Nidderdale AONB. Conversions were often possible within and close to existing settlements subject to appropriate design. The decisions were that the buildings should be allowed to decay and revert to part of the natural landscape. My client and I recognised both sides of the argument but were conscious of the need to support the future needs of the agricultural communities. In later chapters I revisit these core issues facing all our areas covered by national environmental designations.

The use of land to extract minerals and other resources is one area where the need for the control of development, in advance of permission, during and post extraction is now readily apparent. This has not always been the case, and the great scars created on the landscape and the pollution impacts resulting from the extraction of coal and mineral ores in the eighteenth and nineteenth centuries, driven by the first Industrial Revolution, continued into the twentieth century. In Chapter 2 reference is made to the extractive and smelting industries impacts in the Lower Swansea Valley, coal extraction in the Welsh valleys, and to the considerable reclamation efforts which followed many years later to decontaminate and restore these landscapes.

The requirement for controlling mineral extraction and associated development was introduced in the 1947 Act. There are a number of important reasons for requiring planning permission:

1. Minerals can only be extracted where they occur naturally and where their extraction is economically viable and environmentally acceptable. As a result, location options can be limited. Where valuable mineral resources exist, and these are important to the national economy, usually results in the need to protect these from other forms of development via mineral safeguarding areas which are defined in development plans.
2. While the working of minerals is a temporary use of land, it can take place over a long period of time. Once extraction commences it is usually progressive and there is a requirement for regular monitoring and possibly enforcement if planning conditions are not being complied with.
3. The extraction process can have several adverse environmental effects, some of which can be mitigated. The great majority of mineral planning applications will need to include an Environmental Statement setting out the results of environmental impact assessment.
4. Following the cessation of workings the land should be restored to a level which is suitable for a beneficial after-use.[5]

While there are strong and often sustainable reasons to protect various mineral reserves to support parts of the national economy, there are clearly many potentially adverse impacts of extraction to be considered to protect the population of local communities and their surrounding environment. Minerals planning and development control is a specialist area of town and country planning concentrated in County Planning Departments, where two-tier local government exists, and in the designated unitary authority in metropolitan and other areas. The current national planning policies and the derived local policies provide for a sophisticated level of development control. Specialist teams exist, staffed by officers with the blend of skills necessary to assess applications, negotiate planning conditions and restoration schemes, monitor the progress of schemes and their impacts on a regular basis, and ensure that the end restoration scheme is successfully delivered.

The mining of coal by deep mining has been phased out as the result of government policy to cut carbon emissions and move to sustainable green energy sources. The last operational deep mine in the UK, Kellingley Colliery in Yorkshire, closed in December 2015. The same is largely true of opencast coal extraction, though a few private operators remain with significant operations in South Wales and Northumberland.

In March 2019, Cumbria County Council granted planning permission for a new deep mine, Woodhouse Colliery, promoted by West Cumbria Mining. The Secretary of State recovered this decision and granted planning permission in December 2022. This and the other new opencast projects are particularly controversial given the government's climate change and energy policies, with coal-fired power stations to be phased out by 2025. The prime argument in support of these remaining projects is that it is more sustainable to extract these sources of coal than continue to rely on imports from as far afield as Australia.

The extent to which land use should be controlled and public interests protected by the planning system came into sharp focus when considering the extraction of underground mineral resources in the form of coal bed methane gas, by the method commonly known as fracking. In 2001/2 I acted as the planning witness for Stratagas plc in a public inquiry into a pilot scheme for exploratory drilling of a four-well development for coal bed methane (CBM) extraction near Keele in Staffordshire. The County Council, as the mineral planning authority, decided to refuse the application against the clear and comprehensive advice of their officers. Stratagas appealed the refusal and the Secretary of State for Transport, Local Government, and the Regions (including the town and country planning function) then recovered determination of this application under his powers on the grounds that 'the proposals raise important or novel issues of development control'. The Principal Inspector appointed recommended that the Secretary of State should dismiss the appeal.

This inquiry, the evidence considered, and the conclusions reached are an important illustration of the depth of analysis and argument which can be involved in controlling the use of land via specific development, even where that development is of a temporary nature. The important issues under consideration included the relationships between national policy on the future energy mix required to power the country, greenhouse gas emissions and the need for and relative sustainability of the indigenous CBM in the potential mix (all matters under the energy policy remit of the Department of Trade and Industry), and national and local planning policy and material planning considerations. This was one of an initial group of appeals considering pilot well developments with the aspiration of proving the quantum and viability of extraction of CBM in coal mining areas. The novel issues included the potential impacts of the fracking process on the health and wellbeing of the local population and the related material planning issue of public fear of certain adverse impacts (Chapter 8 examines the issues surrounding such risks and the extent to which they and the public concern created are material planning considerations). The other relatively novel issue was the relationship between a planning application for a pilot exploratory development of four wells and the planning implications of its outcome for a much larger production field, in this case consisting of around 200 wells.

The Department of Trade and Industry were supportive of this planning application and had granted several wider license blocks to several private companies, enabling them to apply for the necessary planning permission. The local MP for Newcastle under Lyme, Paul Farrelly, with the support of his neighbouring MP, William Cash, secured a debate in the House of Commons on the specific application and its national and local consequences with responses from the Minister for Industry and Energy.[6]

While the planning process represents the major part of the approval process required to undertake exploratory drilling and fracking, it is important to consider its role alongside the other permits required. The Environment Agency must be satisfied that there is no unacceptable impact to the environment, particularly to a principal aquifer. An environmental permit is required for any borehole drilling and hydraulic fracturing. The Health and Safety Executive will scrutinise the safety of the well design and closely monitor progress on the well should it be approved.

The proposal was opposed by national and local politicians and there were more than 1,000 local objections. I was one of a team of 10 witnesses including specialist consultants and Stratagas representatives. My evidence included assessments against development plan policies, Green Belt impacts, risks to human health (jointly with a pollution specialist), impact on the local economy and need (jointly with a company witness).

The Inspector, in her lengthy report, concluded that 'in spite of the strong support for this proposal from the Department of Trade and Industry, the need for the development of the pilot project in this location is not sufficient to outweigh the environmental harm that would result from this proposal in terms of visual and landscape impact'.[7] In addition, she concluded that the venting of some methane gas would constitute harmful atmospheric emissions. The Secretary of State was in full agreement with the conclusions reached by the Inspector. The impact issues covered in the Inspector's report included noise, dust, air quality, the water environment, visual and landscape impacts, ecology, Green Belt, human health, the risks involved including seismic activity, subsidence and explosions, tourism, the local economy, traffic, need and other benefits.

National policy in setting the context for local policy making and decision taking on applications, recognises that minerals largely constitute a finite natural resource and that a sufficiency of supply must be maintained to enable the provision of infrastructure, buildings, energy, and goods that the country needs. National policy also requires high quality restoration and aftercare of mineral workings, with reclamation occurring at the earliest opportunity.[8]

My final example of land use and development control issues confronting the planning system relates to the degree to which agricultural land should be protected from urbanising development and other changes to this use of land. Seeking to avoid the loss of agricultural land because of its food production potential has been part of the planning system since the introduction of the 1947 Planning Act. The 1942 Scott Report (see Chapter 1 for the foundations of the 1947 planning system) examined the problems facing rural areas and focussed on land use. The Report identified two main problems, the first being the general economic weakness of agriculture and the second being the pace of urbanisation and the issues surrounding building on the best agricultural land. It is significant that, during the same year as the first comprehensive planning act, a new Agriculture Act was introduced, which aimed to achieve low cost and secure food supplies while maintaining price stability and farm incomes.[9] The 1947 Planning Act protected agricultural land via controls on urban development, in particular the general requirement for planning permission for most development. There was a general presumption in the 1947 Act and the related statutory instruments on planning that followed in 1948 (including the first versions of a Use Classes Order and a General Development Order) that agricultural land should remain in agricultural use unless there were strong reasons to permit development.

The effects of the Second World War, including the German naval attacks on Atlantic convoys, greatly affected food imports to the British Isles and emphasised the importance of national food production and the protection of agricultural land. The 1944 White Paper on The Control of Land Use, in its summary case for a comprehensive land use planning system, referred to 'the requirements of sound nutrition and of a healthy and well-balanced agriculture'. The food rationing system introduced during the war continued for several years, with meat being rationed until 1954.

A little-known Act of Parliament, with a key link to the planning system, was the Agricultural Land (Removal of Surface Soil) Act 1953. This brief Act made it an offence to remove, for sale, surface soil from agricultural land. The stripping of soil was defined as constituting development (as now defined in the 1990 Planning Act) as this could be either an engineering operation or a change of use of land from agriculture. This Act, which remarkably is still in place (as amended up to December 2020), makes it a criminal offence to remove more than 5 cubic yards of soil in any three-month period without the grant of express planning permission. Quite how this obscure Act has been enforced over the years is something of a mystery and I suspect that many planning enforcement officers and their legal advisors are not aware of its existence. This legislation seeking to protect productive soils does have a link to current national policy.

The Reading University Report[10] quoted earlier charts the decline in the amount of agricultural land lost to development and concludes that the 1947 Planning Act system had a positive impact: 'Planning has not only retained much more land in agricultural use than would probably have been the case without restrictions on development – it has also done much to preserve the visual qualities of the constructed rural environment.'

The Report does, however, conclude that planning failed to prevent and probably 'exacerbated social and environmental damage'. The social harm was focussed on rising house prices in rural areas, pricing many employed in low-income agricultural jobs out of the housing market in many rural settlements, particularly in southern England. This is a trend that has continued to the present day and spread to high-priced rural areas in the Midlands and the North. The lack of industrial development added to the problem and alternative jobs became scarce. Several attractive villages became commuter settlements, especially where they were located close to good transport communications into the large commercial centres of the main city regions.

The environmental damage occurred because those who promoted the case for a comprehensive town and country planning system argued this on the basis that the main threat to the rural environment came from urban development, and there can be no doubt this was the obvious main threat given the largely uncontrolled urban and highway development of the 1930s. Agriculture was predominantly exempted from planning controls, with permitted development rights covering many engineering and development proposals on farms provided they were for an agricultural purpose. As farming was incentivised to increase yields this led to hedges being removed to create larger fields for mechanisation, moorland being ploughed to expand production, marshland and other water bodies being drained, and buildings being constructed. Local planners had no powers to control these changes and several conservationists became concerned at the level of habitat loss and environmental degradation.

Marion Shoard, the author of *The Theft of the Countryside*,[11] followed her passionate defence of the characteristics of the countryside with an article in *The Planner*.[12] In the introduction to *The Theft of the Countryside* she argued, in forceful language, that:

'The English landscape is under sentence of death. Indeed, the sentence is already being carried out. The executioner is not the industrialist or the property speculator, whose activities have touched only the fringes of our countryside. Instead, it is the figure traditionally viewed as the custodian of the rural scene – the farmer.'

She goes on to catalogue the lost environmental elements of the English landscape as the result of agricultural improvements, including a quarter of our hedgerows, twenty-four million hedgerow trees, a third of our woods, and very many water bodies and flower rich meadows. Much of our agricultural land was being converted to prairie farming for cereal growing or to a grass monoculture to support more intensive livestock farming. In her article in *The Planner*, she argues for a more central role for planning to control environmental change through agriculture by using the means which have controlled urban development. Others sought to argue the need for some greater planning control over development work affecting what should be regarded as environmental assets. John Bowers and Paul Cheshire[13] recognised that the basic economic cause of the destruction of the countryside was what they referred to as 'a subsidy induced prosperity'. They argued that conservation policies, with a restructuring of farm support policies, would achieve a more efficient use of scarce resources and that 'the beauty and natural fecundity of the countryside is a real resource'. In Chapter 13 I return to this issue by examining how comprehensive environmental planning, future land management, and stewardship linked to progressive development and master plans can preserve and enhance natural capital (environmental assets that are given economic values).

An agricultural land classification (ALC) system was devised and introduced in the 1960s with its prime purpose being to protect what became known as the best and most versatile agricultural land from development via the town and country planning system. Technical Report 11, produced by MAFF in 1966,[14] set out the national ALC system for England and Wales, providing a five-grade classification for the subsequent mapping of agricultural land quality. The gradings are based on the extent to which the physical or chemical characteristics of the soil impose long-term limitations on the agricultural use of land for food production. Criteria for the sub-division of grade 3 were provided in 1976 and grades 1, 2 and 3a comprised what was defined as the 'best and most versatile land' for production and defined as worthy of protection in various planning circulars, guidance and policy documents produced in subsequent decades.

In 1987 there was a big shift of emphasis away from the need for the preservation of agricultural land for production purposes. Previously, local planning authorities had been required to consult MAFF on applications that did not accord with the development plan (i.e., the local development plan for an area should have considered agricultural land quality as one of the selection criteria prior to allocating sites for development) and involving the loss of more than 10 acres (4 hectares) of agricultural land in the three grades comprising the best and most versatile land. By the time of the General Development Procedure Order 1995, the threshold for consulting MAFF on applications for development that was not for agricultural purposes had increased to the loss of not less than 20 hectares of grades 1, 2 or 3a land or the loss of less than 20 hectares where this was likely to lead to a further loss amounting cumulatively to 20 hectares or more.

In the 1990s I was appointed by MAFF to act as their planning consultant at a public inquiry into a major employment park application just off the M62 motorway near Goole. I was advised that this was the first time that MAFF had appointed a planning QC and a planning witness to defend the loss of a large area of best and most versatile agricultural land. Most of the land was grade 2 in the ALC. Part of my role was to evaluate alternative sites in the wider locality that could accommodate this development without losing such a large quantity of the best and most versatile land. Prior to and since this inquiry, MAFF rarely took the role of lead objector at public inquiries into the loss of the best quality land.

The current ministry with responsibility for agriculture, food and fisheries is the Department of the Environment, Food and Rural Affairs (DEFRA), which was established in June 2001 when MAFF was merged with other environmental policy areas. Significantly, Natural England, which is part of this large government department, has been given the responsibility for the protection of soils and agricultural land quality and is now the relevant consultee for local planning authorities. This responsibility clearly has strong linkages with their other focal areas of responsibility, which include biodiversity and habitat protection and enhancement.

In 2018, Natural England produced a new guide for assessing development proposals on agricultural land, which retains the aim of protecting the best and most versatile agricultural land 'from significant, inappropriate or unsustainable development proposals'.[15] The NPPF (Feb 2019) now contains a very succinct policy statement requiring local planning policies and decisions to contribute to and enhance the natural and local environment by protecting and enhancing: i) soils and ii) the economic and other benefits of the best and most versatile land.[16]

The current NPPG for the Natural Environment explains why planning decisions should take account of the value of soils and best and most versatile agricultural land.[17] The figure of 20 hectares of best and most versatile land remains as the threshold for consultation on planning applications. The emphasis on protecting the value of soils is new and states that: 'Soil is an essential natural capital asset that provides important ecosystem services, for the production of food, timber and other crops, as a store for carbon and water, as a reservoir for biodiversity and a buffer against pollution.' This new emphasis demonstrates some joined-up thinking between the ministries responsible for planning and the environment, with greater emphasis to a sustainable approach taking account of several of the beneficial properties of soil.

The introduction of these sustainable qualities of soil into planning policy and guidance is a clear link to major legislation passing through Parliament because of the UK's exit from the European Union. A new Agriculture Act became law on 11 November 2020, followed by the Environment Act in November 2021. While the nature of this new relationship has yet to be established, great clarity will be required on the linkages and the responsibilities of the various bodies who will need to be involved. These statutes will have a close link with the planning system over the coming years, and together they represent a radically new approach to managing land and the environment. The Agriculture Act provides for a seven-year transition period in which farmers and land managers will move from gaining subsidies via the EU's CAP system to a new system of payments under the Environmental Land Management Scheme for their stewardship of the land and the public goods which it contains/supports, these being defined in the Environment Act, the main contents of which are outlined in Chapter 13.

* * *

What should be the extent of planning controls on built development and what constitute material considerations for planning decisions?

How far should planning policies and the control of built development go in terms of restricting individuals' or companies' rights to use land which they own or otherwise control. As we have seen, the 1947 Act removed individual landowner's rights to develop their land in favour of a system that generally required planning permission. The earlier part of this chapter concentrated on planning controls over the use of land and the more general controls on the extent to which development is restricted by the current planning system, in particular by specific land use designations.

We now move on to look at examples of how specific built development projects are restricted by the planning system in terms of scale, operation, occupation, and use. A national and political debate on the extent of such controls has ebbed and flowed over many years since the introduction of the

1947 planning system. Initially there was a great level of acceptance of the new system, conditioned by the knowledge that great progress was necessary in the post-war reconstruction and renewal effort. Increasingly, in succeeding decades, the controls on built development have extended to cover more and more aspects of our lives and the environments in which we live, work and play. Many of these have been demanded by the public in the public interest, with parliament also adding to the statutes governing planning and development, despite counter arguments from the governing party that planning controls should be cut back. Other changes have resulted from specific developments giving rise to adverse impacts within or near to local communities.

A key example that has emerged and grown in significance over the last thirty years is the need to introduce controls over the occupation of new houses. The levels of housing need and demand have increased over this period with rising house prices being a significant result of supply limitations. This has resulted in a drive to provide affordable housing for those who are priced out of both the rental and purchase markets. The restrictions on council house building and the right for households to buy their rented council property, introduced by the Thatcher Government, began the decline in social housing provision. National governments realised that the twin impacts of an overall decline in housing supply and social housing required policy responses through the planning system. The answer was to require private developers to provide a percentage of affordable housing units in developments above a specified percentage threshold of the total site provision. This resulted in a requirement for planning permissions to be conditioned and restricted by Section 106 legal agreements to ensure the affordable provision in the first instance and then controlled to ensure affordability in perpetuity. This level of control of occupancy has been supported as being in the public interest in order to retain balanced communities in terms of social mix, retention of employees in an area, and widening the opportunities to get into the market and avoid the adverse social consequences that would otherwise arise.

A continued rise in second home ownership in the UK over many years reached levels where certain local planning authorities in leading UK holiday/leisure destinations decided they had to act to protect the housing and employment needs of local communities and avoid any adverse economic impacts. A recent Resolution Foundation briefing report advises that the total number of adults owning a second home in Great Britain for the period 2014–2016 amounted to 1.4 million and the number owning buy to let properties was 1.9 million.[18] The latter category includes an unspecified number of dwellings owned as holiday lets (some being owned by landlords who live in the local holiday destination). The report compares the trends in the rise of multiple property ownership with the decline in the levels of ownership of a primary dwelling in the UK. Home ownership rates peaked in 2003 for all family units, but for those households headed by 25–34-year-olds the ownership peak was reached in 1989 when 50.3 per cent of young households owned their own home. The ownership rate for this group of young people had fallen to a low of 24.9 per cent in 2015 with a slight rise to 26.5 per cent by 2018. These dramatic national background figures form a significant part of the argument for affordable housing and further planning restrictions favouring local residents as part of new housing provision in the main tourism/leisure destinations comprising certain coastal resorts, national parks and other areas of beauty and recreation quality.

The Yorkshire Dales National Park is one of these areas that is experiencing the problems associated with second home purchases. We now live in the village of Addingham, which, although it lies within Bradford Metropolitan District, is, at its northern limit, only 1 km from the National Park boundary via the Dales Way long distance footpath. We are so fortunate to enjoy the beauty of the Park landscape as a near neighbour and frequent visitor, occasionally as guests in a welcoming Yorkshire pub. The landscape is made up of varied river valleys, high Pennine hills, and moorland and a rich variety of settlements. As with all national parks, the many qualities of the area cannot be retained, managed, and enjoyed without strong local communities comprising a good age range across households, a variety of businesses and employee skills, particularly in agriculture and strong support services.

The Park Planning Authority have learned from decades of experience, since the original designation in 1954, how they must plan and manage the delicate balance of social, economic, and environmental factors that are necessary to achieve sustainable communities, industries (including agriculture, tourism, services and manufacturing), landscapes and habitats.

The current National Park Plan was adopted in December 2016 and includes housing policies that directly address the specific needs of the local communities and the requirement to give these some degree of priority over second homes and holiday lets. This also requires a planning balance to be struck by recognising the benefits to the tourism industry of a range of accommodation, including houses that can be rented on a short term weekly turnover basis. Policy C1 of the Plan considers housing provision in the local service centres and service villages where developments of more than 11 dwellings must contain 50 per cent affordable housing or alternatively 33 per cent affordable and 33 per cent local occupancy restricted. In other settlements, developments of between 6 and 10 dwellings require the payment

Development Management and the Scope of Planning Control 155

of a commuted sum in lieu of the relevant proportion of affordable housing, and on sites of up to 5 dwellings new housing is restricted to local occupancy only.

The justification stated for this policy approach is introduced by reference to a significant shortfall of affordable housing provision, which at the time of the adoption of the Local Plan was twice the proposed annual housing provision in the plan. Reference is made to the National Circular produced by the Department of the Environment, Food and Rural Affairs (DEFRA),[19] which advises a focus on locally derived housing needs by releasing a supply of sites that will support the social and economic needs of communities, i.e., for people who live and work in the area. In the Yorkshire Dales Plan policy justification, there is the direct statement that this new housing provision cannot be lost to second homes and that it has the side effect of reducing prices by about 20 per cent.

While working at Dacre Son & Hartley (see Chapter 4) I was asked to be interviewed by Tom Heap, the rural affairs correspondent at the BBC and a regular contributor to the well-loved programme *Countryfile*. The subject was rural affordable housing provision, with the Yorkshire Dales National Park Authority area selected as an example of the problems being faced. Edited versions of the interview were to be broadcast on the national lunchtime and evening BBC news programmes. I researched some background data on affordable provision, house prices, market, and housing association interest. Average house prices in North Yorkshire County Council area, which included the Park, were then circa £175,000 compared with a lower figure of £150,000 in neighbouring West Yorkshire to the south of the Park. Within the Park the average house price was then £225,000, driven by scarcity, incoming retirees from outside the area and second home purchases. This was some 10 years prior to the adoption of the current Local Plan. The next 24 hours involved live interviews with ITV's regional evening news, the Radio 4 Today programme, local radio, and national press. The market figures were stark, and I had argued that the Park Authority would need to work closely with housing associations and local and regional housebuilders to develop a new policy approach. At this point I became convinced that restrictive occupancy conditions were an essential element of any policy.

The current NPPF and NPPG do not have specific policy and guidance regarding second homes and local occupancy conditions. However, the NPPF at section 5 – 'Delivering a sufficient supply of homes' – and paragraph 78, under the sub-heading 'Rural housing', states that 'planning policies and decisions should be responsive to local circumstances and support housing developments that reflect local needs'. The DEFRA guidance[20] is much more specific although it only has advisory status:

'The lack of affordable housing in many rural areas, including the Parks, has important implications for the sustainability of the Parks and their communities. This can have a detrimental effect on the local economy and undermine the social networks that are key components of sustainable rural communities. The Authorities have a key role as planning authorities but are neither housing authorities nor housing providers.

The desirability of the Parks as places to live is one outcome of their successful long- term conservation and promotion. Demand for housing in the Parks has consistently driven up the price of housing and development sites. Combined with relatively low wages in the local economy, a declining stock of council housing and only modest additions of new affordable units over the last 20 years, the consequence is that much of the stock is now beyond the reach of many local households. This can affect the social and economic diversity of rural communities and may, in some circumstances, undermine social support networks and the viability of rural businesses, which are key components of sustainable rural communities.'

The circular refers to the socio-economic duty of the Parks, via their role as planning authority, to foster and maintain thriving rural economies and states that this has been given added weight by the Taylor Report (2008)[21] and the Rural Advocate's Report (2008).[22]

The Yorkshire Dales Local Plan recognises that affordable housing can, at present, only be delivered via cross subsidy from open market housing. Given that the 'locals only' housing policies only tend to reduce open market prices by around 20 per cent, this tends to increase the need for affordable housing. The Local Plan acknowledges that developers have cited these policies as a significant constraint on house building with long delays in achieving sales. The Park Authority responded by widening the definition of local occupancy to include the whole of the district council areas lying outside the Park boundaries. In addition, to help attract new low impact businesses and families, the definition of local occupancy has been further widened to include the self-employed and households whose children attend school in the national park. A further policy supports housing for rural workers outside settlements where this can be justified as part of the functional needs of agriculture or other rural enterprises.

Holiday resorts in Cornwall and nearby villages have experienced the growing problem of second and holiday homes creating a social imbalance within communities. The extent of the problem is demonstrated by evidence produced for the preparation of the Cornwall Local Plan.[23] In five of the county's parishes, where second homes account for more than 35 per cent of all housing, the average house price is 87 per cent above the Cornwall average. This premium falls to 46 per cent where second home ownership is between 20–30 per cent and down to 23 per cent where the percentage of second homes is between 10–20 per cent. Local buyers on local incomes are therefore very often priced out of the market. The evidence base document points out that the Inspector who examined the draft Local Plan directed the Council to plan to meet both county-wide local household growth and an allowance for future growth in second home ownership resulting in a 7 per cent uplift to the objectively assessed housing need over the plan period. The lack of any specific policy in the NPPF has contributed to this surprising outcome.

Communities preparing some neighbourhood plans in Cornwall are considering the impact that large numbers of second and holiday homes are having on the sustainability of their communities. The County Council have committed to supporting neighbourhood planning communities who can provide 'robust evidence of exceptional circumstances and as a consequence prepare enforceable policies that help ensure permanent residents occupy homes and support local facilities'.[24] The Cornwall Local Plan: Strategic Policies Policy 6 on Housing Mix aims to provide the planning context to support any neighbourhood planning group who wish to pursue this issue. Large numbers of second homes in an area can result in community services such as schools being less viable. The large seasonal use of second homes and holiday lets can affect the viability of businesses such as post offices, shops, pubs, bus services and restaurants.

Residents of St Ives in Cornwall had been concerned for some time about the pricing and community services viability issues arising from a high proportion of second and holiday homes in their community. In preparing the St Ives Neighbourhood Plan they included a policy that required new dwellings to be subject to a restriction that they should be occupied as a person's 'principal residence'. Neighbourhood plans are required to go to a local referendum prior to proceeding towards adoption. In the St Ives referendum 83 per cent of those voting supported the plan and including this key policy. A local developer challenged this policy in the High Court, arguing that more market housing should have been provided to accommodate the full housing needs, that there were difficulties of enforcement, and the policy was potentially discriminatory contrary to section 149 of the Equality Act 2010. The judge dismissed the challenge stating that the purpose of the policy was not merely to meet the housing needs of local people but also to safeguard the sustainability of development by reducing the proportion of dwellings not occupied as a principal residence. The judge, and the evidence base from Cornwall County Council, looked beyond the market price implications to the wider community sustainability issues involved.[25]

The use of the example of dwelling occupancy restrictions demonstrates how, over time, problems in the development market (i.e., social, economic, and environmental problems) can become new issues that the planning system must address to deliver development which is sustainable in all respects. The roles of the public, in an era of neighbourhood planning, and that of the courts, are also well demonstrated, as is the ability of an issue of arguable national significance to move ahead of national planning policy and guidance. A further example of the extent of planning controls over built development includes the limitations that can be applied via conditions on planning permissions for industrial and commercial development, which seek to protect the general living conditions of nearby residents. These conditions can typically cover working, opening and delivery hours, noise levels and building design.

The siting and control of tall buildings has been a contentious and much debated planning issue for several years. As land values in our principal cities increase, particularly in premium locations with strong market appeal, and planning policies seek to achieve higher densities of development, especially around public transport nodes, there have been pressures to build upwards. Leading modern architects, as well as some local authorities and private developers, have also wanted to express their vision for the image and identity of certain cities on a world stage. These forces have come together to drive projects for individually designed tall buildings. A number of these projects have caused a clear division in public opinion between those who agree with and admire the project designs and the positive image they can create, and those who recognise that, individually or collectively, tall buildings can cause significant harm to several existing qualities within the city and to the lives of those who live within their direct and indirect influences. These qualities can include the views and settings of key listed buildings, open spaces, skylines, valued public realm as well as impacts on the local environment and microclimate.

London has many tall buildings, which are now generally grouped in the City financial zone and in Docklands, particularly Canary Wharf. This grouping is a result of early planning application decisions leading to the need for a clear policy approach, design, and locational guidance. The protection of views of St Paul's Cathedral, which was not so long ago the tallest building in the capital, have been a classic example of the conflict between the protection of the high-quality heritage, Wren's masterpiece, and the expression of modern architecture, design qualities and construction techniques. Given the extent to which certain views of St Paul's have been compromised and that the London skyline has changed in several views, there can surely be little argument against the most careful control of additional tall buildings in the capital. The policy in Paris has been stricter, with the establishment of La Defense, Europe's largest business district, located at the western end of the Axe Historique (historical axis) of Paris, some 10 kms from the Louvre in the centre of the city. This leaves the open steel form of the Eiffel Tower as the only tall structure within the historic zones of the city.

* * *

Material Considerations

The Town and Country Planning Act 1990 section 70(2) provides the general considerations to be taken into account by a local planning authority when determining planning applications. They are to 'have regard to the provisions of the development plan, so far as material to the application and to any other material considerations.' While certain amendments have been made to this section, including the need to take account of the content of a post-examination neighbourhood plan, none of them alter the basic position on 'other material considerations'. To be material any such matters must be planning considerations and it is for the decision-maker to decide which material considerations apply to a particular planning application and the weight which should be given to them in arriving at a 'planning balance'.

The current planning legislation offers no further guidance as to what considerations might be regarded as material, other than to include in further sub-sections of section 70 matters introduced through other legislation. These include the environmental information included in an Environmental Impact Assessment; development affecting habitats protected under European legislation (Council Directives on Habitats and Wild Birds); objectives in relation to waste; planning applications relating to listed buildings and development in conservation areas; and representations on the application received from consultees, owners of the land and the public.

Essentially, whether a further consideration is a material planning consideration is a matter for the court. Legal guidance on those considerations that can be included in the potential array of matters that are material to the planning decision is provided in the Encyclopedia of Planning Law and Practice[26] and in various specialist planning law textbooks and briefings. My somewhat dated copy of the Encyclopedia (September 2005) has around 20 pages of guidance on what constitutes a material planning consideration, based on many court decisions. Those matters which are material has continued to widen.

Section 38(6) of the Planning and Compulsory Purchase Act 2004 provides that where the development plan has to be considered in deciding a planning application (i.e., it has policy content which is relevant to the development being proposed), then the determination must be in accordance with the plan unless material considerations indicate otherwise. This presumption in favour of the development plan when deciding planning applications is retained in the NPPF and the policy contents of the NPPF are themselves a material consideration. The NPPF from 2012 onwards has introduced the new presumption in favour of sustainable development. The three overarching and interdependent objectives of sustainable development, which have to be pursued in mutually supportive ways, have the effect of extending those matters that can potentially form material planning considerations. This additional and extended presumption has not yet been enshrined in law. The 1990 and 2004 Acts referred to in this section remain the statutory basis, together with the decisions of the courts.

Writing about the economics and financial implications of planning and other law, FH Stephens provides some important summary points regarding the significance of the courts decisions as to what constitutes a material consideration. Many planning decisions have been quashed by the courts due to one or other material consideration not having been taken into account when arriving at a decision on a planning application. He also points out that there have been very few occasions where the courts have decided that a particular consideration is not relevant to planning. He concludes that what constitutes material considerations frames the limits to public intervention in the development control decision-making system.[27]

The main starting point adopted by the courts is the decision of Cooke J in Stringer v MHLG [1971]:[28]

'In principle it seems to me that any consideration which relates to the use and development of land is capable of being a planning consideration. Whether a particular consideration falling within that broad class is material in any given case will depend upon the circumstances.'

The broad nature of this approach has generally been considered as appropriate by planning lawyers and they have generally regarded any attempts to restrict this as bringing the courts too far into matters of planning policy rather than law.

Several more specific general principles have emerged through decisions of the courts over the years. A local planning authority or an inspector dealing with a subsequent appeal can consider purely private interests. They may refuse planning permission to protect such interests provided that there is a planning purpose or other special consideration involved. Personal circumstances are not often treated as matters carrying much weight in arriving at decisions. The deciding authorities can consider matters regulated by other statutory codes provided that a consideration is material in planning terms. When considering a planning application for a change of use the value, in planning terms, of the existing use of the land or building can be material. A good example is the change of use of a pub into residential accommodation, where the pub is regarded as an asset of community value.

Financial considerations can be material in a number of situations. For example, the profits from the development of school playing fields were held to be material where those profits were to be applied to the restoration of the main school buildings, which were listed. Enabling development can be a material consideration where this directly facilitates provision of other beneficial outcomes on property within the control of the applicant. This usually requires a section 106 planning agreement to tie the applicant into providing the beneficial uplift being promised. The precedent effect of a decision can be material when considering whether it should be supported. The planning history of a site can be material in deciding whether a new development proposal should proceed.

Safety is a material consideration when considering planning applications for development which may be hazardous, or where any type of residential development proposal is in proximity to existing defined major hazards.

The complexity and scale of development proposals clearly affects the number of other material considerations which may need to be weighed in the balance having already taken account of relevant policies in the development plan and the NPPF. If we take the example of new residential development, there are many material considerations that are now accepted as relevant to the decision, some of which will be covered in policies expressed in the local and neighbourhood plans. The following list should be included when dealing with residential planning applications:

- Road access, traffic generation and highway safety issues.
- The layout and density of the development and its fit with the urban grain and character of the neighbourhood.
- Design, appearance, materials.

- Disabled persons access and connectivity to essential services in terms of pedestrian, cyclist, and public transport provision.
- Nature conservation and biodiversity (now including the national policy requirement for biodiversity net gain from development).
- The character of the development site in terms of land use, trees, the presence of any contamination, drainage, and flood risk.
- The existing landscape character and visual amenity, the landscape proposals forming part of the development and the resulting impacts of the development.
- The capacities of existing community and social facilities including schools, GP and healthcare provision, access to public open space and greenspace in general.
- The capacity and siting of utility services.
- The proximity of incompatible uses, nuisance potential from sources of noise, smells or fumes, including air pollution.
- Overlooking and loss of privacy; loss of light and overshadowing.
- Previous planning history of the site and immediate locality.
- Recently added to the list is the requirement for development to be beautiful.

While this list is relatively comprehensive, it should not be regarded as exhaustive.

Issues that are frequently raised by objectors to development proposals and have been dismissed by the courts as not being material, include an individual's right to a view, the effect of a development on the value of a property, restrictive covenants, boundary, ownership, and other neighbour disputes. More weight will generally be attached to issues raised by objectors where there is planning evidence supporting them rather than mere assertion.

The extent to which personal considerations can be material in determining a planning application were considered in a landmark case – Westminster City Council v Great Portland Estates in 1984/5.[29] In his judgement, Lord Scarman determined that in certain specific cases:

'... the personal circumstances of an occupier, personal hardship, the difficulties of businesses which are of value to the characteristics of a community are not to be ignored in the administration of planning control. It would be inhuman pedantry to exclude from our control of the environment the human factor. The human factor is always present, of course, indirectly as the background to the consideration of the character of land use. It can, however, and sometimes should, be given direct effect as an exceptional or special circumstance. But such circumstances, when they arise, fall to be considered not as a general rule to be met in special cases. If a planning authority is to give effect to them, a special case has to be made and the planning authority must give reasons for accepting it.'

In Chapter 8, Planning and Risk, I consider the nature of public concern (anxiety and fear) as a material planning consideration and how this should be treated in arriving at a balanced planning decision.

The up-to-date position on what are material considerations and how they should be applied is usefully summarised in a 2019 decision of the Supreme Court.[30] The case concerned an application for planning permission for making a material change of use of land from agriculture to the erection of what was described as a community scale 500 kW wind turbine. The Council, as planning authority, had been persuaded by its planning officers and the applicant to take into account as a material consideration the applicant's offer to commit to make an annual payment of 4 per cent of the income generated by the turbine to a local community fund. This was to be secured via a planning condition attached to the planning permission. Mr Wright, a local resident, challenged the decision and succeeded in the High Court and the Court of Appeal.

In the Supreme Court, the judgement of Lord Sales was unanimously agreed by his fellow judges. He clearly reaffirmed that a material consideration must serve a planning purpose and that this can only be the case where the consideration relates to the character of the use of the land. It was self-evident, when applying this test, that planning permissions cannot be bought and sold by making payments or the provision of other benefits that do not have a sufficient connection with the proposed use of the land. In addition, 'policy documents cannot make "material" that which would otherwise be immaterial'. Lord Sales held that what is "material" is, in the end, a question of law and if Parliament wished to expand the range of relevant factors in section 70 of the Town and Country Planning Act 1990, it could do so.[31] In arriving at this decision the Supreme Court resisted the invitation made by counsel for the Secretary of State to 'update' the leading case law (Newbury 1981) 'to a modern and expanded understanding of planning purposes'.

In this key decision, the Supreme Court accepted that the criteria established by case law for identifying valid material considerations are the same as those for the imposition of planning conditions when approving a development. They must be for a planning purpose and must fairly and reasonably relate to the development permitted.

* * *

Speed versus quality in decision-making/Development Management or Development Control

As we have already seen there have been regular calls for the system of development control to be speeded up and leading members of the government, including prime ministers and the secretaries of state responsible for planning and business, have called for changes to cut the time taken to determine applications and reduce the alleged adverse economic impact of slow decision making. Margaret Thatcher and David Cameron were amongst these critics during their terms of office.

George Dobry QC led an inquiry commissioned by Geoffrey Rippon, the then Secretary of State for the Environment, and his final report was published in 1975.[32] As a result of a property boom in the early 1970s the number of applications made increased by 35 per cent to 600,000 in 1973 and the backlog of appeals at the Planning Inspectorate rose to 17,000 by the end of that year, from an annual average of 7,000 between 1963 and 1971. The situation arising from this increase in workload was compounded by the effects of local government reorganisation in 1974. In producing his report, he recognised the competing influences of calls for more effective public participation and better protection of the environment. One of his main recommendations was to split applications into minor and major categories. This was a distinction we were to rapidly take up at Leeds by instituting a householder application system with simpler procedures and a much-reduced timescale for decision. The minor/major application distinction remains in the current system.

In 1980 the Thatcher Government produced the Circular Development Control- Policy and Practice aimed at streamlining the system.[33] While the contents were toned down from earlier ministerial statements, some of the wording was still rather aggressive compared with previous and subsequent statements. The introduction to paragraph 5 for example states that 'Promptness, relevance and efficiency are characteristics of good planning. The benefits to the economy and to the individual from the business-like handling of planning applications are very substantial.' This paragraph goes on to state that local planning authorities have a clear responsibility to minimise delay in determining planning applications.

In December 2001, the then Department for Transport, Local Government and the Regions published a green paper entitled 'Planning – Delivering a Fundamental Change'. This document, produced under the Labour Government of Tony Blair, criticised the current town planning system for its complexity, its lack of speed and predictability of outcome, its failure to engage with communities, its lack of customer focus and weak enforcement.

High level proposals for reform were put forward, but the document did not use available evidence and did not offer a detailed prescription for change. The Parliamentary Select Committee responsible for scrutinising the affairs of the Department believed that the Green Paper had exaggerated the weaknesses of the system and played down the strengths of existing practices. 'It had largely overlooked the central value of the planning process as a brokering mechanism between competing interests in deciding how land and buildings should be used.'[34] In their report, the Committee refer to 'the immense positive contribution which planning can make to public life'.

In September 2011 the RTPI produced a short paper responding to the 'Top Five Planning Myths'.[35] The first of the myths addressed is that the default response to a planning application is 'No'. The Plan for Growth (March 2011), produced by HM Treasury, had given rise to this myth. The Government's own statistics showed that, for at least a decade, 80 per cent of planning applications had been granted, with a higher figure of around 90 per cent for those major commercial applications that were critical for economic growth. The second myth was that planning is slow. In the planning debate around this issue reference had been made to 'a slow and prescriptive planning regime'. The then Planning Minister, Greg Clark, had committed to 'speeding up applications that get stuck in the system'. The RTPI paper responds that councils are meeting or exceeding the 8- and 13-week targets set for them by the government despite a reduction in resources. Only 0.7 per cent of applications were taking longer than 12 months to reach a decision. What appear as delays in the planning system are not always within the control of the planning authority. I have come across several cases where interventions from the public have sought to delay a decision with a view to securing an eventual refusal of a particular development. Mechanisms used have included seeking village green status for some of the land involved, high court actions and unofficial referenda (i.e., a poll within a settlement outside the formal procedures associated with the preparation of a neighbourhood plan. The development industry often conflates planning with other consent regimes such as licensing or environmental permits.

As stated in Chapter 5, the Building Better Building Beautiful Commission recognise the importance of arriving at quality decisions and outcomes and that these considerations should be balanced against speed. They also recognise that there is a major resourcing issue in terms of staffing levels and the range of skills required in fairly determining applications.

The term 'Development Management', insofar as it relates to the determination of planning applications and related decisions, appears to have been introduced somewhere between 2004 and 2008, although there is no clarity on this. Wikipedia provides an interesting start by stating that its introduction 'was an attempt to change the culture of planning from the previous process of "development control"'.[36] This entry provides further definition by referring to development management as 'the process of proactively managing development in a local area to achieve the local planning vision and objectives. It has an emphasis on the pre-application stage and delivering sustainable development and includes the principles of "place-shaping" as recommended by the Lyons Inquiry into local government.' Sir Michael Lyons' Report[37] identifies place-shaping in Chapter 2 as capturing the central role and purpose of local government and defines it as follows:

> '... the creative use of powers and influence to promote the general wellbeing of a community and its citizens.'

His report also advises the reduction of central role and prescription, which he concluded would enable local government to better respond to local needs and to better manage expectations regarding the delivery of local services. In order to step up to the challenge of place-shaping, he considered it necessary for local government to develop its style, skills, and behaviours. He further defines place-shaping as covering a wide range of local activity and embracing anything which affects the wellbeing of the local community.

Having carried out further web research, I have been unable to verify the direct link to the Lyons Report. The Planning Advisory Service (PAS) appear as a further major promoter of Development Management as a replacement for Development Control and they are strongly promoting a cultural change in the planning system and the profession via their activities. PAS is a government grant supported team within the Local Government Association (LGA) who provide consultancy and peer support to local planning authorities, learning events and online resources to help them understand and respond to planning reform. The LGA website identifies 'effective development management and good decision-making' as one of the main aims. PAS produce occasional and unofficial blogs on topics including Development Management. One of these addresses what the term actually means, its role within a renewed planning system and the ways in which it affects the traditional divisions in planning departments between the policy and development control teams.[38] This blog suggests that 'the new approach has been driven by the implementation of the 2004 Act', though it is acknowledged that there is no direct reference to Development Management in the Act or in associated guidance.

The blog refers to Development Management as 'the culture change from reactive assessment of others' proposals to seeking and shaping developments'.

The use of the word 'shaping' may be more of a passing reference to Sir Michael's Report, as the blog goes on to state that under the new approach developments will need to be assessed in relation to the way in which they contribute to the outcomes needed by the community. While the 2004 Act does not refer to these community outcomes, the government did give some clear guidance in the first of a set of new Planning Policy Statements (PPS1 Delivering Sustainable Development, 2004), which referred to the requirement to consider the needs and problems of the communities covered by local planning authorities, how these interact, and then relate them to the use and development of land. Current NPPF policy on decision-making makes no reference to Development Management, although it requires local planning authorities to approach decisions in a positive and creative way.

The Planning Advisory Service are the ones taking the lead, with the encouragement of the LGA, in promoting Development Management by providing seminars, guidance documents and individual reviews of local authority procedures. Several authorities have now produced separate development plan documents setting out their development management policies, which are framed in a much more positive way than the old development control policies. A recent report produced by consultants with PAS, on behalf of the LGA and MHLG, entitled Good Development Management[39] 'investigated how local planning authorities across England are designing and implementing projects to improve the efficiency, quality and delivery of Development Management'. Leadership issues and the availability of a talented pool of officers were key matters identified in the 13 projects assessed and the 30 examples of good practice identified in a survey.

Key stages of the revised process identified in the cumulative work carried out by PAS are the pre-application consultation including design advice; facilitating validation, making sure that all the required documents are in place and adequately describe and assess the development proposed; full and clear engagement with internal and statutory consultees and a clear decision-making and authorisation process.

My own interim conclusions are that the Development Management approach is well suited, and indeed needed, for complex and major planning application proposals (with 'major' being based on the current size thresholds applied in the Glossary of the NPPF). However, the question arises as to how the categories and scale of applications, which fall below the major thresholds, are to be handled together with the other sub-functions of development control, including, for example, enforcement, advertisements, listed building and conservation area proposals. The PAS blog author suggests that this will become a diminishing role and questions whether it might be better integrated into other regulatory services. This approach would, I believe, unnecessarily diminish the roles of those officers assigned to the 'smaller' cases. In my experience, a single new dwelling proposal can raise a range of material planning considerations requiring skilled assessment, including the incorporation of design advice. The management of the overall system of decision making, liaison with policy teams and consultees needs review and there is no one approach fits all answer. The need for quality outcomes is not restricted to major and complex proposals. The recommendations in the current Government's White Paper and the Building Better Building Beautiful Commission's Report need to be considered in a national review of the development control/development management process. There is scope for some streamlining, simplification, improved guidance, and innovation within the overall process. These efficiency improvements must be considered alongside the greater time and resource which will be needed to achieve sustainable development, enhanced community engagement and quality and beauty in development outcomes. Job satisfaction, staff motivation, good training and reskilling will be essential to the design of a renewed system. The involvement of national government in the process redesign should be limited, with the emphasis on working with recent practical research by PAS and others, recognising existing best practice, customer satisfaction reviews, reports from the RTPI and the TCPA's Raynsford Review.

* * *

Development Control/Development Management and their impacts on Health and Wellbeing

In this brief review of the scope, content, and operation of the development control function, aspects of the health and wellbeing of individuals and the communities in which they live frequently appear as a material consideration to be taken into account when making decisions on development proposals. It is noteworthy that there are two strong 'bookends' that support the centrality of this key aim in plan-making and decision-taking. The records in between present a patchier emphasis and performance. Throughout the full period of operation of the planning system under review, the underlying theme of planning acting in the public good is present to some degree.

The left 'bookend' is the 1944 White Paper 'The Control of Land Use', which had as its core planning aim the harmonisation of competing land use needs and demands in a way that achieved the greatest possible measure of individual wellbeing and prosperity.

The right 'bookend' is the current version of the NPPF (July 2021), which places the health and wellbeing of communities at the heart of the definition of sustainable development, which all planning policies and decisions are expected to achieve.

For many years the central 1947 Planning Act produced control over development to the advantage of the protection of agricultural land and the countryside. The associated National Parks and Access to the Countryside Act 1949 added further positive leisure and recreation benefits for health and wellbeing via the designation of national parks, AONBs and nature reserves alongside improved rights of access to open countryside.

Within the fields of land reclamation and mineral development, considerable progress was made on retrospective restoration schemes in both urban and rural areas. Over time, the control of mineral developments and their relationships with communities, landscapes and habitats has become more sophisticated by careful site selection and planning, and the incorporation of strong restoration and after-use management conditions in planning permissions. Previous adverse impacts on communities and individual health have been greatly reduced. The protection of the public is a key material consideration in minerals planning policy and decisions on applications including human health. This must be carefully balanced against the need for the mineral within the context of national planning and energy policies, its scarcity and occurrence and the policies of mineral planning authorities.

Public fear and anxiety of potentially adverse impacts from several types of development have increasingly been accepted as material considerations in decision-making. The courts have determined that, in considering such material issues, they should be clearly supported by evidence. Armed with this evidence, it is then for the planning authorities to weigh these considerations in the planning balance. Such fear and anxiety supported by evidence can have an adverse impact on the physical and mental wellbeing of individuals. In Chapter 8, specific examples of risks associated with land use and development are examined and their potential impacts on health and wellbeing, particularly in the form of fear and anxiety.

The example of protecting land in agricultural production, specifically the areas of best and most versatile land for food production had a strong basis in the 1947 Planning Act and the subsequent 1953 Agriculture Act, whose main purpose was the prevention of the removal of surface soil. These controls also had a strong foundation in the clear and purposeful 1944 White Paper. The protection of good agricultural land and soils was a major contributor towards the health and wellbeing of the population as the UK strove to escape from the limitations of years of food rationing. While the 1947 planning system had a positive impact for several years, the gains in food production eventually came at a price in the form of environmental degradation, caused by the intensification of farming methods and the amount of land used for agriculture, leading to the loss of hedgerows, trees, water bodies and habitats. Many advocated the introduction of planning controls to regulate these environmental changes, but this did not occur.

The extent of planning control over the protection of the best and most versatile agricultural land was relaxed in the 1990s, with the threshold for consultation being raised to a potential loss of at least 20 hectares. The latest government planning policy in the NPPF, and allied policies in the new Agriculture Act 2020 and the Environment Act 2021, in a post Brexit context, contain a new emphasis on home food production. Alongside this is a new system of farm support, where farmers and other landowners will eventually be paid for their stewardship of the land in Environmental Land Management Schemes (ELMS). This creates increased competition for the use of the UK's rural land resources and will require a greater involvement of planning, at both policy and land use management/control levels to achieve the best balance of outcomes. These issues are considered in the last three chapters of this book and have been well illustrated in October 2022 by the Truss Government's plans to limit the scope for solar farms on best and most versatile agricultural land by widening the grades of farmland that fall into this category.

National planning policy in England and in similar documents in the devolved nations seeks the approval of applications for sustainable development at every level of the decision-making process (planning officers, planning committees, government inspectors and the Secretary of State). All planning decisions are to be approached in a positive and creative way. The adoption of a 'development management approach' by several local planning authorities signals that an effort is being made to take the required positive, if not always creative, approach. The lack of a comprehensive description of this concept, the extent of its coverage and method of operation, particularly in relation to the implication that it will in some way replace 'development control', are currently obstacles to its universal adoption. The absence of any clear policy direction from the government beyond generalised high- level statements provides no assistance in this situation.

The operation of a positive and creative approach is necessary, especially if all developments are to achieve truly integrated sustainability as defined in section 2 of the NPPF. In my introduction I state that the bar has been set very high, and rightly so. The problem is that the proportion of developments achieving this essential outcome is currently too small. The potential to achieve integrated sustainable development across a wide range of application projects

Development Management and the Scope of Planning Control 171

exists but requires resourcing and reskilling planning departments and clearer policies at the national and local level. The few exemplar projects that have achieved truly sustainable development demonstrate the enhanced health and wellbeing benefits that can be achieved. These should be integral to project evolution, public consultation, and negotiation across a full range of projects, from the small scale to larger and more complex projects.

In this chapter I have tried to demonstrate that the development control system has already protected residential communities from the adverse impacts that new development can bring to the everyday lives of people and communities, which, if they were not subject to adequate control, would affect their health and wellbeing. This has been a gradual evolution where public health, public anxiety and related issues have entered the growing list of material considerations to be considered in reaching planning decisions. As discussed in Chapter 5, and to be considered in subsequent chapters, much more needs to be done to enhance health and wellbeing benefits through development. The greatest potential lies in future housing and mixed-use developments which can benefit both the existing and the new residents.

The provision of sufficient new housing to meet both the overall and specific sectoral needs of all households is a major aim of government planning policy. At last, there is a welcome recognition that quality of design, layout and the living environment are as important as the quantum of provision for a wide range of reasons, a number of which relate directly to improvements to health and wellbeing. The basic satisfaction of a housing need can be a huge boost to the health and wellbeing of households. However, if that initial boost effect dwindles because there are deficiencies in the design of individual dwellings, in the layout of the development, there is a lack of open space and other amenities, and poor connections to services, then the health and wellbeing benefits can be lost.

The provision of affordable housing remains a key and pressing requirement across all local planning authorities and development controls are needed to ensure the provision and retention of this part of the overall housing stock. In certain parts of the country, and particularly those settlements in rural and coastal areas designated, or otherwise recognised, for their beauty, there is a need for additional occupancy controls giving more local households the comfort and security of their own home.

Development control already provides protection to the health and wellbeing of communities in various ways as shown in the examples considered. There is a great deal more that needs to be achieved, whether this is via an expanded and repurposed development control system or a new comprehensively devised and implemented development management process.

Chapter 7

Green Belts – The Need For a New Approach

The topic of Green Belts within the UK probably ranks as one of the most publicly recognisable and appreciated within the town and country planning system. The policies that govern their designation, review and alteration and their role in controlling development are subjects that give rise to strong debates and passionate stances by the public and the politicians who serve them nationally and locally. This public/political stance is invariably to oppose applications for development within the Green Belt, and any changes proposed to the adopted boundaries when they come up for review in the production of the local development plan.

Many of my colleague members of the RTPI regard the designation of Green Belts, and their generally successful control of urban sprawl, as one of the greatest achievements of the planning system that evolved from the 1947 Act. Many countries have adopted Green Belts, including those in the Commonwealth, who have based their planning system on the UK's 1947 model, and others such as the USA, South Korea, France, Germany, and Brazil.

The development constraining effect of Green Belts, over many decades in the case of those in the UK, requires reappraisal because of the clearly adverse impacts that have been apparent for at least the last thirty years, particularly in relation to the housing market in terms of supply and affordability. The cases for and against change have been strongly made by policy institutes, political and economic think tanks, and academics. Practicing planners have been generally too busy and too constrained to actively engage in this debate. After many years, practical involvement with Green Belt policies, boundary reviews and a wide variety of development projects, I have come to the firm conclusion that there is now an overwhelming case for changing the current planning approach. My experiences, often very frustrating ones, have not led me to argue for the wholesale abandonment of Green Belts or for very large deletions from their current extensive land cover. My proposals, which align with those of other advocates for change, are for a more sophisticated planning approach to the land use and management of all areas of our Green Belts.

This new approach will need to take full account of the ways in which truly sustainable development will be achieved, how the newly emerging environmental policies can be fulfilled, in particular those linked to climate change adaptation and mitigation. As we have seen, the health and wellbeing of our people and communities have, in policy and decision-making terms, been placed at the heart of sustainable development. At the same time, environmental policies, programmes, and projects have become central to the achievement of all the improvements required for our physical and mental health. Given the extent and the proximate urban location of Green Belt land, it is now essential that we make the very best use of this land. All the key overarching government policies are in place in the NPPF to support this new approach with the sole exception of Green Belt policy. As well as the presumption in favour of sustainable development, policies exist to enhance the environment, habitats, public recreation, well-designed places, walking and cycling, and meeting the challenges of climate change.

In addition to the above listed policies, national planning policy supports 'Making effective use of the land'.[1] This short but highly significant component of planning policy is to encourage the multiple benefits to be gained from both urban and rural land including via mixed-use schemes and realising opportunities to achieve net environmental gains. The examples given for these environmental gains include new habitat creation and the improvement of public access to the countryside. The English government state that it is necessary to 'recognise that some undeveloped land can perform many functions, such as for wildlife, recreation, flood risk mitigation, cooling/shading, carbon storage or food production'.

There is clearly a substantial disconnect between these enhancement policies and the basis they provide for decision-taking, which form the overarching strategic planning approach of the government, and the policy and decision-taking requirements contained and advocated in section 13 of the NPPF 'Protecting Green Belt land'. This disconnect forms a key part of the argument for a change in Green Belt policy and practice, together with fully readdressing the adverse housing supply and affordability impacts, the condition of the urban-rural fringe and unsustainable travel impacts.

The introduction of well-designed enabling development, particularly but not exclusively residential, into discrete parts of this multiple land use mix are part of the way forward. This approach is not the 'evil' envisaged by large sections of the public and, in my experience, a large proportion of national and local politicians. Such an approach can have many benefits, which need to be addressed and logically and practically weighed in the ongoing debate.

Prior to examining these arguments and providing some examples of the way in which these multiple use benefits of Green Belt land can and should be achieved, it is helpful to briefly review the history and purpose of

Green Belts, the extent of their coverage and some key public and political misunderstandings associated with their functions.

The establishment of certain forms of rural exclusion zones around towns and cities has an ancient base. In the Bible reference is made to God asking Moses to command the Israelites to give towns to the Levites to live in from their land inheritance and to include pasture lands around the towns.[2] The prophet Mohammed, in the seventh century, decreed that a wide tree-planted belt be established around the city of Medina. In 1580, Queen Elizabeth I banned all new building in a three-mile-wide belt around the city of London. This was primarily established as a cordon-sanitaire to prevent the spread of plague and other contagious diseases. In modern times the broad concept of protected zones around large urban areas re-emerged in continental Europe, the most notable example being the Ringstrasse in Vienna. The London Society and the Council for the Protection of Rural England put forward a proposal for a London Green Belt, up to two miles wide, in the Development Plan of Greater London 1919.

It is widely acknowledged that Ebenezer Howard is the founder of the comprehensive socio-economic and planned Green Belt, which formed a key part of his Central Social City concept. The central city was to be limited in scale by protected surrounding countryside used for agriculture, forestry, and recreation, with smaller planned satellite towns beyond the green girdle. This concept gave birth to the Garden City Movement and the case for establishing new towns to house overspill from large conurbations. One of the most notable proponents of integrated town and country planning was Patrick Geddes, a biologist. He placed great emphasis on the environment in combination with the social needs of city-dwellers. A further proponent of integrated town planning and the establishment of Green Belts was the architect of Hampstead Garden Suburb, Raymond Unwin. He became advisor to the Greater London Regional Planning Committee from 1929 and regarded the Green Belt's chief role as being the provision of recreation land on the edge of large urban areas as compensation for the lack of open space. He went on to propose that land should be purchased by local authorities to form a 'green girdle' around London, preferably in the form of a continuous tract of land three or four kilometres wide.[3/4]

In 1944, Patrick Abercrombie produced his Greater London Development Plan, which had been commissioned by the wartime government and was published the following year. He was strongly influenced by the work and writings of Howard and Geddes and by the Barlow Report. The Plan was derived on the basis that the population of London and its surrounding ring of settlements could be contained at its then current level by looking at the problems on a regional and strategic basis. His approach therefore covered a radius approximately 30 miles from the city core. Within inner London the density of the population and the prevalence of slum development required reduced densities to be introduced to accommodate open space and improved infrastructure. The resulting overspill population amounted to some 600,000. Abercrombie calculated that a further 400,000 people should be added to this total to deal with urban problems in the outer areas. The Green Belt incorporated into the Plan was to be some five miles (8 kms) wide on average. The selected width of this belt was primarily on the basis that the existing towns and the proposed new towns beyond the Green Belt would accommodate the overspill accommodation and the jobs they required. This followed the principles of Howard's earlier model of balanced and contained communities linked by a transport system.[5] Abercrombie also included what he described as 'a family of Green Belts', with local Green Belts proposed around the existing outer towns such as St Albans and Harpenden.

In the years prior to and following the Second World War, other city authorities had made plans for Green Belts and some, for example Sheffield, had purchased land on their urban periphery with recreation provision and landscape preservation aims. We have seen how the Restriction of Ribbon Development Act (1935) had sought to control urban sprawl along arterial highways. The more widespread issues of urban sprawl continued in the post-war years prior to new development plans under the 1947 Act coming into operation. The Barlow Report had concluded that the need for population and industrial dispersal and the improvement of inner-city areas would result in significant overspill problems in most major cities. Harold Macmillan's incoming Conservative Government in 1951 embarked on a large house building programme, which resulted in a total of 340,000 dwellings being constructed in 1954. Few local authorities outside London seemed to be minded to halt the peripheral expansion of their main settlements.

The TCPA and a few individual lobbyists expressed their concerns regarding the pressures of continuing expansion of urban areas and the need for better proposals to deal with overspill development. This helped to persuade the Minister of Housing and Local Government, Duncan Sandys, to make a speech in Parliament on 26 April 1955 calling for restrictions on the growth of large urban areas. This led to the publication of the MHLG Circular 42/55 issued on 3 August 1955. In his speech the Minister stated:

'I am convinced that, for the wellbeing of our people and for the preservation of the countryside, we have a clear duty to do all we can to prevent the further unrestricted sprawl of the great cities.'

He goes on to refer to the Home Counties local planning authorities having submitted development plans for a 7–10 mile wide Green Belt around Greater London and congratulates them for this progress. He refers to other local planning authorities elsewhere in the country as endeavouring to restrict further development around their large urban areas but goes on to regret that no formal Green Belts had yet been proposed. In the Circular he asks local planning authorities to consider establishing a Green Belt. The introductory paragraph of the Circular stresses 'the importance of checking the unrestricted sprawl of the built-up areas and of safeguarding the surrounding countryside against further encroachment'. The Minister is stated as being satisfied that 'the only really effective way to achieve this object is by the formal designation of clearly defined Green Belts around the areas concerned'.

Paragraph 3 of this landmark Circular states the three purposes of Green Belts as being 'a) to check the further growth of a large built-up area; b) to prevent neighbouring towns from merging into one another; or c) to preserve the special character of a town'.

Paragraph 5 of the Circular contains the national government's development control policy, which authorities were to apply within designated Green Belts. Approval is not to be given, 'except in very special circumstances, for the construction of new buildings or for the change of use of existing buildings for purposes other than agriculture, sport, cemeteries, institutions standing in extensive grounds, or other uses appropriate to a rural area'.

The implied urgency of the situation is delivered in the final paragraphs of the Circular by advising that those authorities wishing to establish Green Belts in their area should forward a sketch plan of the approximate boundaries to the minister as soon as possible. A detailed survey would then be required to define precisely the inner and outer boundaries, which would subsequently form part of the development plan.[6]

It is important, in this context of the formal introduction of Green Belts into the planning system, to emphasise the dissent of the senior civil servants involved as the Minister had issued the Circular against their advice. Elson[7] notes that Dame Evelyn Sharp, the Ministry's Permanent Secretary at this time, responded to supporters of the proposals, in particular the planning lawyer Desmond Heap, then President of the Town Planning Institute, by stating that it was easy to define Green Belts but where was the implied overspill residential development to be accommodated. Peter Self in his book *Cities in Flood*[8] provides one of the earliest critiques of the Green Belt approach in his book dealing with 'the problems of urban growth'. He concluded that however Green Belts were defined there were limits to what they would be able to achieve. Continuous protection would be required but he considered that this could only be done if reasonable alternatives were available to accommodate urban overspill.

Self goes on to set out how the higher income households will be able to acquire their new house beyond the Green Belt and easily afford the longer commute into their city centre jobs, while others' wishes would be penned up inside the constrained urban area. He supported a regional planning approach like that of Howard and Abercrombie, which could accommodate new towns and town expansions to take overspill with jobs to settlements beyond the core city and its Green Belt.

The 1955 Circular and Duncan Sandys single-strand strategic policy approach led to a flood of sketch Green Belt proposals being submitted by local planning authorities. By the early 1960s a total of 69 sketch Green Belt proposals had been made and were at various stages of evaluation/approval by the Ministry. The original London Green Belt was approved in 1959 with further extensions approved in 1972/3. The West Riding of Yorkshire Green Belt was approved in 1966 and was the subject of my geography degree dissertation submitted in the early summer of my final year. My analysis was not very deep or profound and impacted my final honours degree award. The first Green Belts were not approved for the West Midlands and Manchester conurbations until 1974 and further expansions followed. Elson's provides a detailed explanation of the proposed conurbation and historic city Green Belts.

By 1979 the area of established Green Belts in England extended to 721,500 hectares. In 1993 this had expanded considerably to 1,652,310 hectares and by 2003 to 1,671,580 hectares. At the end of March 2020, the area covered by the Green Belt in England was estimated at 1,615,800 hectares, which covers some 12.4 per cent of the land area.[9] This is a remarkable total when it is compared with the percentage of the total land area which is urban and developed which totals 9 per cent (the figure for the UK as a whole is 6 per cent).[10] The extent of Green Belts should also be generally considered alongside the extent of other nationally designated areas including the national parks, AONBs and national and international nature reserves, which unlike Green Belts are designated and protected for their environmental and recreation qualities. In 2020 these areas covered a total of 3.444 million hectares and 26.4 per cent of England's land cover and 6.789 million hectares and 27.8 per cent of the UK's land cover.[11]

The change in the area covered by Green Belt designation in England over the last 17 years (2003–2020) is a loss of 55,780 hectares or 3.34 per cent of the land area subject to designation. The total loss over this period includes the redesignation of 47,300 hectares of Green Belt land as part of the New Forest National Park in 2005. The actual loss to residential and employment development is further reduced by changes to urban open space, the removal of areas of land which are already developed (e.g., 28.78 hectares at Biggin Hill Airport in the London Borough of Bromley) and land used in major highway schemes and subsequently deleted. The land lost to residential and employment development over this 17-year period is of the order of 0.5 per cent of the 2003 maximum annual coverage. This does not support the widely held public perception that there has been extensive erosion of the Green Belt and that this is primarily a result of residential development. This is just one of the myths which emerge in the on-going debate.

It is important to remove some further myths and misunderstandings regarding the nature and role of Green Belts in UK and English planning policy and decision-making prior to assessing the case for change. The second myth is created by the mis-selling of Green Belt designations as permanent. Many of the Green Belts established following the issue of Circular 42/55 have changed very little in recent decades, as the preceding national statistics demonstrate. In the early decades of the designation and approval of Green Belts the predominant changes were to expand their extent particularly around London and the other main conurbations. The amount of land in the Green Belt more than doubled in the fourteen-year period to 1993. Since then, the amount of change to well established Green Belt boundaries has been limited. The public then can perhaps be forgiven for this misconception of permanence. It is indeed the government who are most to blame for this confusion, as is demonstrated in the latest NPPF. The opening paragraph of section 13 states that: 'The Government attaches great importance to Green Belts' and 'the essential characteristics of Green Belts are their openness and their permanence'.[12] Paragraphs 135 to 139 inclusive then go on to refer to 'exceptional circumstances' where Green Belt boundaries might need to be reviewed. Such circumstances are not defined or described in this policy document or in the NPPG, which is supposed to provide more detailed guidance. Insofar as a list of what might constitute exceptional circumstances exists, this has been achieved via the courts and includes meeting the objectively assessed housing needs of a local planning authority and matters that relate to this such as sustainable development. It is for decision-makers to decide the weight to be given to such circumstances.

It is clear from the current NPPF and from court cases that Green Belt boundaries can be reviewed and that such reviews can identify safeguarded land between the urban area and the revised Green Belt boundary 'in order to meet longer term development needs stretching well beyond the plan period' (paragraph 139 of the NPPF). Essentially, therefore, government policy relates the longevity of Green Belts to the development plan cycle of preparation, adoption, and subsequent review. Typically, a local plan now establishes a 15-year plan period and seeks to provide enough land for housing and other land uses for that period and for a longer period of about 20 years, where it is considered, safeguarded land should be identified in the plan. The fact that the boundaries of most Green Belts have not been changed for several decades, and in some cases from their first establishment (e.g., the West Riding of Yorkshire Green Belt established in 1966), means that pressures for change have been building up over a long period.

The third myth is that Green Belt is an environmental policy rather than simply a policy to control the spread of urban development. Many members of the public, and even a few planning authorities, consider Green Belt designation has an environmental basis or provides a direct recognition of the environmental qualities of the land included in the designation. There is a strong tendency in planning discussions and public consultations, and in public examinations and inquiries, to align Green Belts with environmental policies. Environmental designations and the policies that protect them have one thing in common with Green Belts and that is they are very often a constraint on development. Any similarity ends there as environmental designations are selected and defined based on their qualities, such as landscapes of character and visual beauty (national parks and AONBs) and areas protected for their biodiversity and habitat qualities (Nature Reserves, Ancient Woodlands). The land to be included in Green Belts is not selected on any qualitative basis though it may include local nature reserves, attractive woodland, and other environmental features. Green Belt boundaries have historically been selected based on the general width of the land area required to constrain urban growth and to keep towns separate from one another. Most planning authorities have extended this purpose of separating settlements to include villages, an approach to which national government has acquiesced, without giving specific policy support. The only environmental factor included in the five purposes of Green Belt, as defined in the NPPF (paragraph 134), applies to the preservation of the setting and special character of historic towns. The number of towns/cities meriting this level of protection are limited in number and include York, Chester, Bath, and Oxford.

I have experienced planning authorities applying the presence of a Green Belt as a site-selection criteria in combination with environmental designations in a discrete stage of the selection process rather than treating this designation and its purposes as a wholly separate and subsequent stage where it is the sole criterion to be applied.

The fourth misconception is that the designation of Green Belts has in some way provided much greater access to recreation for urban dwellers. While Howard and the early proponents of Green Belts envisaged that public recreation would be both a main role and a benefit of designation, they generally allied this aim with a programme of public acquisition of the land to facilitate this. The great majority of Green Belt land is privately owned, and the major land use is agriculture. Local planning authorities have generally not sought to negotiate access agreements with landowners. However, this position could and should be radically changed as a result of the introduction of the new statutes covering agriculture and the environment. The introduction of these statutes leads to a need to adapt planning policy and development management relating to land use, including land in the Green Belt.

When Green Belts were first established by Circular 42/55, they had three stated purposes:

'(a) to check the further growth of a large built-up area.
(b) to prevent neighbouring towns from merging into one another; or
(c) to preserve the special character of a town.'

It is important to note the emphasis on towns and large built-up areas, although several the designated Green Belts had already included medium to large sized villages and small towns into the types of settlements that should remain separate. Green Belts are consequently generally more extensive and complex for multi-settlement authorities in the conurbations. In the West Yorkshire conurbation Green Belt, the metropolitan districts of Leeds and Bradford each contain several principal free-standing towns and large villages.

A full Planning Policy Guidance document on Green Belts was first published in 1988 and a revised version issued in 1995.[13] These versions of the policy guidance provided the extended five purposes of Green Belts. The 1995 version, for the first time, included a set of positive objectives covering provision for recreation and outdoor sport, the improvement of damaged and derelict land around towns, retaining and enhancing attractive landscapes, securing nature conservation interest and the retention of land in agriculture and forestry. An important introduction was the statement regarding the most important attribute of Green Belts as being their 'openness'. The five purposes of Green Belts defined in 1995 remain in the NPPF as do the main defined characteristics of openness and permanence. The NPPF wording is as follows:

'(a) to check the unrestricted sprawl of large built-up areas.
(b) to prevent neighbouring towns from merging into one another.
(c) to assist in safeguarding the countryside from encroachment.
(d) to preserve the setting and special character of historic towns; and
(e) to assist in urban regeneration, by encouraging the recycling of derelict and other urban land.'

The first two purposes are essentially the same as those in the 1955 Circular. The retention of the word 'large' in the first purpose is significant. The third purpose of the Circular has been refined to make it clearer that this relates to historic towns and their settings. This strengthens the argument that this purpose relates to a select group of historic towns with a wide historic core and listed buildings of great heritage significance. These historic buildings can still be viewed from the surrounding landscape that forms the setting of the defined historic core, including the landmark building(s). An individual example is York Minster and the outer boundary of the York Green Belt, and the development control policies in operation are strongly influenced by the vistas of this great church. Examples of grouped heritage assets are the spires and towers in the historic skyline of Oxford and the Georgian crescent and terraces, which together with earlier historic buildings are key features in vistas from the countryside surrounding the city of Bath. In some recent Green Belt boundary review studies, I have examined, local planning authorities and their consultants have included a criterion which seeks to weight the significance of the relationship of a tract of Green Belt with the small historic core of a town, where that is not intervisible with any viewpoints due to surrounding modern development. In one case the authority sought to stretch this purpose to include relationships with the setting of listed and scheduled heritage features outside an urban area.

Regarding the alleged permanence characteristic of Green Belts, the PPG2 1995 document introduced an interesting qualification by stating at paragraph 2.1 'their protection must be maintained as far as can be seen ahead'. There is no such qualification in the current policy. The introduction of the key attribute of openness in 1995 caused some significant head scratching amongst planning professionals and lawyers. It is not untypical of civil servants and ministers, in drafting policy and guidance, to introduce key terms without any definition or attempt at clarification as to what is meant. It was left to three initial public inquiries, following the publication of PPG2, to arrive at a determination of what was meant and what factor(s) should be assessed to determine the extent of a development's impact on the openness of a particular tract of Green Belt land. My friend Andrew Williamson and I acted as advocate and planning witness in one of the three landmark cases. The project involved a scheme for the residential conversion and redevelopment of a largely redundant psychiatric hospital in the Green Belt adjacent to the village of Burley-in-Wharfedale. Prior to the opening of the inquiry, we had compared our interpretations of the term openness and agreed that it was distinct from the visual qualities of the area of Green Belt in which the proposed development was to be sited and related to the amount of urbanising development and activity that would be introduced. Helpfully, PPG2 distinguished the impact of a development on the visual amenity and landscape of an area as a separate assessment. Andrew produced a simple clarifying example. He drew a footprint of a small, new dwelling set in the middle of a woodland and the same dwelling footprint and massing set in an open moorland location, visible from a wide radius. He concluded that the urbanising impact on the openness of each area of Green Belt was the same or very similar, while the visual amenity impact was substantially different. The inspector accepted this interpretation and concluded that impact on the openness criterion was related to the amount of urbanisation comprising the new built development and the traffic activity associated with it.

The latest NPPG about openness (as published on 22 July 2019) for the first time provides some clear guidance on the factors to be taken into account when considering the potential impact of development on the openness of the Green Belt. The examples included are those identified by the courts. These include both spatial and visual aspects and the degree of activity likely to be generated. The spatial can relate to quantitative matters such as the footprint and volume of built development. The visual aspects are largely covered by the visual impact of the proposed development.

While Green Belt policy and guidance, post the 1955 Circular, has consistently referred to the leading purpose of Green Belts as being to check the further growth or unrestricted sprawl of 'large' built up areas, this has not been followed in policymaking and decision taking at national and local levels. The designation of the larger Green Belts covering multiple settlements, have defined settlement boundaries for cities, large towns, and a range of villages. While some of the smallest villages have been 'washed over' to include them in the Green Belt, allowing for only small infilling developments, other medium and larger villages have had their developed areas tightly defined by the adopted Green Belt boundary. It is often the case that these villages have no brownfield land or other land suitable for development. As a result, local housing need, particularly of a low cost or affordable nature, is very difficult to meet and is either exported beyond the Green Belt boundary or remains unmet.

The inner boundaries of Green Belts have generally been defined in a simplistic single purpose manner related solely to constraining the further growth of the urban areas which they surround. The great majority of these inner boundaries have remained unaltered over several decades and most of them are unchanged since the original adoption of the boundaries in the development plan. This means that there is little or no 'wriggle room' to accommodate development. I have come cross several examples where the inner boundary does not follow a well-defined on-site feature and, in a few cases, cuts across the garden areas of existing properties, or places part, or the whole, of a residential curtilage in the Green Belt when its closely associated neighbours are within the urban boundary. Many boundaries, both inner and outer, were defined in a desk-based exercise deploying ordnance survey maps rather than via an original or follow-up on the ground survey.

The alteration of Green Belt boundaries can only be achieved via the preparation of a new local development plan or the review of the existing adopted plan. A process of Green Belt review has been established with local planning authorities often needing to employ a large planning consultancy to resource this very significant piece of work as part of the preparation of their new development plan. There is an overriding national policy requirement when undertaking such a review and seeking to establish a case for changing boundaries. It is necessary to establish that there are exceptional circumstances for altering Green Belt boundaries. Paragraph 136 of the current NPPF states that such exceptional circumstances must be fully evidenced and justified. As part of this high-level justification a local planning authority needs to demonstrate that it has fully examined all other reasonable options for accommodating its development needs, including making as much use as possible of brownfield and underutilised land and optimising densities of development. The need for consultation with neighbouring planning authorities to see whether they could accommodate some of the need is also included as part of the national policy approach. This latter approach is under threat as the government have recently indicated that they are reviewing the need for the duty of groups of neighbouring authorities to cooperate on several aspects of development plan preparation. This is somewhat ironic as the alternative approach of preparing a regional or city-region plan was formally abandoned in 2010 by the then Conservative Government. Many argue that the most appropriate way of reviewing Green Belt boundaries for the conurbations should be based on a regionally prepared plan. Howard, Unwin, and Abercrombie would have readily agreed as will many urban and regional planners in other countries. It is both logical and sensible.

The courts have considered several cases that have involved the determination of what constitutes exceptional circumstances. The level of housing need in a particular planning authority has been accepted as one example. While the determination of what can constitute exceptional circumstances is a matter of law, it is for the individual planning inspector examining the submitted draft development plan to weigh and determine whether the evidence exists to justify the application of that exceptional circumstance.

Gaining planning permission for development in the Green Belt remains very difficult, even though the NPPF policies have introduced some limited relaxations. Paragraphs 143 to 146 inclusive of the NPPF define what are now considered to be appropriate developments. The great majority of development proposals are inappropriate, and it is necessary to establish very special circumstances to justify such development. The central test stated in paragraph 144 is as follows: 'When considering any planning application, local planning authorities should ensure that substantial weight is given to any harm to the Green Belt. "Very special circumstances will not exist unless the potential harm to the Green Belt by reason of inappropriateness, and any other harm resulting from the proposal, is clearly outweighed by other considerations." From my experience it is extremely difficult to establish that very special circumstances exist. High Court judges have remarked that this test sets a higher bar than the 'exceptional circumstances' test required to support boundary changes through the development plan. Very special circumstances can include the ability to meet housing need, if it can be established that the development proposed cannot be accommodated elsewhere in the general locality where the need arises, and there are other benefits which cumulatively meet the test.

* * *

The case for and against change

There is absolutely no doubt in my mind that a very serious and well-informed debate needs to take place, as soon as possible, on the future operation of Green Belts in the UK. My own investigations and extensive experience have focussed on the situation that has emerged in England. In the preceding parts of this chapter, I have attempted to provide an adequate background for this debate, but I would strongly encourage those who have engaged in the arguments, or wish to do so in the future, to read articles and assessments from both sides of the divide.

Journalists producing headline articles need to carefully review the factual basis as well as considering both sides of the argument, alongside related sustainable development policies. Those relying on such articles and broad total figures for deletions also need to carefully review and fully understand the evidence before they make pronouncements. The Council for the Preservation of Rural England (CPRE) provided a highly negative commentary in the following article and a number of their publications defending Green Belt policy restrictions are based on generalisations.

The Times Environment Editor produced a news review under the following headline: 'Green Belt is eaten up to meet appetite for new homes.'[14] The headline was based on the accurate fact that 5,070 hectares of Green

Belt land had been deleted in the year ending 31 March 2018. This was the highest total change since comprehensive records were first collected from English local planning authorities in 2003/4 by the Ministry of Housing, Communities and Local Government (MHCLG), who are responsible for national Green Belt policy. The report produced by the National Statistics section of MHCLG provides a much clearer and balanced position, which includes a very significant change in the mapping system that the English local authorities were asked to follow for the 2017/18 mapping of changes. The following quotations from this report, under the sub-heading 'Revisions to 2017/18 estimates' clearly demonstrate the need for accurate and considered journalism and the avoidance of headlines which stoke the arguments of the 'no change' camp:

'Revisions are made each year to the published estimates for the previous year in order to accurately calculate the net change in Green Belt area.'

'As part of an ongoing scheme of general improvements to mapping accuracy and boundary definitions, local authorities have been asked to map their Green Belt areas against the ONS Mean High Water Mark boundary. This results in greater accuracy in the delimiting of designated Green Belt land where it meets coastal or estuarine areas and is to ensure a consistent national approach in boundary definitions.

Thirty-five local authorities reported revisions with the majority due to ongoing mapping improvements. Sixty-seven percent of the overall revision in Green Belt area for 2017/18 is attributed to West Lancashire, which was reduced by eleven percent following the exclusion of land below the high tide mark.'[15]

At the 31st of March 2017, West Lancashire was ranked first amongst English local planning authorities regarding the proportion of its total boundaries covered by a Green Belt designation. The proportion was a staggering 99 per cent. This had reduced to 90 per cent in 2018/19 following the change referred to in the previous paragraph. The Green Belt had previously been generously drawn to include a large area of the inter-tidal flats of the Ribble Estuary. Table 4 of the Technical Notes to the quoted MHCLG document provides a further breakdown of the revisions made by the other 34 authorities and leads to the clear conclusion that only a small proportion of the total removed related to development.

The following are the broad arguments in support of an early national review of Green Belt policies, their purposes, objectives, and performance.

These can be distinguished from further, more specific, arguments relating to the nature of the changes required.

The English Green Belts collectively comprise a large land resource totalling 12.4 per cent of the land area of England that has relatively restricted access for recreation purposes and has considerable potential for realising multiple uses. During recent years a whole suite of policies have been introduced that are of direct relevance to land use in the Green Belt. Further positive changes to these policies were proposed by the Johnson Government in the August 2020 planning White Paper. This group of policies relate largely to environmental issues but also include related approaches on health and wellbeing and the achievement of sustainable development. The topic policies include habitat improvements and biodiversity enhancements, adaptation to climate change and the mitigation of its adverse impacts, the provision of more publicly accessible and better-connected greenspace and improvements to the landscape. Health and wellbeing policy and improvements to the physical and mental health of the public are strongly related to extended access rights and better management. The positive environmental improvement and stewardship roles planned for farmers and landowners in the Agriculture Act and the Environment Bill will have a major impact on the use and management of Green Belt designated land. This is a great potential force for beneficial planning outcomes and must be positively planned for.

The decades of operation of the predominant singular purpose and focus of Green Belt policy has resulted in the very successful containment of both major urban settlements as well as the unintended containment of many smaller settlements. It is however this continuing success of the main purpose which has created a group of problems.

This primary group of problems, which have grown over the last thirty years or so, relate to housing need and supply, continuing rises in market prices and expanding affordability and market access issues for younger households. Related to this group are issues associated with longer commuting distances and the sustainable location of development. The success of the containment policy has not been sufficiently counter balanced by long term, effective and imaginative regional plans in the form of those envisaged by Abercrombie's Greater London Plan. The lack of regional coordination and planning started at an early stage of the generally negative containment process. The post-war government had set up regional offices of the Ministry of Town and Country Planning to play a monitoring and coordinating role. These were abolished by the incoming Conservative government in 1951. Peter Hall recalls this in his book on urban and regional planning and goes on to criticise the approach of introducing Green Belts in a planning vacuum.

There were no real initiatives to introduce strong regional planning to fill the vacuum until the late 1980s. The emphasis was on regional economic policy with the introduction of grant systems and development area status to address the socio-economic deprivation that existed in those regions outside the South East and East Anglia. In Chapter 9 I examine various attempts at regional planning post-1947. A notable recent attempt to resolve the accommodation of urban growth and extensive housing need occurred in the South West region in the form of the West of England Joint Spatial Plan (a city region plan for Bristol and Bath and the surrounding hinterland of North East Somerset and South Gloucestershire). The consortium of the four local planning authorities broke up in March 2020 following a very difficult Examination in Public. The draft plan's proposals had included 12 strategic development areas and some elements of development by changes to Green Belt boundaries. In the early stages of the preparation of this plan I acted for one of the country's leading strategic land developers to review potential areas and sites for development. It was clear that there was a major problem to be resolved if the collective identified housing growth was to be met without unnecessarily adverse impacts on the extensive coverage of primary environmental constraints. When the Green Belt areas around Bristol and Bath were overlayed on the GIS mapping of international, national, and regionally protected environments and the capacity for brownfield land was assessed it was very clear that certain areas of Green Belt represented the leading sustainable choices for growth. These areas emerged as prime choices because they were in or near sustainable travel corridors and they avoided adverse environmental impacts.

This city-region example fully illustrates the stark choices that need to be evaluated when seeking to accommodate urban growth in an appropriately comprehensive manner. These choices need to be faced in a brave and positive way.

The strong supporters of the Green Belt containment policy frequently refer to the large amount of brownfield land available within urban areas, which they argue is capable of meeting all or most of the future housing needs of a local planning authority. The Council for the Preservation of Rural England (CPRE) are one of the main proponents of this view.[16] This is an argument that has been put forward on a fairly consistent basis over the last thirty years and in many areas, it is wearing very thin to the point of becoming threadbare. Brownfield or previously developed land is a relatively fixed resource and the take-up of brownfield sites for housing has for many years provided the great majority of housing completions in the larger urban areas. New brownfield sites come forward on a very limited basis and do not constitute a sufficiently large number which can replenish the diminishing supply. The CPRE Report clearly disagrees with my conclusion and the experience of others who have considerable market and research experience. The summary states 'that brownfield is a perpetually regenerating resource with the potential to provide a steady pipeline of development opportunities'. They also claim in the Report that the Brownfield Registers of local planning authorities show that enough suitable land is available to build more than one million homes across 18,000 sites comprising 26,000 hectares. This finding has been criticised by several sources. Even if this statement was anywhere near being accurate the total potential quoted would only equate to approximately three years' supply based on the government's annual target of building 330,000 homes in England. Local Plans under production or recently adopted must look forward over a period of 15 years when assessing their housing requirement.

Considerable emphasis has been placed on brownfield land delivery by successive governments in the provision of policy led targets, special funding schemes and the statutory registration of brownfield land that is both suitable and deliverable for new housing. The delivery of urban brownfield land for housing has been a major policy objective since the late 1990s. In 1998 the UK government set a national delivery target of 60 per cent of all new development to be on brownfield land. In June 2014 the government introduced Local Development Orders and Pre-Planning Approvals to support housing development on brownfield land. In January 2016 a £1.2 billion fund was announced to prepare brownfield sites for the construction of starter homes. In 2017 the Town and Country Planning (Brownfield Land Register) Regulations were published, requiring all English local planning authorities to keep, and annually update, a register of deliverable sites. The CPRE in 2019 were continuing to lobby government to introduce a national 'brownfield first' policy. It is significant that when the first version of the NPPF was produced in March 2012, while continuing to encourage the use of Brownfield land, it did not set a target threshold.

The Government's 60 per cent target set in 1998 was consistently met in the period 2000–2006 at an average of 2,774 hectares/annum, which was only marginally above the annual figure of 2,644 hectares achieved between 1989 and 1998. When preparing planning case for residential planning applications and appeals I frequently researched the Annual Monitoring Reports of local planning authorities to ascertain their past performance on the delivery of housing on brownfield land. During the period from 1995 to 2019, Bradford Metropolitan District has consistently exceeded a total of 80 per cent delivery of housing completions on brownfield land within the urban areas. When preparing their Core Strategy part of the development plan (adopted in July 2017) they provided a detailed analysis and justification of the strategic sources of supply to meet their housing requirement for the period 2004 to 2030. Of the total requirement of 42,000 dwellings, they assessed that 11,000 (26 per cent) would be needed from Green Belt releases. A local MP formally objected to the adoption of the plan despite the clear support given by the Inspector who led the Examination in Public. The Secretary of State recovered the draft plan and following a lengthy review finally agreed that the plan could proceed to adoption as he had accepted the Council's justification. This represents one of several clear illustrations that reliance on a high proportionate delivery of new housing on urban brownfield land is becoming increasingly difficult to justify.

This conclusion has been echoed by well-respected sources. The Joseph Rowntree Foundation, in one of their many well researched reports on the housing market, consider some of the impacts of the successful delivery of brownfield housing particularly in deprived neighbourhoods.[17] While referring to the generally positive outcomes in achieving brownfield targets in urban areas, they raise concerns including that of high-density developments that often do not lead to sustainable development:

'Concerns have been widely raised about the sustainability and appropriateness of continuing such a high-density brownfield regeneration approach to deliver the government's ambitious housing target in the future. At the turn of a new decade, it is, therefore, timely to reconsider housing planning strategy and targets to meet projected housing needs in the most sustainable manner.'

The successful utilisation of brownfield land for housing over many years, together with the high densities often achieved in the core areas and around transport hubs, has the effect of placing pressures on greenfield sites in urban areas. The first of these is the pressure of intensified use of those that are designated as recreation areas. The second is the pressure to develop the remaining undesignated or otherwise unprotected sites. Both groups of greenfield sites are a precious resource, which arguably need greater protection than many areas of Green Belt.

The Town and Country Planning Association (TCPA) has been, and remains, one of the most notable supporters of Green Belt policy since its inception and this organisation, as the successor to the Garden City Association, founded by Ebenezer Howard, have been the main protagonist of his ideas. Some readers may therefore be surprised to learn of the contents of their policy statement on Green Belts issued as early as May 2002. The introduction to this short but succinct policy statement begins as follows:

'Green Belt policy has been one of the most notable achievements of the planning system and has had widespread public support. The TCPA has consistently supported that policy since its inception. However, the Association believes that a reappraisal of the roles, purposes and extent of green belts is now necessary. In some parts of the country, they inhibit the appropriate sustainable development of urban areas, and often they constrain the opportunities for reducing social exclusion.'[18]

It is notable that these conclusions were arrived at following a very lengthy period of support and at a time when the land covered by Green Belt designations had not yet reached its peak. The TCPA remind us in this Statement that they had historically not supported the fifth purpose of Green Belts when it was introduced into national policy in the 1980s because they considered that the links with inhibiting urban regeneration were unproven. In their policy critique they go on to state that their support was on the basis that Green Belt policy 'should be applied in a way which is complementary to a strategic package of policies including new towns and town extensions and conserving the countryside for recreation and agriculture.' In other words, they were holding fast to Howard's integrated policy approach.

The TCPA continue by stating several reasons that support their case for a reappraisal of Green Belt policy. They point out the need for integrated and sustainable urban extensions that are well related to public transport corridors and include a variety of open spaces. They considered that such extensions can be in direct conflict with Green Belts. Extensive Green Belts 'inhibit the scope for encouraging new forms of development in rural areas and for the diversification of their economies and communities'. They also considered that Green Belts should be conceived as 'eco-belts' that include zones for a range of ecological and sustainable uses including allotments and smallholdings, community woodlands and green energy projects. At a local level they considered that Green Belts were too restrictive in their control of small-scale development. They went further by encouraging the identification of green wedges and more limited strategic gaps in some parts of the extensive conurbation Green Belts. Finally, they advocated new sustainable land management and the reinforcement of the positive land use objectives introduced in PGG2.

The conclusions reached by the TCPA and the Joseph Rowntree Foundation, both of whom are well respected organisations in the fields of planning and housing need/supply, and the fact that their conclusions were reached 18 and 10 years ago respectively, represent part of the very powerful case for change that has now accumulated.

A further frequently recurring criticism of Green Belt containment policies are their adverse impacts on out-migration of households from the main urban areas, restricted by the large conurbation Green Belts, in search of suitable housing accommodation that cannot be met within those urban areas due to size requirements, lack of choice, affordability and related amenity issues. This need or desire to 'leapfrog' Green Belt boundaries has existed almost from the period in which the large conurbation Green Belts were first introduced. This is particularly true of the London Green Belt where at first assessed overspill requirements were planned for through the provision of new town and town expansion schemes. However, this regional planning approach has been absent for a large part of the period in which the Metropolitan Green Belt has operated. As a result, the unplanned 'leapfrogging' has expanded in terms of the numbers of households moving to settlements beyond the outer Green Belt boundary and the distances being moved. The extension in distances moved was initially caused by the considerable expansion of the outer Green Belt boundary and subsequently by the widening of search areas to achieve suitable and affordable housing. The expansion of London's socio-economic market influence in terms of both housing and employment provision created the phenomenon referred to by geographers as ROSE (the rest of the South East). This extended the length of commuter journeys back into the London core to unprecedented distances and impacted on the health and wellbeing of employees via daily travel duration and associated stress.

Over the period 1985 to 2015 I took part in many development plan public examinations, particularly for Leeds and Bradford Metropolitan Districts, where the subject of the constraining Green Belt and the annual levels of out-migration of households, especially to the adjacent districts of North Yorkshire, were recurring issues. This year on year 'leapfrogging' of the outer Green Belt boundaries to the north was primarily a result of unmet housing demand and need transferring across local planning authority boundaries. I and fellow consultants persistently argued that housing supply needed to be enhanced in the two exporting districts to meet the needs of households in terms of the number, size, and affordability of dwellings in the future supply. It was clear from estate agents operating in the receiving districts why households wished to make these moves and the impact that this was having, over time, on house prices in the proximate North Yorkshire districts. While annual population statistics provided the basic out-migration figures more specific surveys were required to establish the impacts. Increases in commuting journeys along specific arterial routes were key indicators. The opening of the M1 extension around the eastern side of Leeds to join the A1(M), travelling northwards, had a rapid impact on the housing search area with the result that additional parts of Harrogate District, including Ripon and Boroughbridge, have subsequently been added to the Leeds Housing Market Area. While these impacts do not compare with the extent of commuting journeys and house price rises in the South East, they do not lead to sustainably located development and work journeys.

Two policy institutions, the Adam Smith Institute[19] and the Centre for Cities,[20] produced reports in 2015 arguing for changes to Green Belts to help meet the growing crisis in housing supply. They both referred to the need to make a series of discrete land releases in London's Green Belt focussed on easy sustainable travel distances from selected railway stations. The Adam Smith Institute report considered three different policy initiatives to reform the effect of urban containment policies on house prices, house sizes, costs to businesses and the environment. The third of these initiatives grabbed the headlines and proposed the release of all intensively farmed land within 800 metres of a railway station. They argued that this would provide land for one million new homes for London via the deletion of 3.7 per cent of the Metropolitan Green Belt. The Centre for Cities went further by suggesting a case-by-case review of Green Belt designations within 2 km of a railway station leading in the case of the Metropolitan Green Belt to sufficient building land for around three million homes.

In November 2017 the House of Commons Library produced a briefing paper on the Green Belt and an updated version of this three years later.[21] This paper provides members of both houses and the wider public who care to read it with a balanced and well-informed assessment. Under the sub-heading 'Is the Green Belt hampering growth?' Paragraph 2.2 introduces this part of the debate by stating:

'Whether this level of protection for the Green Belt remains unnecessary or appropriate – or whether, conversely, it places obstacles in the way of providing new housing – remains controversial.'

They go on to refer to several other organisations who argue that changes are now necessary. The first reference is to an Institute of Directors report (February 2011) favouring some Green Belt land release. In its initial analysis of the 2017 housing White Paper, the RTPI suggested that the 'role, purposes and social impact of Green Belts should be revisited and Green Belt boundaries "may well" need to change, albeit with safeguards'. The third body referred to is the Organisation for Economic Co-operation and Development (OECD), who criticised the Government's Green Belt policy for being an obstacle to house building. The strength of the OECD's criticism is particularly surprising given that it comes from an independent body with a world-wide view. In a paragraph on housing supply in their 2011 economic survey of the UK, they refer to the response of supply to demand as being one of the lowest amongst OECD countries over the last 20 years and then go on to state that: 'Green Belts constitute a major obstacle to development around cities, where housing is often needed. Replacing Green Belts by land-use restrictions that better reflect environmental designations would free up land for housing while preserving the environment.'[22]

A small number of those entering the debate seek a more sophisticated and detailed response to the problems arising from the strong Green Belt constraint. Notably the Landscape Institute carried out a survey of its membership to gauge their reactions to the operation of Green Belt policy.[23] This subsequently resulted in a briefing document calling for a review of the Green Belt.[24] This paper, produced in 2018, calls for a re-purposing of Green Belts, the transformation of its landscape and a restructuring of its land uses to deliver several social and environmental objectives including:

- Encouraging healthy lifestyles and contact with nature.
- Reducing flood risk and pollution.
- Improving water and air quality.
- Promoting social cohesion.
- Facilitating sustainable new development.
- Enhancing biodiversity.
- Mitigating climate change.

This is a positive and enlightened contribution to the debate, which forms a good basis for a reformed policy and a practical approach. However, there are some key factors which need to be added to be able to deliver fully sustainable outcomes with public support and recognition of the benefits that can be achieved. The incorporation of economic factors and the delivery of clear economic benefit is necessary to achieve the three interlocking objectives of sustainable development. Many of the papers advocating change give insufficient recognition to the need for the full involvement of landowners and the public throughout the process, from policy formulation to the production of zonal master plans, planning application schemes, management plans and ongoing stewardship responsibilities. Howard and his successors had argued for public ownership of Green Belt land and wide recreation access. This did not transpire and there will consequently need to be processes that ensure that private landowners are persuaded to be at the heart of the delivery process. My recent experience is that this can be achieved, and multi- use/multi-benefit projects conceived and realised.

In Chapter 5 I refer to the example of my work on an Area Plan for the town of Ilkley in the context of the emergence of a new relationship between planning and design at the national level and the actions necessary to ensure that planning contributes to good design and beauty. Ilkley is a small but very attractive town with a population of 14,500 and a very good range and quality of services, including rail and bus connections. It has an attractive rural and urban environment. The population is skewed towards the age groups in retirement and there is a need to retain and attract younger working households. All these factors in combination create need and demand for market and affordable housing. The town was voted as the best place to live in the UK in the 2022 Sunday Times awards, a somewhat ironic outcome given the very limited provision for new housing development at this time.

Following GIS mapping of primary and secondary environmental constraints, and analysis of the very limited number of potential development sites remaining within the urban area, it became apparent that some release of Green Belt land would be necessary to meet the future housing requirement. Any development sites within the existing urban limits and currently in the Green Belt, at the eastern and western edges of the town, would have to meet the mitigation policy requirements arising from the protection of internationally designated habitats on the moorlands to the north and south of the town. The Green Belt zones/sites to the east and west of the town, where we have been acting for landowners, were free from primary environmental constraints except for an area of protected river floodplain, which would need to be retained and used for complementary purposes.

The early area master plans have been developed on a multiple land use basis with the involvement and agreement of the landowners. These sites and their land ownerships can provide the required contributions to the total residential requirement distributed to this Principal Town in Bradford MD, as defined in the currently adopted development plan, the Core Strategy (July 2017). The Core Strategy policy proposes that 1,000 new dwellings should be provided in and adjacent to the urban area of Ilkley. The newly emerging Local Plan now proposes a much-reduced future supply of 300 dwellings. The only justification for this reduction of 700 dwellings is a desire to avoid the loss of too much land from the tightly designated Green Belt surrounding the town. The urban area has no sites identified in the Brownfield Register, the importance of protecting the list of urban greenspaces identified in the Neighbourhood Plan and the existence of three conservation areas covering large parts of the town all verify that the number of housing completions within the urban area will be restricted. Additional evidence charting the development of brownfield sites and large gardens over the last thirty or so years further underlines this point.

The level of residential development on both sites, being promoted through the new development plan, would cover a net area substantially less than that which will be devoted to a mixture of environmental uses. The latter will include habitat provision and enhancement, new recreation areas, footpath and cycleway connections, plus the creation of newly landscaped corridors and the retention and management of existing landscape features. Enhanced connectivity is provided to neighbouring communities and the town centre and transport hub. The multiple use and environmental enhancement of the land holdings will require more detailed master planning and new management plans to ensure the long-term future of these areas for the benefit of existing and future residents and visitors. Effective management plans and stewardship will ensure that the health and wellbeing benefits realised are maximised and retained in the long term.

It is ironic that the *Sunday Times* announced Ilkley as the 2022 winner of the best place to live in the UK on 10 April. The many positive qualities of the town and its surrounding countryside were highlighted. However, there was no acknowledgement of the current absence of land formally designated for market and affordable housing other than that of one online commentator who added the words 'and, importantly no significant planned development' to his list of positive characteristics.

Projects of this type can make a major contribution to the repurposing of the Green Belt, with clear land use and management plans being prepared. The question of long-term ownership and stewardship of large tracts of Green Belt land needs to be resolved together with the preparation of land use mosaic master plans for their future multiple use and management plans covering a wide range of matters. Given that some 65 per cent of the 14 English Green Belts are in agricultural use, the new Agriculture and Environment Acts will help to guide and facilitate many of the positive changes required. The Environmental Land Management Schemes (ELMS), which will progressively replace the existing CAP payments scheme, align extremely well with these proposals for repurposing Green Belts, as do the policies in the Environment Act.

Professor Sir Dieter Helm, in his 2019 book *Green and Prosperous Land*,[25] provides his positive economic blueprint for the future of the UK environment. His emphasis is on all forms of natural capital and the need to recognise and revalue the economic benefits they can, and indeed must, provide. While we may differ to some degree in our views on the need to allow some residential development on land currently designated as Green Belt, I fully support his approach to natural capital and his main conclusions. We also agree on the need for more detailed planning in terms of both future land use and management of Green Belt land. My approach is to see the great majority of residential development on former Green Belt land as enabling development that will bring with it a raft of environmental and public benefits in the form of recreation and enhanced access provision. The amount of Green Belt land required for residential development would equate to between 1–2 per cent of the current nationally designated land.

Planning authorities will need to prepare or commission proactive plans for their Green Belts to promote and accommodate the new approach. Some overview plans should be prepared on a cooperative basis, either at a city-region level or a smaller grouping of planning authorities. This can be done relatively quickly as most of the information is available from existing plans and databases, held either by the planning authorities or other public bodies such as the Environment Agency, Natural England, and the Water companies. These wider plans would cover such matters as extensive cross-boundary woodland planting, large recreation areas, major habitats, the management of the upper catchment areas of river systems and the creation/enhancement of wider corridors/networks for biodiversity, landscape, recreation, and sustainable communication. Local planning authorities, in cooperation with neighbourhood planning bodies, will need to prepare more detailed plans, which can either be incorporated into newly emerging development plans or produced as separate link documents within the development plan system. Interim measures should be put in place to provide for the early incorporation and approval of projects that can demonstrate the provision of a wide range of public and sustainability benefits and the capacity to integrate into wider sub-regional proposals.

The public need to be engaged in this process and to approach the formulation of proposals at all levels with an open mind assisted by education, mutual learning and clear presentations of the many benefits and opportunities arising. These proposals do not constitute an approach that seeks to progressively dismantle or remove large areas from the Green Belt. Many of the environmental benefits and the proposals that deliver them will need to be delivered on a truly permanent basis, such as the provision of public open spaces, community, and other woodlands, protected habitats and rights of access. This multi-use and multi-management approach should involve bodies such as the Woodland Trust, the County Wildlife Trusts, and the National Trust, who are well able to take on ownership and management roles where landowners, including farmers wish to share or relinquish these.

Sustainable residential developments will need to demonstrate how they can provide multiple environmental and public benefits while also creating high quality design. This approach can solve or mitigate many of the problems existing within urban fringe landscapes, including poor land management, degraded landscapes, fly-tipping and poor public connectivity with the wider countryside and the existing residential communities. This is not an approach that simply seeks to detach areas of land from the Green Belt for more housing. Instead, it is an integrated land use and design approach that aims to repurpose areas of Green Belt to provide a range of health and wellbeing 'services' to existing communities and future residents.

This approach is more agile and sophisticated than the prime containment purpose of existing Green Belt policy. The existing Green Belt policies in the NPPF lack the injection of any strong objectives for Green Belt land use and their environments. This approach by the national government is totally out of step with their call for a positive and proactive approach to plan-making and decision-taking. It also forms a very poor fit with their progressive environmental policy approaches, contained to some degree in existing planning legislation and policy and in recent thinking, as contained in the White Paper 'Planning for the Future' (August 2020) and the recent Agriculture and Environment Acts.

After many years of containment, without any other policy outlets as envisaged by Howard and his successors, the time has arrived for radical changes to Green Belt policy and planning. These changes will readily align and integrate with other planning policy and practice changes that are seeking environmental enhancements, the creation of beautiful places and major improvements to the health and wellbeing of the nation.

Salt's Mill: Saltaire World Heritage Site.

Saltaire worker's housing: remains highly marketable today

Three Cliffs Bay, the Gower Peninsula: The first Area of Outstanding Natural Beauty (AONB).

Beamsley Beacon, North Yorkshire: The Yorkshire Dales National Park meets the Nidderdale AONB.

Ilkley – Panorama of the town environs.

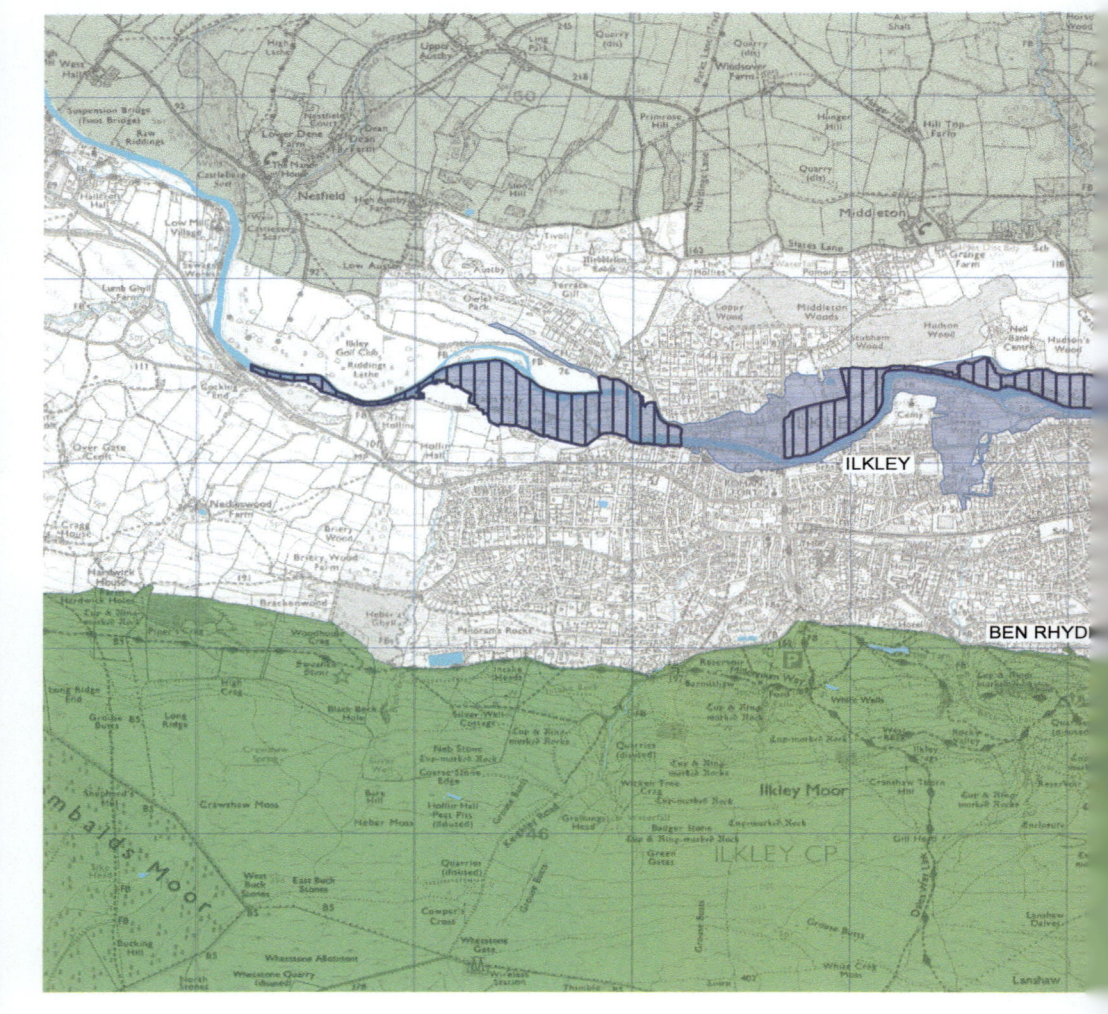

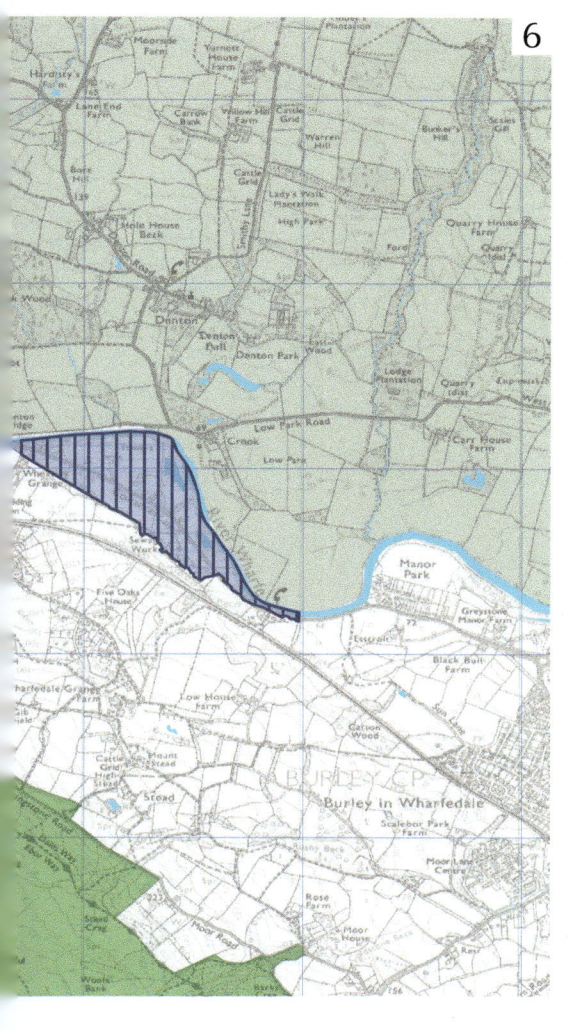

Key

■ Ilkley Moor Designations - SSSI, SPA, and SAC

□ Nidderdale AONB

▨ Environment Agency Ilkley Flood Zone 3a

▦ Environment Agency Ilkley Flood Zone 3b - The Functional Floodplain

Ilkley – Primary environmental constraints (GIS mapping).

Key

- Ancient Woodland — Ancient and semi natural woodland
- Ancient Woodland — Ancient replanted woodland
- Woodland undesignated
- Registered Park and Gardens
- Bradford RUDP OCT 2005 Urban Greenspace
- Bradford RUDP OCT 2005 OS2 Recreation Open Space
- Bradford RUDP OCT 2005 OS3 Playing Fields
- Bradford RUDP OCT 2005 OS6 Allotments
- Bradford RUDP OCT 2005 SEGI/RIGS Site of Ecological or Geological Importance
- Bradford RUDP OCT 2005 Site of Local Conservation Importance
- Local Nature Reserve
- Watercourse

7 Ilkley – Secondary environmental constraints (GIS mapping).

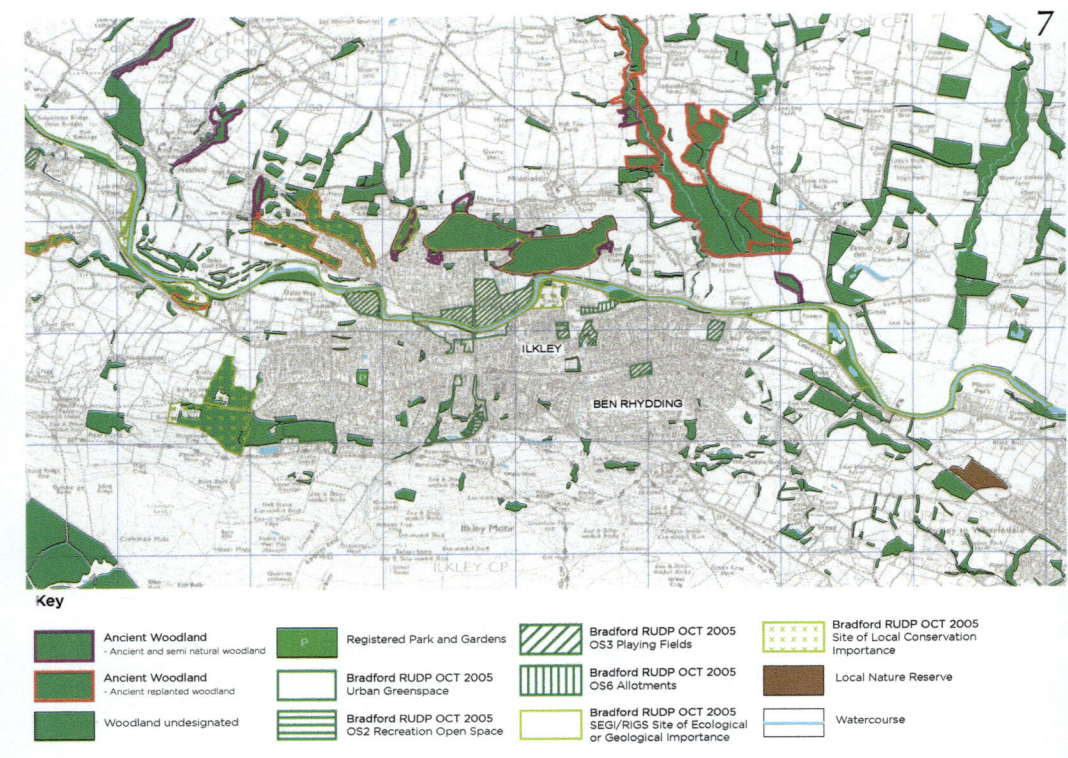

8 Ilkley – North and South Pennine Moors Special Areas of Protection (SPA) & Special Areas of Conservation (SAC) – habitat mitigation zones.

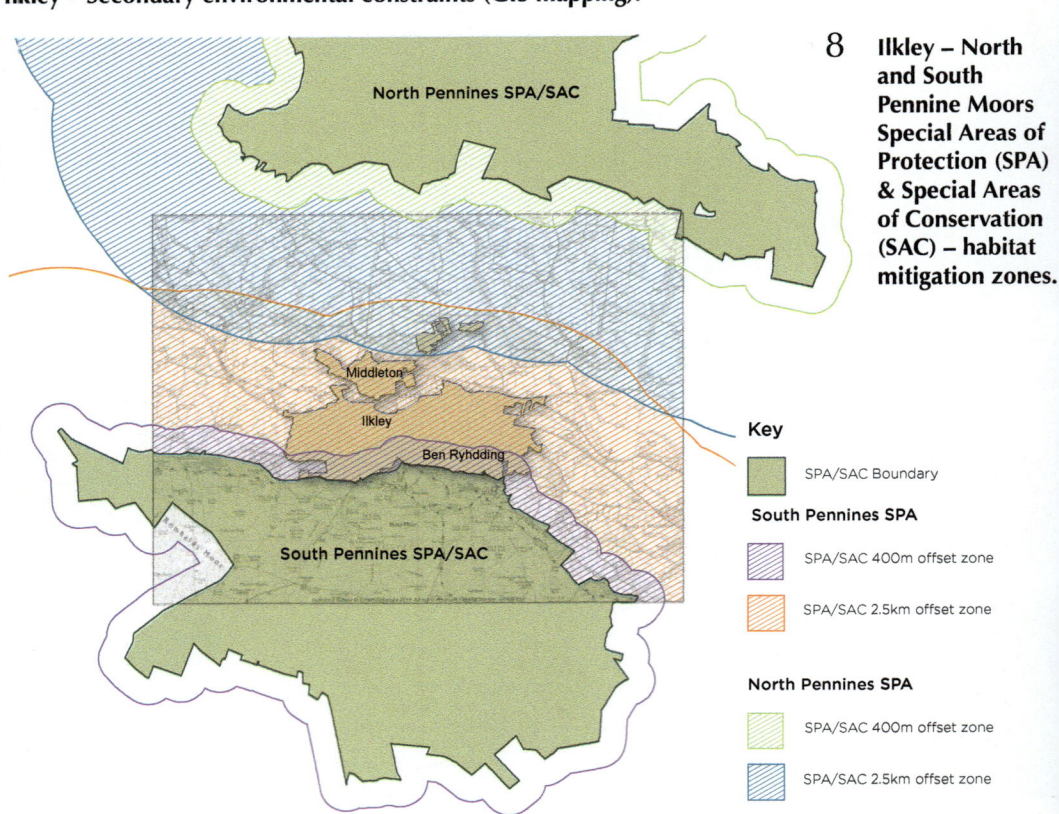

Key

- SPA/SAC Boundary

South Pennines SPA
- SPA/SAC 400m offset zone
- SPA/SAC 2.5km offset zone

North Pennines SPA
- SPA/SAC 400m offset zone
- SPA/SAC 2.5km offset zone

9

A	Ilkley Golf Club	E	Castle Road Allotments	I	Ilkley Cemetery
B	Ilkley Lawn Tennis ans Squash Club	F	East Holmes Fields (east of Middleton Avenue)	J	Ben Rhydding Sports Club
C	Ilkley Park	G	Ilkley Pool and Lido	K	Ilkley Moor
D	East Holmes Fields (west of Middleton Avenue)	H	Middleton Woods		

Key

- Public access land
- Existing strategic greenspaces
- River Wharfe Corridor
- Existing footpath links connecting green spaces
- Proposed footpath links connecting green spaces
- Proposed network of green spaces creating green infrastructure gateways to the Warfe Valley creating significant new public open spaces that connect the river valley to the moor and provide robust boundaries capable of absorbing new development that book end the settlement
- Strategic green hubs as park of new green spaces network
- Key vistas guiding network of green spaces
- Gateway feature to be safeguarded (view line to Cow and Calf from river side)

Ilkley – Landscape Strategy Plan.

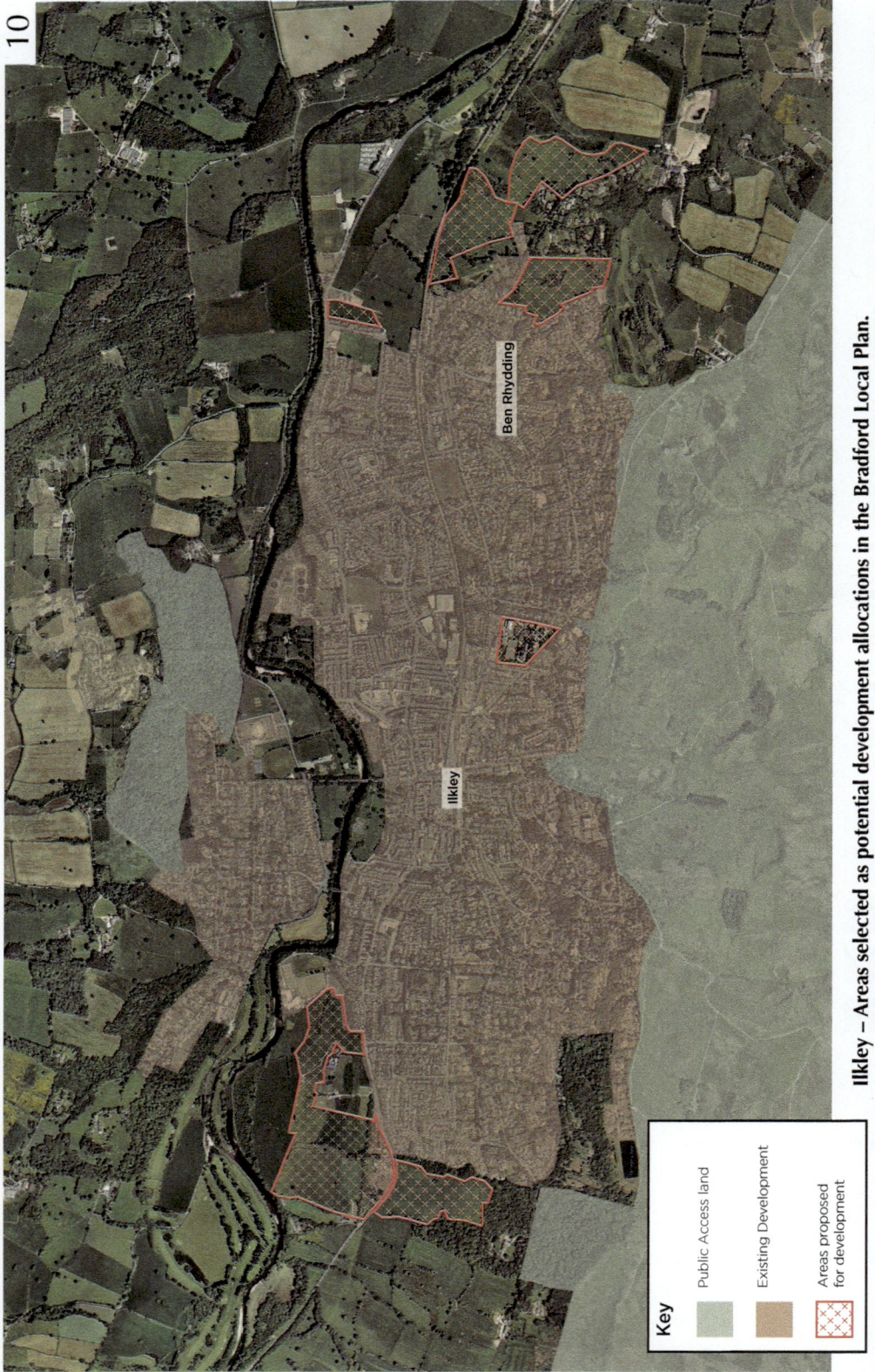

Ilkley – Areas selected as potential development allocations in the Bradford Local Plan.

Ben Rhydding

Ilkley

Key
- Public Access land
- Existing Development
- Areas proposed for development

3D model of Ilkley Roman Fort – scheduled ancient monument plus assemblage of listed buildings.

All Saints Church Ilkley development project – external panorama.

Internal photographs of the refurbished Church.

15 Multi award-winning sustainable residential development by Citu – alongside the River Aire Leeds City centre.

16 Marmalade Lane, Cambridge – multi award-winning cohousing community.

17

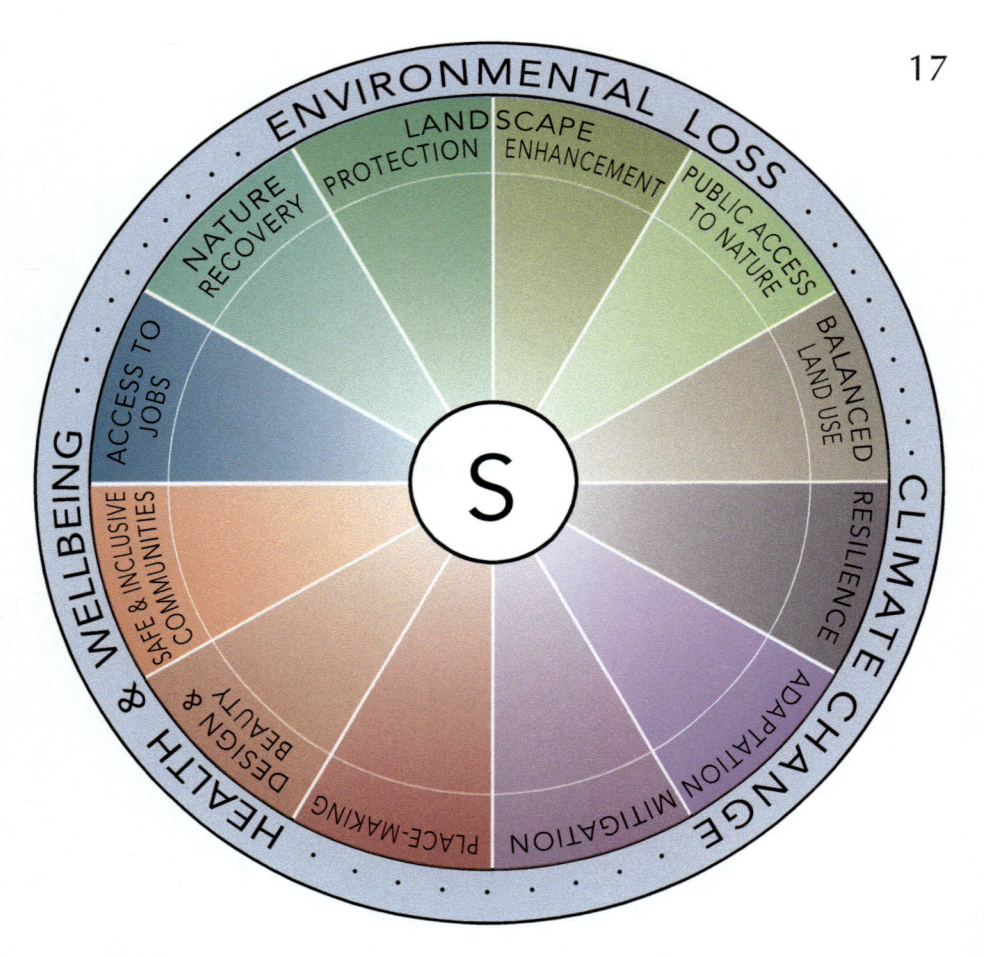

The Wheel of Sustainable Development & the three interlocking challenges.

Competition for land-use, solar farm near Bentham, North Yorkshire.

The Leeds–Liverpool Canal south of Skipton: access to heritage, nature, recreation and symbolising hope for the future.

Chapter 8

Planning and Risk Management – Consideration of Public Anxiety and Fear

In Chapter 6 I examined the operation of the development control part of the UK planning system, including the consideration of what constitute material planning considerations that can be taken into account in arriving at planning decisions on applications and appeals. Using examples of the control of certain types of land use and development, I sought to demonstrate how these decisions can relate to the protection of individual and community health and wellbeing. It is valuable to extend this investigation to a specific sub-set of material planning considerations that relate to public concern and the public perception of risk. These risks and consequent concerns are associated with potential adverse environmental and related impacts that can directly or indirectly affect human health and wellbeing. In Chapter 6 I briefly introduce the example of the extraction of coal bed methane by the process known as fracking. The public have raised their concerns regarding the perceived and actual (i.e., they have occurred elsewhere when using this process) environmental harm that could occur as the result of such development.

As summarised in Chapter 6, it is for the courts to decide what constitute material planning considerations and for the decision-maker to decide which established material considerations apply to a particular decision on a planning application or appeal and what weight should be given to those selected in applying the planning balance to that decision. Any consideration that relates to the use and development of land is capable of being material to a planning decision. Issues of public safety are relevant considerations as are public concern, including anxiety and fear, regarding perceived risks arising from particular land uses or types of development. The extent to which there is objective evidence available to support public concerns is, as we shall see, an important consideration.

The main reason for choosing to examine this area of planning practice and decision-making is that public concern, in particular anxiety and fear, can have a deleterious effect on the health and wellbeing of individuals, groups, or a whole community. These experiences can be sharp and acute or prolonged over months or years. They can affect the mental or physical health of individuals and in some cases both. To demonstrate the increasing significance of this subset of material planning considerations and how they have influenced decisions I have used planning project examples, including some where I have been the planning consultant helping to promote the case. In addition, I have reviewed legal articles that provide the most helpful commentaries on specific project examples and the case law arising from them. These illustrate the evolution of the law and some of the difficulties that are faced by the decision-makers when seeking to understand and give appropriate weight to particular public concerns.

Before looking at these examples and legal reviews, it is helpful to briefly set out the current national planning policy and guidance in England that are directly relevant to the considerations of public safety and public concerns. An overview of the public perception of risk is also presented.

As previously stated, the health and wellbeing of individuals and communities lie at the heart of the national planning policy with its focus on the delivery of sustainable development. When public concern is expressed in relation to the potential impacts of specific development proposals on health and wellbeing the policy approach contained in the NPPF can at best be described as patchy. The lead relevant part of the NPPF is section 8 entitled Promoting Healthy and Safe Communities. Paragraph 91 introduces the current policy approach by stating:

> 'Planning policies and decisions should aim to achieve healthy, inclusive and safe communities.' Reference is made to creating safe places so that crime and disorder and the fear of crime do not undermine the quality of life or community cohesion. Paragraph 95 requires planning policies and decisions to promote public safety in the context of 'possible malicious threats and natural hazards'. Policies for the layout and design of developments need to be informed by the most up-to-date information available from the police and other agencies regarding the nature of potential threats and their implications. The information to be provided by these agencies includes the 'appropriate and proportionate steps that can be taken to reduce vulnerability, increase resilience and ensure public safety and security'. There is an emphasis here on crime, terrorism, and defence and only a brief reference to 'natural hazards'.

Section 10 of the NPPF – 'Supporting High Quality Communications' – requires planning policies and decisions to support the expansion of electronic communications networks. Paragraph 112 introduces this approach by stating that:

'Advanced, high quality and reliable communications infrastructure is essential for economic growth and social well-being.'

This means that planning decision-makers will have to consider the weight to be attached to this social benefit and the public concerns regarding potential adverse health impacts. These latter considerations have arisen in decisions in the courts. It is interesting that the NPPF provides specific policy instructions to local planning authorities on the health-related issues. Paragraph 116 of section 10 of the NPPF states:

> 'Local planning authorities must determine applications on planning grounds only. They should not seek to prevent competition between different operators, question the need for an electronic communications system or set health safeguards different from the International Commission guidelines for public exposure.'[1]

Such specific statements are lacking in other areas where public concerns arise on health, environmental and related issues associated with various categories of development. Searching the National Planning Practice Guidance (NPPG) provides very limited content on how public concerns in these categories might be considered. Development on or near unstable land is covered in the NPPG, while guidance on planning applications for waste disposal facilities are contained in a separate document.[2]

The guidance in the NPPG refers to the important role of the planning system when considering land stability in minimising the risk and effects on property, infrastructure, and the public, and helping to ensure that development does not occur in unstable locations or without appropriate precautions.[3]

The National Planning Policy for Waste requires waste planning authorities, when determining planning applications, to consider likely impacts on the environment and on amenity against criteria which are set out in Appendix B of the document and the locational implications of any advice on health from the relevant health bodies. 'Waste planning authorities should avoid carrying out their own detailed assessment of epidemiological and other health studies.' Air emissions including dust should be considered in relation to the proximity of sensitive receptors including ecology and human settlement.

An understanding of how the public perceive risk and how this gives rise to specific concerns when confronted with development applications, within or in the general locality of their community, has become of increasing importance to planning decision-makers. This understanding should include an appreciation of how the public's increasing awareness of environmental issues has expanded, aided in some respects by the greater media attention given to these issues and the increased reporting of development proposals that impact on people's lives. Social scientists and others have studied how the public perceive risks of various kinds, what influences them and how people react in different ways to the risks posed. Writers from various professional backgrounds have identified a basic division, or in one commentator's view 'a battle', between those members of the public who are termed rationalists (or realists) and populists (or constructionists). The Health and Safety Laboratory (HSL) in the UK have produced a 'Review of the Public Perception of Risk, and Stakeholder Engagement'.[4] The authors state in the Executive Summary that:

> 'There is a division between approaches to risk that reflects the debate within the social sciences between the realist and constructionist perspectives, about what is knowable in some "final" sense.'

This HSL document was commissioned specifically to inform the Health and Safety Executive (HSE) regarding the public perception of risk in relation to hazardous industries. A legal article by Chris Hilson contrasts the 'battle' between the rationalists and the populists. His article opens with a key question:

> 'If majority scientific opinion has stated that a proposed incinerator's emissions are safe, or that a proposed mobile phone mast poses no risk to human health – but local residents nevertheless continue to perceive that there is a risk – should planning authorities be able to deny planning consent based on this public concern?'[5]

The main findings of the HSL research include the conclusion that risk perception at the individual level is a result of many factors as opposed to those rational judgements based on the likelihood of harm. The legal and other articles I have considered appear to be agreed that the decision-maker examining a planning application or appeal needs to gain an understanding that combines the positions taken by the rationalists and the populists. Reflecting on my own involvement in planning appeals involving public concern as a key consideration, I agree that this more comprehensive understanding incorporating both viewpoints is necessary.

The following example of building houses above land, previously subject to the mining of sand in Victorian times, illustrates a number of the issues involved in assessing public concern for the local planning authority, consultants involved in putting forward the case and the subsequent decision maker, a planning inspector.

I was involved as the planning consultant for the housebuilder promoting this unusual scheme from the initial pre-application stages through to giving evidence at the public inquiry. The scheme at Glasshoughton, part of the urban area of Castleford in Wakefield Metropolitan District, involved the full remediation of the site and its shallow mining workings and the development of 62 houses. The site extended to 2.3 hectares and consisted of three field units mainly covered by scrub vegetation plus hedgerows and rough grassland. The site was surrounded by existing residential development. The shallow mine workings, extracting sand for the local glass industry, had left a network of underground tunnels. As part of preparing the application plans the tunnel network and indications of collapsed sections were mapped by local cavers. The collapse of some of the tunnels had created surface subsidence and 'crown-hole' depressions with clear on-site evidence of this disturbance and the migration of voids. Specialist geotechnical, noise, dust and vibration reports were prepared by members of the consultancy team.

The planning application was submitted on 12 August 1998, following pre-application meetings in which the planning officers had confirmed that the site was brownfield and was on the register of land suitable for residential development. No points of principle were raised against the proposed residential use. A public meeting was arranged in April 1999 and attended by some 70 residents from the immediate locality, together with local councillors and planning officers. The technical experts within our team presented details of the proposed remediation method, which at this point in time comprised the extraction of sand and limestone with on-site crushing, backfilling, and compaction of this material. This process was estimated to last for a total of 20 weeks. Various public concerns were expressed at this meeting, including some relating to the potential noise, dust and vibration impacts of the remediation proposals and others to the longer-term impacts on their nearby properties. This latter group of concerns included problems associated with the insurability of their properties, declines in property values and the ongoing risk of subsidence. Primarily because of the expressed public concerns, our clients changed the proposed remediation scheme to one involving grouting the tunnel areas beneath the proposed dwellings and access roads.

The application was considered at planning committee on 3 December 1999, with an officer recommendation of approval. In short, the planning officers and their expert consultees considered the proposals to be acceptable and technically sound subject to compliance with a series of conditions. The application was refused on three grounds. The first of these was the effect of the remediation works on the amenities of adjacent residents via noise and dust impacts; the second related to the loss of open undeveloped land, which would result in an unacceptable impact on the amenity of the area. The third reason is the main one that is relevant to our consideration of public concern as a material consideration:

> 'The proposed development due to its location in an area of unstable land, would give rise to justified anxiety and fear for public safety resulting from potential off-site effects including subsidence and disturbance as a result of the proposed site remediation works and subsequent development.'

The level of concerns of local councillors and the public supported by the planning committee covered both short and longer-term impacts. Having taken legal and planning advice, our housebuilder client decided to appeal the Council's decision. In my planning evidence I was asked to deal with the planning aspects of the third reason for refusal while a colleague dealt with technical considerations. I was guided in my planning assessment of the public concern expressed in this case (i.e., the fear and anxiety arising from certain perceived impacts of the development proposals) by Andrew Williamson, acting as the advocate for the appellant. Three leading cases and two articles in the *Journal of Planning and Environmental Law* were identified and I was asked to distil the relevant findings for this appeal and how they should be applied by the inspector in arriving at his decision.

The first of the two articles is by Andrew Piatt, published in May 1997,[6] and briefly concerns itself with whether public opposition to a proposal is a material consideration is in its own right and, if so, what are the consequences for a particular decision. The second article, by Neil Stanley, followed in October 1998.[7] This was the leading legal commentary at the time and remains a highly relevant and valuable contribution in the current context. This article deals primarily with the decision-maker's dilemma, which has, if anything, intensified over the period since this was published. The three leading cases at the time of preparing my evidence were all related to waste handling systems (Gateshead MBC v Secretary of State for the Environment, Newport CBC v Secretary of State for Wales and Envirocor Waste Holdings v Secretary of State for the Environment).[8] Neil Stanley considered that the Gateshead and Envirocor cases provide useful illustrations of the nature of the evidence required to assess the impact of a proposed development upon the public and the environment. He contrasts the technical approach to risk assessment of the experts with the criteria used by the public to assess exposure to development related hazards.

In my proof of evidence, based on the cases and learned articles, I agreed that fears and apprehension (anxiety) brought about by the particular use of land can be a material consideration. Such public concern is a material consideration provided that the concern is brought about by the proposed development, which again I agreed applied in this case. The question then arises as to what weight is to be applied to any such material consideration. To help the planning witness and the decision-maker the types of public fear in such planning cases can be quite simply categorised in the following way:

- Fears which are justified by evidence.
- Fears which are not justified by evidence but are genuine and understandable.
- Fears which are both unjustified and unreasonable.[9]

Research referred to in Neil Stanley's article shows that the public use 'rules of thumb' to describe and express their concern. 'Inferences are drawn from what the public can remember hearing or seeing in order to calculate the likelihood of an event occurring, or reoccurring.' From the public meeting and written submissions from individuals, I was able to conclude in my evidence that residents were identifying past occurrences of property damage and linking this to the likelihood of a future event(s) associated with the proposed residential development.

From his extensive research, Stanley identifies 13 factors that can contribute to the public's assessment of risk in a particular case and how this is translated into varying degrees of public concern. Under the sub-heading 'Conclusions from public concern research' he refers to psychologists and decision-researchers forming two 'clusters' of factors that are especially important to the assessment of the degree of public concern. The first cluster of factors is described as 'dread risks' and the second cluster are described as 'unknown'. Industrial developments, including the incineration of waste, which exhibit dread risks will generate very high levels of public concern. In the planning inquiry into the Gateshead clinical waste incinerator a substantial amount of time was devoted to the consideration of the environmental impact of emissions upon humans and the surrounding environment. Public concern focussed on the toxic impact of dioxin and cadmium emissions and the frequency of plant accidents.

In the Glasshoughton appeal case I stated in my evidence that the type of hazard raised by public concern tended to fall more towards the dread risk cluster but not to a clear extent. In my evidence I examined the public's concerns by assessing them against the main characteristics of dread risks identified by Stanley and others. In this case the public were mainly concerned that the drilling, traffic, and other actions associated with the grouting and construction scheme would trigger off-site tunnel collapses leading to structural damage to property and financial loss for individual households. An additional concern related to potential damage to the approach highways to be used by construction traffic. The public were concerned about the uncertainty of the situation based on past events and they had created a causal link between the development and any future structural damage that may occur. A major distinguishing factor in this case was that the risk to property above and adjoining the sand workings already existed as a result of the former mining and the incidence of subsidence, which has created varying degrees of collapse and some structural damage.

I went on to argue that the core issue was whether the grouting remediation works, which would stabilise the site ready for residential development, would have any identifiable off-site impacts on land stability. The public concern was centred on past events, uncertainty, and the potential for future linked events. The appellants on the other hand were relying on extensive technical evidence to substantiate their firm contention that there would be no structural damage resulting from the remediation and construction works. This conclusion was supported by the Council's expert geo-technical consultants and by the experience of grouting schemes for shallow mine workings in other parts of the country.

The general position arising from the case law relevant at the time of this public inquiry is that public concern becomes a material consideration in its own right, but to gain weight in a particular case it must be justified or substantiated in some way. The decision-maker must consider the weight they apply to the public concern against that of the appellant's technical evidence plus the other material considerations that are applicable. I argued in my evidence that there were a number of factors that should reduce the weight to be given to public concern in this case, including the clear national and local support for the reclamation and re-use of derelict land, the tried and tested nature of the grouting solution and the extent of the investigative work carried out, which included two sets of test drilling, geological investigations, the mapping of the tunnels by underground survey and the large measure of agreement between the technical experts employed by the planning authority and the appellants.

Approximately one third of the Planning Inspector's decision letter of nine pages covered the issue of public concern. He makes an interesting comment on the lead court cases: 'Apart from clearly establishing that this subject (public concern) is a valid material consideration and a possible reason for the refusal of planning permission, irrespective of whether the fear and anxiety is warranted, they are somewhat inconclusive.' He picks out a key comment from the judgement of Lord Justice Glidewell in the Gateshead case as the most important to emerge from all the cases:

> '... if in the end ... public concern was not justified, it could not be conclusive. If it were, no industrial development – indeed very little development of any kind – would ever be permitted.'

The inspector goes on to add his own opinion by stating that: 'It would indeed be perverse if any public concern, however irrational or unjustified, could prevent development. It is necessary, therefore, to consider whether the genuine fears of residents are justified or not. Such a consideration requires a thorough assessment of all relevant evidence presented at the inquiry.' The inspector concluded, having summarised the evidence, that the likelihood of off-site subsidence and damage to neighbouring property being caused by the remediation works and housing development was negligible. He further concluded that the fear and anxiety for public safety, while understandable, should be regarded as unjustified. The appeal was allowed, and planning permission was granted subject to several conditions.[10]

I have experienced first-hand the public concern on health and wellbeing grounds regarding waste incineration and treatment projects. In the 1990s I was appointed by International Technology Europe (ITE), an American company who were trying to establish large industrial waste incinerators on a regional basis in the UK and Europe. They had been unsuccessful in two appeal cases in the North East region of England. Their third attempt at success was on North Humberside within the general locality of a small village, Paull, but located within an industrial zone. I was one of 14 witnesses, 9 of which were American technologists/scientists. On the opening day of the inquiry, a substantial band of protestors marched down the village main street to the inquiry venue dressed in white boiler suits and gas masks, bearing stark warnings on placards. The inquiry opened a couple of weeks before the Christmas holiday and the locals turned out to be generally friendly, with some of them providing cakes and coffee during breaks. The need and technical/scientific case were very strong indeed, but the appeal was 'called-in' for determination by the Secretary of State and turned down on development plan policy grounds that had little to do with the balance of evidence given at the inquiry. Our leading counsel, Andrew Gilbart QC, advised the company to appeal to the High Court providing strong grounds for challenge. However, ITE decided that three straight defeats were enough and did not pursue the challenge. The public opposition and concern in all three cases had a determining influence on the outcomes, but it was clear that in the Paull case political influence swung the decision. I simply introduce this to demonstrate the power of public opposition and concern in some cases, sometimes with justification and in others totally lacking any evidence base. The encouragement of opposition to incinerator projects is well demonstrated by the publication of a campaign guide for opposition groups by Friends of the Earth.[11]

The next group of development proposals that I briefly consider are telecommunications masts. We have seen how the NPPF seeks to guide, one might say restrict, public opposition to planning proposals for new mast facilities. A short review of the court cases is interesting in the context of whether there is evidence of adverse impact on the health and wellbeing of the public who live close to a proposed installation. The Government have consistently stated in current planning policy and previous planning guidance, going back to the revised version of PPG8 – Telecommunications,[12] that local planning authorities should follow the ICNIRP guidance and not seek to set any health safeguards that are different to that guidance. These guidelines relate to the transmission of electromagnetic fields, from mobile phones telecoms base stations, a form of non-ionising radiation. The International Commission's guidelines are based on the avoidance of excessive localised and whole-body heating. Public Health England have recently produced an updated version of their guidance on the subject.[13] This is a somewhat bland document that supports the government's planning policy approach and the updated guidelines of the ICNIRP. While the support for using these guidelines is very strong, some counter evidence has emerged over recent years. It is not the purpose of this chapter, or indeed this book, to explore in any detail the extent of evidence for and against certain actual, or alleged, health impacts of particular types of development. It is relevant to broadly assess whether public concerns are understandable and may be justified. On this basis I only refer to one example from many documents that seek to establish that there are real health risks. This is an objection letter produced by the co-founder and Charity Director of the Radiation Research Trust (RRT), Eileen O'Connor, addressed to the members of Ofcom in March 2010.[14] The writer had been living 100 metres from a mobile phone mast and for a number of years suffered sleep problems, headaches, dizzy spells, and vertigo. Subsequently she developed breast cancer and reports that she was living in a cancer cluster. Her GP had not been able to recognise the earlier symptoms, but her condition was later recognised as 'electro sensitivity'. Her research and campaigning are based on her personal experiences and wide-ranging meetings and reviews of scientific opinion. In September 2008 the RRT brought together leading experts in the fields of science, politics, and regulation at a major international conference – 'Electromagnetic fields and health – a global issue' held at the Royal Society, London.

The Chairman of ICNIRP made the following clear statement: 'The ICNIRP guidelines are neither mandatory prescriptions for safety, the "last word" on the issue nor are they defensive walls for industry or others.' Sir William Stewart, who had chaired a government established group of experts to investigate and report on the possible health effects of electro-magnetic waves, also spoke at the 2008 Conference. In his report published in May 2001 it was concluded that there was no evidence of health risks but advised caution until more research was done. In his conference address he made the following statement:

> 'Since 2000 there has been a mass of publications, reports, observations, and views purporting at the very least to implicate phones/base stations as a cause of adverse health effects. At a time of uncertainty when more information is required, non-peer reviewed articles should not be ignored. Doing so is ridiculous. They may be right but unproven and/or offer pointers to be thought about and followed up.'

Sir William, the RRT and others called for the application of the precautionary approach in the UK. Several European countries, regions and a few city authorities have adopted their own standards regarding public exposure levels below those advised by ICNIRP. Finally, amongst growing scientific evidence on the health impacts of electromagnetic radiation, it is worth mentioning the 2007 BioInitiative report produced by an international working group of scientists, researchers, and public health policy professionals. The report provides detailed scientific information on health impacts when people are exposed to electromagnetic radiation many times below the ICNIRP guidelines.

In this context it is, to say the least, interesting that the 2020 Public Health England guidance refers to the existence of health-related evidence, reviews and research including ongoing studies in the UK and elsewhere plus the support of the World Health Organisation and a European Commission scientific expert committee as supporting the ICNIRP exposure levels guidance. Following this summary review of the evidential position on the health and wellbeing impacts, I now turn to examine relevant judicial decisions and legal commentaries on the issues these considered, to see if they assist planning practitioners in their difficult decision-making role.

The comparison of the Gateshead and Newport court cases on waste treatment revealed a division between support for the rationalist approach to public concern and the populist approach. The more recent case law relating to telecoms masts in England has again revealed inconsistency in court decisions with the same division in the approach to public concern between the rationalist and populist. In Phillips v First Secretary of State [2003][15] there was a challenge to the siting and appearance of a mobile phone mast on a site close to the claimant's home. This is predominantly a decision involving the need to consider alternative sites. The judge also identified that there was an important public concern element to the case. The then government policy contained in PPG8 2001 recognised that public concerns about the health implications of a development can be a material consideration. The judge concluded that the existence of such concerns was one of the reasons why the location of telecommunications structures is such a sensitive issue. He then went on to state: 'It seems to me to follow, again as a matter of principle, that if there were two alternative sites each of which was otherwise acceptable in environmental terms, it would be open to a decision-maker to refuse approval for one of those sites if the location of a mast on that site would give rise to substantially greater public concerns than its location on the alternative site.' The judge made it clear that he was not saying that this was the way in which a particular application ought to be decided, but it would be lawful for a decision-maker to approach the matter in that way. The claimant's contention in this case was that she had been denied the ability to present evidence on the suitability of an alternative site that the appeal inspector would have considered. The judge supported the claimant's case that the inspector's conclusions may have been different if an alternative site had been available.[16]

A second case, Trevett v Secretary of State and others,[17] makes some important points, particularly regarding the way in which government guidance should be considered by the decision-maker. At the time of the Phillips and Trevett decisions, government guidance was contained in the revised version of PPG8 and specifically refers to the way in which health considerations should be taken into account when taking decisions on telecoms development. Having stated that these can constitute a material consideration and it is for the decision-maker to determine the weight to be given to such considerations the guidance then states:

'However, it is the Government's firm view that the planning system is not the place for determining health safeguards. It remains central Government's responsibility to decide what measures are necessary to protect public health. In the Government's view, if a proposed mobile phone base station meets the ICNIRP guidelines for public exposure it should not be necessary for a local planning authority, in processing an application for planning permission or prior approval, to consider further the health aspects and concerns about them.'

In the Trevett case the judge, Sullivan J, citing a Court of Appeal case where Schiemann LJ had spelled out the status of government guidance, re-stated that 'guidance is no more than that: it is not direction, and certainly not rules. Any appeal panel which, albeit on legal advice, treats the Secretary of State's Guidance as something to be strictly adhered to or simply follows it because it is there will be breaking its statutory remit in at least three ways: it will be failing to exercise its own independent judgement; it will be treating guidance as if it were rules; and it will, in lawyers' terms, be fettering its own discretion. Equally, however, it will be breaking its remit if it neglects the guidance. The task is not an easy one.'[18]

The Trevett case relates to an appeal decision to grant conditional planning permission for three communication masts at three separate sites in the Stroud District Council area of Gloucestershire. The Council had refused planning permission on all three mast proposals. Sullivan J concluded that the inspector had acted properly in all respects in relation to the way in which the PPG8 guidance should be considered particularly given the somewhat contradictory nature of the content of that guidance. The local planning authority had however:

'... failed to examine the basis for the public's fears to see what, if any, weight should be given to them and whether they justified a refusal of planning permission, given the benefits of the proposals. Just as it is erroneous to proceed on the basis that perceived health risks cannot justify a refusal of planning permission unless they are objectively justified, so it is equally erroneous to assert, as was the Council in effect, that merely because there are perceived risks to health, that justifies a refusal of planning permission without any regard to the extent as to which those fears are objectively justified in the circumstances of the particular case and given the particular characteristics of the site in question. Those are the factors which the Inspector considered in concluding that the perceived risks as to health, whilst material considerations, did not justify a refusal of planning permission in these three cases.'

It is difficult to draw clear conclusions from these different case decisions. Since these court decisions were made, increasing scientific evidence has emerged suggesting that there are actual adverse health outcomes linked with the electromagnetic waves. As a planner who has had to make, or advise on, many difficult decisions, I can only conclude that there is now scientific evidence on both sides of the argument and that public concern including fears and anxiety are more likely to have some justification in certain cases. The evidence on both sides will need to be weighed very carefully indeed in arriving at planning decisions. The government have not shown any sign of changing their firm policy approach of sticking to the ICNIRP guidelines, which in the context of the Chairman's comments at the 2008 London conference, looks more like a political balancing act given the policy support expressed in the NPPF for the 'economic growth and social well-being' benefits of high quality and reliable communications infrastructure (paragraph 112). At the time of drafting this Chapter, a legal challenge is being prepared under a public funding appeal that seeks to change the current Government's policy approach and total reliance on the ICNIRP's guidelines.

The public's perceptions of the potential and actual risks to the environment, and to health and wellbeing as a result of certain types of development project, are growing. This growth is fuelled by the heightened level of understanding of the issues and the actual impact outcomes that might arise, by wider media reporting, by the existence of well organised national and local action groups who offer their support and by actual evidence of specific events and impacts. I have described the Government's policy approaches to the issues raised on developments that present some risk and produce public concern as at best patchy. The example relating to the health hazards of mobile phone base stations now requires urgent review, which might come more quickly if the proposed legal challenge is progressed and succeeds. The same criticism applies to national planning guidance, which does not address many of the issues and complexities giving rise to the decision-maker's dilemma.

Both Stanley and Piatt, in their commentaries (1997/1998), concluded that new government policy guidance was necessary to ensure that the planning process can manage development related risks effectively. Stanley argued that 'there was an obvious need for a judicial decision of the House of Lords to clarify what is required to "justify" or "substantiate" public concern.' Piatt reminded us that the purpose of town and country planning legislation was to put the protection of the public interest at the forefront of decision making. We can now add to this the emphasis throughout the NPPF to achieve development which is truly sustainable. While this approach has not yet been enshrined in English law it does give added weight to the protection of the public interest. The problem is that the government do not place this approach at the centre of all their topic-based policies.

The lack of clear national policies and guidance on many types of risk-based development does not help the decision-maker. However, there are several principles that emerge from the cases examined, the commentaries referred to and from individual appeal decisions:

- Public concerns and public perception of risks go well beyond mere expressions of public opposition, provided that there is a direct link between the proposed development/land use and impacts on neighbouring land users.
- Public concern can often comprise fear and anxiety and those fears can be of different types/origins. It is important for the decision-maker to understand these to attribute appropriate weight to them.
- The decision-maker must also be able to review and understand the nature and content of the evidence presented to substantiate public concern and the evidence presented by the applicant/appellant to arrive at a balanced decision.
- The public perception of risk or concerns regarding potential adverse impacts can be a basis for refusing planning applications/appeals, even though these are not supported by scientific or logical reasoning. Such concerns can tip the balance of considerations against the proposal. A good example of this type of decision is the consideration of alternative sites.

These principles need to be demonstrably followed by decision-makers at all levels of the system which includes those ministers responsible for planning in the MHCLG. Public concern and the fear and anxiety, which is often apparent in the assessment of application projects, are not only material planning considerations, but they are also at the heart of sustainable development in the form of the 'social objective' that directly embraces health and wellbeing. This central objective of the three interlocking objectives includes the delivery of a safe built environment and supporting the health, social and cultural wellbeing of communities. Section 8 of the NPPF – Promoting healthy and safe communities – has an emphasis on public safety, but this is mainly focussed on crime and disorder. Public fear of crime is referred to but there are no references to the ways in which planning has to consider other sources of public concern and fear arising from developments that pose specific or potential risks. There is one major area of public concern that spans land-use and development planning and is clearly linked to all aspects of town planning and specifically to risk management. The topic of climate change and the role of town planning in mitigation, adaptation and future policy formulation is considered in Chapter 12. The impacts of climate change pose current and future risks and there is little doubt that these are matters of great, though not universal, public concern. These concerns include fears and anxieties regarding current and yet to be realised impacts. Flooding and the increased frequency of flood events is the most apparent among the current concerns and in impacted communities the anxiety and fear are readily apparent.

There can be little doubt that public concern regarding potential health and wellbeing impacts from certain types of development will appear more frequently as a material consideration in many application decisions. Without clearer and more explicit national policy statements and improved guidance for decision-makers further planning appeals and high court challenges will follow.

Chapter 9

Regional Planning – Its Rise and Fall and the Case for a Resurgence

Regional Planning in the UK has had a chequered history, with England displaying the most volatile position since the adoption of the 1947 planning system. The devolved nations, and in particular Scotland, have displayed a greater attachment to some form of regional planning brought about by a mixture of clear regional/national identity and culture, distinct socio-economic problems, fewer urban concentrations, and large rural areas. For a large part of the history of the modern UK planning system, regional planning was predominantly regional economic planning with little emphasis on regional spatial planning. The origins of regional economic planning in this country lie in the great socio-economic inequalities arising mainly from the Great Depression of the early 1930s existing between Scotland, Wales, the North of England, South West England and the more prosperous areas of South East England, East Anglia and to some extent the West Midlands.

The emphasis on the need to provide economic support to certain regions starts with the 1934 Special Areas (Development and Improvement) Act, with four areas designated for assistance: South Wales, Scotland, North East England, and West Cumberland.[1] In Chapter 1 we identified the significance of the wartime Barlow Report (1940) and its radical policy initiatives to redistribute population and industry in order to reduce unemployment and congestion mainly in Greater London. Barlow continued to influence post-war planning and the 1945 Distribution of Industry Act passed by the wartime Coalition Government. The 1934 Special Areas were enlarged and redesignated as Development Areas. A system of loans and grants were introduced alongside the financing and construction of new trading estates. The system was somewhat weakened by the incoming Conservative Government. Throughout the 1960s there were various changes to the extent and definition of the assisted areas, and the Labour Government launched a National Plan in 1965 together with Regional Economic Councils in 1965/66. This national/regional system has been compared to the 'French Model'.[2]

In the late 1970s the Inner Urban Areas Act of 1978 introduced a major shift away from the more peripheral assisted areas to the inner cities of the major urban areas. The Thatcher governments of 1979–1990 demolished much of the regional economic planning structure and financing, arguing that unemployment had worsened in the South East and the West Midlands and that the great majority of grant monies had been targeted at capital intensive industries including heavy engineering, shipbuilding, iron and steel, coal mining and heavy chemicals, with a poor job creation result.

This background on regional economic planning forms important contextual material for the main subject of this chapter, which is regional spatial planning. As we will see there have been examples, in geographical and organisational terms, where regional economic and spatial planning have been aligned or integrated. A brief review of the history of spatial planning at the regional level in the UK also forms useful context in developing the case for its reintroduction. The example of French regional planning assists this case.

The work of Ebenezer Howard and his blueprint *Garden Cities of Tomorrow* is the first holistic model for city-region spatial planning born out of a recognition of the inequalities and health and wellbeing problems associated with the growth of London in the latter part of the nineteenth century. Part of Howard's 'grand plan' for the Greater London area was realised in the form of Letchworth and Welwyn Garden Cities and, subsequently, in the regional plans of Patrick Abercrombie in 1944 and 1946. These plans for the London area were the first attempt to establish a framework for the dispersal of the urban population from the overcrowded and still unhealthy inner urban areas of London and the other major conurbations. These advisory regional plans were prepared by distinguished planning consultants either on behalf of central government or an advisory committee of local authorities.[3] The question of where overspill from the conurbations should be accommodated could not be decided within the framework of local government, even though the 1947 Act had led to the rationalisation of the number of local planning authorities. For example, there were some 67 local authorities operating within six miles of the centre of Manchester. Fifteen local authorities administered the area covering the Clyde Valley to the east and west of Glasgow.

The various regional plans that emerged contained proposals for new towns and town expansion schemes and green belts. The central aims of these plans were to improve the conditions in the overcrowded central areas and to provide the out-migrants with new homes and better living conditions and working opportunities. These early examples of regional framework plans, centred around population overspill schemes, were in no sense holistic spatial plans that looked at all the inter-related topics that need to be covered to achieve successful outcomes, including large scale countryside, environmental and transport planning. There were also problems associated with poor relationships between the exporting and importing local authorities and the lack of drive and purpose from central government.

In the early 1970s I became a member of the Nottinghamshire/Derbyshire sub-regional planning team, with responsibilities for economic development alongside my main role as Economic Development Officer for Nottinghamshire County Council. This was an integrated macro plan approach, which included the establishment of the Mansfield–Alfreton Growth Zone in the declining areas of the coalfield. This was relatively successful in achieving some intra-area transfer of existing industry, together with the attraction of new companies from outside the sub-region to newly established industrial estates with good access to junction 28 of the M1 motorway.

It is useful at this point to introduce a comparison with the French approach to regional planning. The primacy of Paris, the national capital, emerged strongly during the nineteenth century when its population increased threefold, and it became the dominant centre for government, industry, finance, culture, and learning. In the twentieth century the rise of Paris 'to a position of unchallenged supremacy within the French urban system' continued.[4] Regional planning commenced in the mid-1950s, largely as a response to the over-dominance of Paris, with the establishment of 22 planning regions concerned at first with economic development. These were formed in the context of a national planning system with national plans being produced every five years from 1946 onwards. The first three of these national plans concentrated on economic development and attempts to curb the industrial expansion of Paris by decentralisation to the other regions. The fourth national plan, introduced in 1962, was the first to widen investment planning to cover housing, health, education, and transport and to attempt to change the urban system and the dominance of Paris.[5]

In 1964 the French government introduced the concept of the 'balancing metropolis' or the 'metropoles d'equilibre', which recognised both the potential of the French provincial capitals as well as their very weak roles in the national system. They considered that this was the only way of reducing the pull of Paris. There was also an additional incentive arising from France's membership of the European Economic Community, which gave rise to the desire to put the main provincial cities on an equal footing with those of other member countries of the Common Market. This had the aim of integrating them into the evolving European network of transport and other communication services. Eight cities, or groups of cities, were designated as metropoles d'equilibre, these being considered as the areas most

Regional Planning – Its Rise and Fall and the Case for a Resurgence   229

likely to produce a positive response to this national policy initiative. Those chosen included the single cities of Strasbourg, Bordeaux and Toulouse, and groupings including Lyon–St Etienne–Grenoble, Lille–Roubaix–Tourcoing, Nancy–Metz–Thionville, Marseille–Aix, and Nantes–St Nazaire. In 1966 urban planning regions were established for the eight selected metropolitan areas. Over the following decades the metropoles were connected to the Paris transport hubs by the TGV high speed rail system. New motorway connections were also introduced. This national policy has seen the growth of these selected metropolitan areas, their specialisation in certain industries (e.g., aviation in Toulouse), relative improvements in higher education, health and service provision and greatly improved communications. The movement of industry and headquarters from Paris has been limited. The creation of directly elected regional governments in 1984 has strengthened the position of the eight metropoles as well as smaller cities. The regional economic and spatial planning systems are further strengthened by political and other linkages between the 36,000 French communes, regional and national government.

This brief examination of the French regional planning system is a very useful comparator for the current position in England, with its lack of a national plan, the relative dominance of London and South East England and the strong north-south socio-economic divide. The government are seeking to address this divide via a process of 'levelling-up' which lacks clarity and purpose, though some improvement has subsequently occurred with the publishing of a White Paper (presented to Parliament 2 February 2022).[6] The position is still relatively weakened by the absence of formal systems of regional economic and spatial planning. Prior to addressing the current need to reintroduce regional spatial planning and how this can play a major role in the so-called 'levelling-up' process, I examine the emergence of Regional Planning Guidance and the subsequent English Regional Spatial Strategies.

Non-statutory Regional Planning Guidance (RPG) was introduced in 1986 and 13 of these documents were issued in the period up to 1996. The guidance was designed to inform the preparation of structure plans by the local planning authorities but was prepared with very little public engagement and had little standing in the decision-making process. The RPGs were progressively replaced by Region Spatial Strategies (RSS), which were much more comprehensive documents and were introduced as part of major changes to the planning system by the Planning and Compulsory Purchase Act of 2004. An RSS was required to be prepared for each of the nine English regions, which had been defined in the Regional Development Agencies Act of 1998. This Act, introduced by Tony Blair's incoming Labour Government, was the first step towards realising their vision for an integrated form of regional government and planning in England. The next stage was reached in 2003 when the Regional Assemblies (Preparation Act) was passed. This gave the deputy prime minister, John Prescott, the power to call regional referendums asking voters whether they would support directly elected regional assemblies. Three were planned for 2004 and these were all in the North of England (the North East, North West and Yorkshire and the Humber regions). Greater London, at this time, had its own regional assembly, which had been established in 2000 following a local referendum. The Greater London Assembly (GLA) has 25 members and, together with the London Mayor, continues to provide an elected upper tier of local government in the capital. The reasons that this regional tier of government was established and has survived for the last twenty years are, in my view, a result of the complexity of the urban issues faced by London, the metropolitan outlook of its voters and the need for a coordinating strategic plan to provide a clear framework within which the 32 borough planning authorities can work.

Part 1 of the Planning and Compulsory Purchase Act 2004 in requiring the production of Regional Spatial Strategies made them part of the statutory development plan system for England. While this was not accompanied by the regional government model desired, due to the lack of a public appetite for this in the trial referenda, it was the first time that a comprehensive two tier regional and local development plan system was introduced. The Regional Development Agencies Act of 1998 had made provision for nine regional areas including London. Had the Government decided to proceed to elected regional assemblies they would have had political independence but ironically would have been implementing the Secretary of State's policies.

Regional Planning Bodies (RPBs) were established in the place of formally elected assemblies and at least 60 per cent of their membership was made up of elected councillors from the constituent local authorities of each region. Where a region contained a national park the Park Authority was allowed to have nominated members on the new organisation. The RPBs were required by section 3 of the 2004 Act to consult a broad range of regional stakeholders through focus groups and planning forums. The RSS for each of the nine regions was to be kept under review and the implementation of policies and proposals were to be monitored with an annual report on implementation. Each RSS was to include a formulation of appropriate policies and then make a statement of those policies. These would be the policies of the Secretary of State. While the RPBs were independent in political terms the role of the Secretary of State was extensive and covered the provision of the national policy context, approval of the regional policy content, consideration of the full draft

RSS and the need to hold an Examination in Public (EIP) and subsequently to approve any revisions recommended by the appointed inspector.

Collaborative working with the constituent local authorities was required by section 5 of the Act to ensure that their local knowledge and advice were considered. Section 6 embedded community consultation in the statutory process of preparation and approval. Each RSS, as a major strategic spatial plan, was required by European law to be subject to Strategic Environmental Assessment (SEA). The English approach was to subsume SEA in the production of a wider Sustainability Appraisal. Not surprisingly there was a tension between the Secretary of State's policy-making powers and the delegated roles of the constituent local authorities.

My recollections of working with the draft and final versions of a number of the RSS and attending EIPs representing groups of private sector clients was the relative thoroughness and comprehensive nature of the process and admiration for the knowledge, evidence and creativity that went into the making of these regional spatial plans. They were truly comprehensive and combined strategic land use, spatial and transport planning in a clear and effective way. The requirement to work with other regional bodies ensured that infrastructure planning was appropriately integrated. Utilising specialist consultants in demographics, economic forecasting and related subjects enabled the production of housing and employment forecasts for each of the constituent local planning authorities across a common plan period. This 'top down' approach was not seen as a problem by my colleague practitioners and my housebuilder and commercial developer clients. While the preparation and approval process were quite lengthy, the end products provided a reliable, integrated, and comprehensive basis for the preparation and review of local development plans. The EIPs were, in my experience, well conducted by the inspectors appointed by the Secretary of State. There was little, if any, of the tensions and disagreements now experienced between the participants at EIPs into draft local development plans, though I acknowledge the different level of the subject matter and the fact that the public did not have the same level of engagement with, and understanding of, the regional/sub regional material as they do with local policies and proposals.

The RSS process was, in a word, a success. My experiences and those of my contemporaries strongly support this conclusion as does my subsequent reading on this subject. The strength of these views is heightened by the rapid removal of this upper tier of planning in England. I recall the shockwave amongst the planning profession, which accompanied the announcement from the new Secretary of State, Eric Pickles, immediately following the election of the May 2010 Conservative/Liberal Coalition Government. Pickles was a bluff Yorkshire Conservative and former leader of Bradford Council. His immediate action in May 2010 was to revoke the application of the system and to announce its forthcoming abolition. The abolition was formalised by statute in July of that year. This was in my experience one of the biggest, if not the biggest, acts of political vandalism against the planning system since its introduction in 1947.

Guidance to local planning authorities followed and sought to provide some justification for the dramatic change. In his letter the Chief Planner at the DCLG[7] stated that:

> 'In the longer term the legal basis for Regional Strategies will be abolished through the Localism Bill that we are introducing in the current parliamentary session. New ways for local authorities to address strategic planning and infrastructure issues based on cooperation will be introduced.'

The derivation of housing need and requirements was to be the responsibility of the individual English local planning authorities 'identifying a long-term supply of housing without the burden of regional housing targets'.

One of the main drivers for the Secretary of State and his colleagues in abolishing the system was their determination to remove the 'top down' prescription of housing numbers and to replace what was regarded as an overly bureaucratic tier of planning with a more localised system. The logic for the replacement sub-system was, as we shall see, somewhat lacking. However, the planning profession were by no means alone in their dismay at the untimely passing of the regional spatial plans. In March 2011 the all-party House of Commons Communities and Local Government Committee published its report on the implications of the abolition of the RSS.[8]

The Committee launched their inquiry in the summer of 2010. Their terms of reference included the following:

- The implications of the abolition of regionally derived housing building targets for levels of housing development.
- The likely effectiveness of the Government's plan to encourage local communities to accept new housing development using financial incentives.
- The level of incentives that will be needed to ensure an adequate long-term supply of housing.
- Consideration of the arrangements that should be put in place to ensure appropriate cooperation between local planning authorities on matters formerly covered by RSSs, such as waste, minerals, and flooding, as well as housing.

- They also questioned how the data and research previously collated by the Regional Local Authorities Leaders' Boards should be made available to local authorities and how that research would be updated.

The Committee were assisted in their task by written and oral evidence from the planning sector and from community and environmental groups, the housing industry, local authorities and representatives and Ministers from the DCLG.

One of the witnesses quoted in the introduction made this telling statement:

'Having taken 30 years to build up the strategic planning system, and perhaps three years to prepare each Regional Plan, it has taken literally 3 months to abandon the whole process and create a situation of complete paralysis in the planning system.'[9]

The Committee heard evidence for and against RSS, but in their conclusions and recommendations there was a clear emphasis on the need for some form of regional planning to plug the definite gap appearing in the summer of 2010 between increased localism and national policy and decision-making. There was clear concern regarding this planning vacuum and the absence of any transitional provisions, which had always been provided in the past when major changes to the planning system were being progressed through the legislative system.

The Committee placed emphasis on the wide reach of the RSS and its ability to provide a clear, evidenced base for taking strategic planning decisions on wide reaching, difficult and complex projects that had impacts beyond one local planning authority's jurisdiction. Several private sector bodies and regional infrastructure providers gave evidence that demonstrated how RSS had helped them in their regional planning and decision making. The Committee summarised the origins of the main objections to RSS as coming from those who were concerned with protecting the Green Belt and land that was seen as having local values, from the pressures of housing development. The RSS were seen by these objectors as not reflecting the aspirations of local communities. The reality is that the much debated 'top-down' housing targets were arrived at following a considerable amount of negotiation with the local planning authorities.

A group of some 29 organisations from a diverse range of sectors wrote to the Secretary of State at the end of July 2010 advocating some form of strategic planning which it defined as 'larger than local' planning:

'We wish to ensure that the larger than local planning and investment – which we term strategic planning – is carried out to address the most pressing issues facing the nation such as economic recovery, meeting housing need and demand, sustainable transport, regeneration, sustainable development and growth, investment in our infrastructure, biodiversity loss, climate change and reducing inequality.'[10]

The Committee went on to recommend that the Government include provision for effective strategic planning arrangements in the Localism Bill. In December 2010 the Coalition Government published their 'essential guide' to Decentralisation and the Localism Bill.[11] The Executive Summary of this document outlines the plan for the shift of power from Big Government to the Big Society. The Coalition set out their commitment to decentralisation and devolution of power with six essential actions ranging from lifting the burden of bureaucracy, through empowering local communities to do things their way to strengthening accountability to local people. The Guide included notes explaining the changes proposed to the planning system in part 5 of the Localism Bill. Chapter 1 of part 5 included powers to abolish the regional strategies, places a duty to cooperate on local planning authorities in relation to the preparation of development plans and makes changes to the processes for adopting, examining, and publishing development plan documents. Chapter 3 made provision for the preparation of neighbourhood development plans.

The Government decided that Local Enterprise Partnerships (LEPs) would not be defined in legislation and would not therefore have a statutory purpose. LEPs were to differ across the country in both form and functions to best meet local circumstances and opportunities.

The RTPI in a paper on the planning content of the Localism Bill expressed a number of concerns with regard to the content of the Bill regarding the Duty to Co-operate, including that it does not require joint working, there was no requirement to reach agreement on policy and it does not set out what would constitute a failure to cooperate and no sanctions are specified for such failure.[12] The RTPI had earlier provided a briefing note on strategic planning that quoted a number of examples of local planning authorities undertaking joint working in spatial planning, including the preparation of joint Local Development Framework core strategies, sub-regional working and sub-regional strategies.

* * *

A review of what happened regarding regional planning pre and post the Localism Act (2011): The example of South West England

The South West of England is the largest of the nine English regions and is larger than Wales. Inward migration is the key element of population growth. Major urban growth has occurred around Bristol, Bath, and Weston-Super-Mare (the West of England sub-region), in south-east Dorset and in other key towns. Growth has also impacted on the character of smaller market and seaside towns. Need and demand for new housing is only being partially met and this is placing a strain on the physical and social infrastructure, with all the main urban areas of the region experiencing road traffic congestion and pressures on open spaces and countryside.[13]

Two further rounds of local government reorganisation in 2009 and 2019 reduced the number of individual planning authorities from 53 to 35. The Counties of Cornwall and Wiltshire became individual unitary authorities with the abolition of their 'second tier' districts/boroughs. The County of Dorset and its seven second-tier councils became two unitary authorities, six of these councils forming the new County and the other Christchurch Borough being added into a new Bournemouth/Christchurch /Poole urban area and known as the South East Dorset conurbation. In 2020 the region's local government structure comprised 12 unitary authorities, 3 County Councils – Devon, Gloucestershire, and Somerset with a total of 18 districts and boroughs in a two-tier arrangement – and two national park authorities – Dartmoor and Exmoor.

The situation in the rest of England adds to the complex array of alternative governance systems. Provision was made in the Localism Act 2011 for directly elected mayors, subject to confirmatory referendums in local authorities specified by the Secretary of State. The West Midlands and Greater Manchester are examples of Metro-Mayors, as are the city-regions of Liverpool and Sheffield. The Metro- Mayors were elected as part of the devolution deals provided for in the Cities and Local Government Devolution Act 2016. West Yorkshire, in recent times, has had a Combined Authority for certain limited areas of government, including transport and economic development, but in May 2021 elected its first Metro-Mayor under a devolution deal and assumed enhanced devolved powers and funding from central government.

Returning to the South West Region, the sub-area we can describe as the Bristol City Region, but confusingly known as the West of England in local government terms, is made up of four unitary authority areas – the City of Bristol, Bath and North East Somerset, North Somerset, and South

Gloucestershire. This largest urban agglomeration in the South West also elected a Metro-Mayor in May 2017 with responsibilities for strategic government including planning, transport, and skills. The City of Bristol also has its own elected Mayor.

All the local authorities in the South West Region first came together for regional planning purposes in 1990 via the establishment of the South West Regional Planning Conference (SWRPC), which comprised elected members from each authority. They worked together to assist the government in the production of the first Regional Planning Guidance in 1994. In 2000 the South West Regional Assembly succeeded the SWRPC and worked with the Regional Development Agency on the production of the Regional Spatial Strategy (RSS) in common with the other English regions. It then took many years of hard work by planners and politicians to produce an agreed draft spatial strategy document across this large and diverse region. The draft plan then spent four years in the central government processing stage, longer than for any of the other draft RSS. By the time the fate of revocation befell all the 9 RSS, the South West RSS had the unfortunate distinction of never being put in place as a finally approved plan. All those bodies which had any form of regional planning remit in the South West were abolished including the Regional Assembly, the Regional Development Agency, and the Government Office for the South West.

Following the Localism Act, a number of the South West planning authorities took the Duty to Co-operate requirement in the Act seriously and decided to produce joint Core Strategies or Local Plans. All these joint plans have responded positively to the wider realities of socio-economic planning issues, which encompass areas wider than the administrative boundaries of an individual local planning authority. This is particularly true of housing market and journey to work areas. These factors along with many other transboundary issues comprise the case for some form of regional or sub-regional spatial planning.

This case was nowhere more apparent than in the West of England subregion. Geoff Walker, in his excellent review paper, charts the origins of attempts at city-region spatial planning in the West of England and the most recent attempts at producing the West of England Joint Spatial Plan.[14] I choose to highlight these attempts to demonstrate some of the complexities involved, and my personal experience of seeking to utilise the draft plan on behalf of one of my strategic land developer clients.

The West of England, sometimes known as the 'twin city region', is based on Bristol and Bath, with North Somerset and South Gloucestershire forming its more rural hinterland. The sub-region has one of the fastest growing economies in the UK and a population of 1.1 million (at the 2018 mid-year estimates). Early attempts at sub-regional planning go back as far as 1930 when Abercrombie and Brueton published their 'Bristol and Bath Regional Planning Scheme', covering an area very similar to the current West of England. In 1971 'Severnside – A Feasibility Study' was published, which looked at the planning issues and opportunities that needed to be dealt with and identified the accommodation of major growth as a leading planning challenge. In the 1974 local government re-organisation the new County of Avon was created with the aim of achieving a strategic level of administration and planning of services in the sub-region. The new County Structure Plan was to provide the necessary strategic planning framework. However, the Structure Plan was many years in preparation and, by the time it was approved in 1985, the last of the English structure plans to achieve this goal, the plan period end date of 1991 was getting close, and alterations were necessary. The County of Avon was abolished in 1996 following a further local government reorganisation in England and the four unitary authorities replaced it. They quickly accepted, with some government prompting, that full joint-working was essential for the future of this sub-region. The Joint Replacement Structure Plan reached deposit stage in 1998, but the Examination in Public went through a number of phases trying to reach agreed total housing requirement numbers and their distribution.

The short-lived RSS followed and included a sub-region-specific set of proposals. Further complications followed the establishment of the West of England Combined Authority in May 2017 as there was not political agreement across all four councils and North Somerset did not become a member of the Combined Authority. By this point the Replacement Structure Plan was more than 12 years old and the framework provided by the RSS sub-regional plan content had been abolished. Once again, the four authorities decided that a joint approach to strategic planning and transportation was essential. A Joint Planning Executive Committee was established, made up of elected members from the four authorities, and work on the West of England Joint Spatial Plan (JSP) commenced in 2015. A Joint Local Transport Plan was prepared in parallel with the JSP.

The JSP had to deal with some controversial issues, with the most significant being the need to accommodate 105,000 new homes in the plan period to 2036 within a sub-region exhibiting a particularly high coverage of primary and other environmental constraints. A total of 12 Strategic Development Locations (SDLs) were identified and allocations from existing local plans were being reviewed as these could provide more than half of the total housing requirement. Some of the proposals involved development of land designated as Green Belt. At the beginning of August 2019, the Examination in Public into the submitted draft JSP commenced, but after just seven days of hearings the two inspectors wrote to the four councils expressing very significant concerns regarding the soundness of the JSP. In particular there were concerns regarding the basis for selection of the 12 SDLs and the extent to which real alternatives had been considered. The inspectors requested more evidence on this process and other aspects of the plan. Having received further evidence their concerns remained, and they formally wrote to the Councils advising that future hearings were being cancelled and matters could not be further progressed without 'a very substantial amount of further work', which included new rounds of public consultation and policy decisions.

Prior to the submission of the Draft JSP I had been engaged by a strategic land developer client to review the ongoing suitability of a portfolio of potential residential allocations, on which they had development options with landowners, in the context of the newly emerging plan. Having examined various evidence base documents, the draft plan, and the draft Local Transport Plan, and in particular the detailed GIS mapping of constraints for the total plan area, I could readily appreciate the problems facing the JSP planning team. While none of the potential allocations I was examining were identified SDLs, it quickly became apparent how the options for major developments were limited to certain less-constrained strategic transport corridors and urban expansion schemes if extensive primary environmental designations were not to be adversely impacted. I regard Green Belt designation as wholly subsidiary in importance for reasons already explained in Chapter 7.

The West of England authorities (with the continuing exception of North Somerset), including the Combined Authority, to their credit have not given up on the great need to produce a strategic plan. In November 2020 they commenced work on the preparation of a new Spatial Development Strategy with a Future of the Region Survey. The timeline to produce a draft plan envisaged reaching the examination stage in 2022.[15]

* * *

Where does this leave regional planning in England and what is its future?

The true strategic planning vacuum created by the abolition of RSS and the relatively weak elements introduced in the Localism Act, which were declared by the former Secretary of State, Eric Pickles, to be some kind of replacement, were clearly not fit for the purpose of strategic planning in England. The Duty to Co-operate was weakly based, and many local planning authorities merely paid it lip service, as did some inspectors at local plan examinations. The move to neighbourhood plans and greater local determination clearly had merits but was never going to contribute to strategic planning; taken on its own it would be likely to increase the vacuum.

Greg Clark, who had been appointed Minister of State at the DCLG in May 2010, was as we have seen, instrumental in establishing the NPPF, became Secretary of State for this Department in May 2015. His understanding of planning and policymaking was undoubtedly superior to that of his predecessor. I think he recognised many of the problems confronting the planning system, though his initial appointment at DCLG was as the Minister in charge of decentralisation, a central policy approach of the incoming Coalition Government. He therefore had a key role in the formulation and content of the Localism Bill. However, by September 2015 he was clearly persuaded that there were significant problems with the current local plan making process. Together with his Minister of Housing and Planning, Brandon Lewis MP, he established the Local Plans Expert Group (LPEG) with a remit to consider how local plan-making could be made more efficient and effective. LPEG's report was published in March 2016 and identified that the problems surrounding the production of local plans were significantly greater than they had first anticipated.[16] At this point in time less than one third of England had an up-to-date local plan and LPEG estimated that less than half of the country's housing needs were currently provided for in local plans. The key problems identified that affected and delayed plan-making included: the agreement of housing needs; difficulties surrounding the Duty to Co-operate, including the distribution of unmet housing needs; a lack of clarity on key issues including Strategic Housing Market Areas (SHMAs), strategic planning, Green Belt, and environmental constraints.

The LPEG report identified the challenges surrounding strategic planning as a key barrier to local plan preparation and the accommodation of growth. The report recommends that the Duty to Cooperate should be given 'more teeth'. The 2017 Housing White Paper included proposals for more and better joint working and promoted the option of collaborative working on strategic spatial plans.[17] This was taken further in the Neighbourhood Planning Act 2017, which included new powers for the Secretary of State to direct the preparation of a joint local plan where this would 'facilitate the more effective planning of the development and use of land in the area'. Joint working on strategic areas became a key criterion for the allocation of funding for housing infrastructure and planning delivery. The significance of strategic planning to the allocation of funding was further emphasised by the Minister for Housing, Kit Malthouse, in September 2018 when he stated:

> 'Our general thrust is for groups of local authorities to come together to form a kind of strategic partnership and vision for a particular region or area, fundamentally so that we can fund the infrastructure that's related to it. We are unable to put the infrastructure that's required through the [Housing Infrastructure Fund] against proposals unless they have that kind of pan-regional or cross-area coordination.'

Revisions to the NPPF in July 2018 made it a national policy requirement for strategic policy-making authorities to collaborate in order to identify the relevant strategic matters that they need to address in their plans. The government made it clear that effective and on-going joint working between these authorities and other relevant bodies was a key part of the production of a positively prepared and justified strategy, and to demonstrate that this was taking place they were to produce statements of common ground.

All these actions were steps on a path back towards strategic spatial planning, but they fall well short of a universal reintroduction of this into the national planning system and there was no sign of an acknowledgement that the abolition of RSS was a big mistake, but then that's politics. These national government actions were primarily driven by the high levels of housing need and failing delivery, but they were ad hoc in nature.

At the time the West of England JSP Examination opened, four other groups of authorities were preparing the 'new style' JSPs including Greater Exeter, Oxfordshire, South Essex, and South West Hertfordshire. In all cases the individual constituent local authorities were to make the necessary decisions required to take the plans all the way to adoption. The Planning Advisory Service (PAS) organised a strategic planning workshop in February 2019 together with Catriona Riddell Associates. The presentation advised that there was no rule book for the preparation of JSPs and that the outcome of the West of England JSP would be a learning curve for all involved in this new form of strategic plan-making. The JSPs are long-term strategic investment plans steering major growth, which is to be delivered over longer periods than may apply to normal local plans, but they are statutory plans and must be legally and policy compliant.

In addition to these JSPs, which are in various stages of preparation, there are a number of other regional planning initiatives underway with government support. A major example is a growth plan for the Cambridge–Milton Keynes–Oxford Arc (CAMKOX). The National Infrastructure Commission was asked by government to prepare a study that considered how the future growth potential of these three cities could be maximised in the form of a strategically planned knowledge cluster incorporating many new jobs and homes. The Commission's final report was published in 2017[18] and sets out how up to one million new jobs and homes can be created within the Arc, including the provision of a new rail line and expressway connecting Oxford and Cambridge. On 19 February, the Minister of Housing announced the next step in the development of these major growth zone proposals in the form of a strategic spatial plan prepared with the constituent local planning authorities.

There are other examples of quasi-spatial plans being prepared at a regional or sub-regional/city-region level. Where city-region devolution deals have been agreed, and elected mayors appointed, for those sub-regions the need for some form of spatial planning has been recognised locally and usually encouraged by government as the most useful basis for targeting investment and growth. The Sheffield City Region, comprising nine local authority areas, has produced a City Region Economic Plan involving the LEP. More significant in spatial terms is the Integrated Infrastructure Plan commissioned by the Combined Authority and the LEP, which identifies seven growth areas and future transport networks based on initial spatial capacity analysis.

The Greater Manchester Spatial Framework was initially supported by all ten constituent authorities, but recently (December 2020) ran into difficulties when one of them, Stockport, decided to oppose the proposals based on the loss of circa 1.2 per cent of their Green Belt and issues relating to traffic congestion. This again demonstrates the difficulty of holding together JSPs and city-region plans where differences of political approach and lack of real cooperation emerge to spoil years of preparatory work in drafting such plans.

A brief comparison with spatial planning in the devolved nations (Scotland, Wales, and Northern Ireland) illustrates a much stronger emphasis on national/regional planning. Scotland produced its first national spatial plan in 2004 and has had a regional spatial approach for many years. The Wales Spatial Plan was produced in 2014 and Northern Ireland was the first to produce a national spatial framework in 2001. The strong national identities of the three devolved nations in terms of their geography and cohesiveness, culture, and socio-economic conditions, linked with the opportunities presented by the devolution deals, have strongly influenced their approach. England stands out as unusual, not only in the UK context but when compared with developed nations across the world, in not having a regional tier of governance or strategic spatial planning. As we have seen, this was not the case in the period 2003–2010. The RSS were cast aside and there has been no attempt to replace them by an updated or modified version. What has emerged is a patchwork preparation of growth and investment orientated regional spatial plans, which are not comprehensive in their coverage of the issues which need to be tackled, relate to a variety of different spatial and administrative geographies and, in a number of cases, have not yet reached anything like satisfactory outcomes.

One of the greatest arguments for regional spatial planning alongside regional economic plans in the UK and England is the current and long existing mega-region inequalities most notably characterised in England by the north-south divide. The differences between the South East and the North have been the most marked, but there are also significant disparities between the South East and the Midlands and the South West of England in terms of investment and business performance. The Johnson government, in their pre-election manifesto, promised a 'levelling-up' agenda as a prominent part of their economic and investment policies. The Coalition Government of 2010–2015 proposed a Northern Powerhouse as part of their national industrial strategy, focussed on investment in transport infrastructure, education and skills and business growth and productivity. In 2018, Transport for the North (TfN) was established as the first statutory, sub-national transport body in the country. TfN is a partnership between public and private sector bodies working with central government and is a statutory partner of the national transport infrastructure bodies, including the Department for Transport, Network Rail and Highways England, with the aim of delivering the pan-northern strategic transport priorities.

The Northern Powerhouse strategy and the establishment of TfN were encouraging foundations for establishing a holistic plan for 'levelling-up'. They have not yet been followed by a coordinated government strategy with a large investment programme based on a comprehensive evidence base and spatial plan. The approach to date has been selective growth deals for some city-regions, with much lower investment than required, and a scattering of town improvement and other ad hoc schemes aimed at quick wins for a relatively low investment outlay. These fall well short of a coordinated government-backed plan and a consistent single programme of action investment, which focusses on the achievement of major gains in productivity, skills and education, essential support infrastructure and the health and wellbeing of northerners.

The large regional disparities between the North and the South of England (not including the South West Region) in investment in business and infrastructure are a deep-seated problem that goes back to the depression era of the 1930s and these have continued to the current day as London and its widening hinterland have expanded their influence economically and politically in a much-enlarged greater south east (i.e. beyond the formally defined South East Region). These regional disparities and lost opportunities are emphasised when we recognise that the overall investment cake to be shared regionally is smaller than it needs to be as UK investment in business and infrastructure has lagged behind its world competitors in the G7 nations.[19] Figures from HM Treasury for the year 2010/11 help to demonstrate the huge differences in spending per head between London and the South East on those sectors of investment spending that are directed at economic growth including infrastructure and science and technology. In this year spending per head on public transport in London was around 3.5 times greater on average than the three northern sub-regions, while science and technology spending was around half of that spent per head in London.[20]

Having spent my formative years and most of my working life in the North, I feel let down by most governments who have run the country. This was emphasised when I became involved in civic/private sector lobbying groups seeking major transport investment in the Leeds City Region in the form of a new mass transit system and subsequently guarantees that the eastern phase 2b leg of High Speed 2 will be delivered to the Leeds Station Hub. Despite promises and much hype from leading government ministers, the level of investment required for the Northern Powerhouse Rail project and other pan-northern schemes has not been guaranteed. Given the comparative lack of investment in the North over many decades the term 'levelling-up' applied by Boris Johnson, the former prime minister, is somewhat inappropriate. A redistribution of wealth and growth is called for, but I leave it to others to argue the economic case and return to the need and search for a system of spatial planning that will help to both advocate and facilitate the required redistribution.

In spring 2021 there was no sign of the government proposing a universal system of region spatial planning, whether in a city-region or wider format. However, what this search revealed was that leading figures in the private and third sectors have got together to form The One Powerhouse Consortium, which has commissioned major planning/multi-disciplinary consultancies to produce 4 mega-region spatial blueprints for the North, the Midlands, the South West and the South East.[21] The Consortium is partnered in this project by the Royal Society of Arts and supported by The Sir Hugh and Lady Sykes Charitable Trust. The expressed belief of the Commission and their supporters is that: 'a substantial part of the problem of regional inequality can be solved not just by money, but by the transformative potential of spatial planning.' Working with three consultancies, Aecom, Atkins and Barton Wilmore have created the draft spatial plans for the four mega-regions of England 'to sit alongside the existing spatial plans for Wales, Scotland and Northern Ireland'.

The plan for the North, prepared by Atkins,[22] has been developed by utilising the extensive network of economic, social, environmental, and urban plans at different levels of spatial geography and reflecting the foundational work undertaken by Transport for the North and the Northern Powerhouse organisation. Atkins acknowledge that their 'blueprint' for the North is by no means the finished product. Atkins and the Commission conclude that it is now time for spatial planning to help deliver the Northern Powerhouse vision and to facilitate effective and efficient collaboration.

My conclusion is that a universal form of spatial planning in England is needed now more than ever. The post 2010 regional spatial planning vacuum has helped to emphasise the many planning issues we need to confront at the supra-geographic level extending well beyond the boundaries of individual planning authorities. This case is both emphasised and expanded by the post Brexit requirement to create a fairer and more equal country and to enhance productivity across all regions, the so-called 'levelling-up' agenda. The urgency of addressing the crises of climate change and biodiversity decline and the enhanced desire for greater health and wellbeing among the regional populations, impacted to differential extents by the Covid pandemic, come together with the other main factors to create the strongest possible array of arguments for government action on this issue. Truly sustainable development can only be achieved by adding this middle tier to the planning system.

There is a wide recognition in the planning and environmental professions that a coordinated multi-agency, long-term and strategic approach is required for regional spatial planning. While the four mega-region spatial plans could set a very helpful context, I believe that they will be too broad brush and that more detailed and specific plans along the lines of the multi-authority Joint Spatial Plans are needed. The mega-region plans might provide a useful framework for these. The decision that needs to be made is the form and geographical extent of these groupings to ensure that rural hinterlands and residual areas do not fall outside this tier of planning. Flexibility will be necessary to achieve this.

Finally, I select some of the examples that impact on land use, socio-economic and environmental futures, and investment and infrastructure planning to demonstrate the need for regional spatial planning and the great utility of such plans.

The regional water authorities in England have demonstrated why a regional spatial planning tier is directly important to the services they provide and the strategic investments they need to make in the future to maintain and improve those services. Thames Water, in their evidence to the second session of the House of Commons Communities and Local Government Committee on the abolition of RSS, stated that the region that they covered required 'large scale investment in infrastructure to address urgent environmental needs such as river water quality or increasing demand for water'. They pointed out the RSS for the South East contained detailed policies on these issues and 'were essential guides to local authorities when deciding on planning applications'. In my home region, Yorkshire Water own and manage uplands that make up water gathering grounds for a series of large reservoirs. They have a lead and linking role in providing and managing recreation facilities, improving these environments, maintaining/enhancing biodiversity and contributing to climate change mitigation and adaptation. The South and North Pennine Special Areas of Protection (SPAs) are international habitat designations with the main aim of protecting a small number of endangered bird species. The North Pennine SPA overlaps with the Nidderdale AONB, which is popular for outdoor recreation. These areas require integrated management regimes and in the case of the South Pennines SPA cross the boundaries of a number of local planning authorities.

The links between strategic rail infrastructure planning, in the form of new or enhanced rail routes and the creation of new spatial planning opportunities, create choices across several districts. It is important to select and design routes that are environmentally sensitive at a regional and local scale, as well as providing the best opportunities for the selection of growth points around new and enhanced stations.

Situations are occurring in parts of England where a core, highly urbanised local authority can no longer accommodate its full housing needs across normal local plan periods of 15 to 18 years. Bristol in the West of England JSP area is such an example as are Manchester and Birmingham City Councils. We have seen that there has been local authority disagreement in the West of England and Greater Manchester spatial plan areas, where one of the planning authorities selected to meet some of the housing need overflows have refused to do so and withdrawn from the cooperative agreement. Regional spatial planning is essential to achieve a fair distribution within the strategic housing market areas based on a strong and comprehensive evidence base. The existing impact of natural out-migration to adjoining authorities needs to be fully recognised and planned for in the most sustainable way, which can, as I and others have argued, lead to the conclusion that selected parts of a Green Belt need to be released.

The location of strategic, often controversial developments, which serve a wide catchment beyond a single local authority area, often prove politically and publicly difficult to accept when the operating company or organisation choose a site and apply to a single local planning authority. Major waste treatment facilities such as incinerators are a good example of such essential projects. The rationale for such a project and the site selection process can be more easily justified by a regional search for a short list of sites that meet the operating criteria and avoid a wide array of environmental and other constraints.

There are many more examples that make up a compelling case for strategic spatial plans. Amongst these are the socio-economic aspects of the different levels of health and wellbeing experienced by individuals and households such as lack of access to open spaces, gardens, the wider countryside, and full recreation opportunities. Without major adjustments via the planning system true levelling-up or rebalancing of the regions will not occur.

Chapter 10

Planning and Transport – What Kind of Relationship?

The way in which we now need to address the relationships between issues of transport and movement and those of the environment, spatial planning, and land use, have much stronger significance and importance for our future than has ever been the case in the post-war era. Some of the impacts and problems we must now address arise from the way in which we have, in recent history, allowed the private car to become an essential fixture in our lives. The other main issues we now must confront are the linkages of car usage and the travel mode choices we make as we face the major challenges of climate change and individual and collective health and wellbeing. Traffic congestion, the continuing dominance of the internal combustion engine and the consequent air pollution are themselves contributing factors to climate change and to deteriorating health and wellbeing for too many people. These direct effect relationships have been made all too apparent by the onset of the Covid pandemic across the world, with associated lockdowns of society and business as a main protective reaction by governments. These have reduced the ability to travel by both private car and public transport and restricted our movements from home to work, leisure and education outside the main lockdown periods. The result has seen large reductions in traffic on our roads and buses and trains, when running, carrying few passengers. Temporary benefits have included significant reductions in air pollution and some reclamation of road space for residents and active travel modes.

We have much to learn from the period of the pandemic, as well as from the extensive period over which most of us have allowed road transport to increasingly dominate large parts of our lives and living environment. As the problems of road traffic growth intensified and became more apparent attempts were made to make changes which would reduce the impacts of road traffic. These have included early pedestrianisation schemes, congestion charging zones, some public transport improvements, and by-pass construction, amongst many others. Some of the remedies applied have simply diverted traffic away from certain areas, such as town centres. In addition, recent governments have continued to build new roads and improve existing ones so that they have an increased carrying capacity and usually have the effect of generating new trips. In most areas of the UK these additional highway improvements have not been counter-balanced by improvements in public transport with the level of attractiveness required to entice travellers from their personal vehicles. At the same time road freight transport has grown with limited diversion of internal freight to the rail and water transport systems. There are, however, some encouraging signs on the horizon regarding car ownership and use, particularly amongst the younger generation of travellers.

The detrimental impacts of high car usage on our health and wellbeing are evident from road accident statistics involving drivers, their passengers, cyclists, and pedestrians, as a result of air pollution and the mental stress effects resulting from driving in adverse conditions in congested urban areas and on busy motorways. Prior to examining town and country planning's relationship and role associated with transport, now and in the future, I look back at how this has played out since the introduction of the 1947 Act planning system.

There are many stories concerned with the progressive invention of the internal combustion engine, which cover a period extending from the end of the eighteenth century, through the nineteenth and into the early part of the twentieth century. Henry Ford's Model T emerged at the end of this evolution and mass production of motor cars began and made them more affordable for a growing proportion of the American population. The growth of the automobile industry was slower in Europe, but mass production expanded after the Second World War.

In the UK, back in 1952, less than 30 per cent of the distance travelled in the UK was by car/van/taxi, with buses and coaches accounting for 42 per cent and rail 17 per cent (measured in billions of passenger kilometres per year). As household incomes increased and private cars became a more affordable commodity the growth in car usage was rapid and went from circa 60 billion passenger kms/year to 650 billion in 2016. By 1970, 75 per cent of all passenger kms were by private vehicle and this had risen to 85 per cent by the late 1980s. This proportion has subsequently levelled off.[1]

Public policy generally encouraged driving in the post-war period. During the 1950s a typical long-distance journey would involve a route taking you through the centres of several towns and villages. As the number of cars on the roads increased, journey delays annoyed drivers and the urban population became more averse to traffic in their communities, the pressure for urban by-pass schemes grew. Several by-passes were constructed and the first section of the UK motorway system, the Preston by-pass, was constructed in 1958 (now part of the M6 motorway).

Over the period 1952 to 2016, travel by bus and coach suffered a long-term decline to just 4 per cent of total passenger kms/year, which represented one tenth of the share at the beginning of the 1950s. Rail has recovered from a share of just 5 per cent in the mid-1990s to more than 10 per cent. On average the amount of time each person spends travelling per day is about one hour, with little change since the mid-1960s, though prior to the onset of the Covid-19 pandemic a small but growing proportion of UK commuters were spending two to four hours on their daily commute, predominantly into London.

The number of households in England without a car has fallen from 85 per cent around 1952 to some 20 per cent in 2020. The number of vehicles registered for use of the UK road system (cars, vans, trucks, and buses) had risen to 40.4 million by March 2020, of which cars accounted for 35 million.[2] The UK has one of the worst urban traffic congestion problems in Europe based on a comparative assessment of around 100 cities and their urban traffic blackspots. One element of good news emerges from this condensed review of UK traffic growth in that the 40.4 million registered vehicles are now emitting a decreasing level of CO_2 emissions as the proportion of ultra-low emission vehicles begins to rise.

The growth in the number of vehicles and the highway space to accommodate them, while it has, for a long period of time, provided us with new levels of freedom of movement and expanded leisure, shopping, employment, and other opportunities, has progressively created problems, particularly as the expansion in vehicle numbers and trips per vehicle have moved ahead of the available road capacity in our growing urban areas.

In Chapter 4 I made brief reference to the post-war city planners (engineers, architects, and planners) acting in either a chief officer capacity or as appointed consultants to prepare city plans for post-war renewal in many of our major urban areas. Prior to this, Abercrombie had provided for ring roads in some of his city plans and Thomas Sharp's plan for Oxford included a relief road, though these were not based on specific forecasts of future traffic movements. In Chapter 4 I referenced the work of Professor Otto Saumarez Smith, who produced a group of 10 biographies of Britain's post-war urban 'planners'.[3] He collectively refers to them as 'bogeymen', seen by the public as widely responsible for several urban highway proposals that despoiled existing townscapes, broke up existing communities and committed pedestrians to a series of subterranean and elevated walkways. Not all these massive-scale schemes were realised, but those that were substantially implemented left significant urban blight, much of which is still evident today. This gave 'planning' in its widest sense a bad name and town planners have in subsequent decades been tainted with responsibility for such negative impacts. Saumarez

Smith has subsequently written a book that seeks to rebalance the reputations of a number of these city planners, including Graham Shankland.[4]

One of the ten was Stanley Gordon Wardley, who became City Engineer and Surveyor of Bradford. In 1953 he set out his vision for a reconstructed city centre including ring roads, expressways, and pedestrian subways. Much of this system was built and still has a dominant effect on the city centre, segregating land use zones and committing pedestrians to subways or major surface highway crossings. Inner urban motorways were planned and implemented for Newcastle, Glasgow, London (the elevated Westway) and Leeds, amongst others.

In 1963, Colin Buchanan and his team produced their comprehensive study *Traffic in Towns: A study of the long-term problems of traffic in urban areas.*[5] The report was commissioned by the Minister of Transport, Ernest Marples, in 1960 following the Macmillan Government manifesto pledge to improve the road network and relieve congestion in urban areas. Even at this early stage the incoming Macmillan Government had been looking at measures such as congestion charging to solve what was becoming an all too apparent problem. At the time of the Report there were 10.5 million registered vehicles, with projections of 18 million by 1970 and 27 million by 1980. Buchanan made the following prescient statement:

> 'It is impossible to spend any time on the study of the future of traffic in towns without at once being appalled by the magnitude of the emergency that is coming upon us. We are nourishing at immense cost a monster of great potential destructiveness, and yet we love him dearly. To refuse to accept the challenge it presents would be an act of defeatism.'

Having visited the United States, and Los Angeles in particular, Buchanan saw the problems as having great urgency and that if rebalancing steps were not taken then the motor vehicle 'would defeat its own utility and bring about a disastrous degradation of the surroundings for living'. He was keen to see the preservation of the heritage remaining in our towns and cities alongside quality environments for urban centres and residential zones. The Report established the need for comprehensive land use planning and movement studies. The City of Leeds was one of three study areas, together with Newbury and Norwich, and I return to the Report's proposals for the city as part of my main example of the evolution of city transportation strategies, illustrating the relationship of transport and town planning.

Following the publication of *Traffic in Towns*, the separate Ministries of Transport and Housing and Local Government (responsible for town planning) published a joint circular in 1964, which proposed the production of land use transportation studies to achieve a coordinated approach to land use and transport planning. Buchanan, who was knighted, became President of the RTPI in 1963/64.[6]

Land use transportation studies had been developed in the United States and were first commissioned in the UK following the 1964 joint circular. The original study was for the West Midlands and was followed by Greater Glasgow, both starting in 1964, Teesside and Belfast in 1965, Merseyside in 1966 and West Yorkshire in 1967. In Chapter 2 I refer to my brief experience of working with the specialist land use transportation team, which formed part of Essex County Council's large planning department. This team was working on joint computer modelling with Hertfordshire and Kent County Councils as a necessary sharing of cost and expertise on these major studies. I recall a visit to see the large-scale computer being used in the joint modelling operation, occupying a full wall of a large office.

The objectives of these studies were strongly traffic orientated despite the apparent aim in the government circular of achieving a fully coordinated land use and transport planning approach. Michael Bruton, in his book on transportation planning,[7] summarises the purpose of the West Midlands Transport Study as being a comprehensive survey of all forms of traffic in the conurbation and an analysis of their relationship with different land uses and the demand factors for movement. This analysis would then be used to make forward projections to assist the planning of future road and public transport networks.

Bruton and other authors make various criticisms of these major studies. They were too concerned with the technical problems of estimating future traffic flows and network planning to accommodate these. There was too little public involvement and the discernment of their transport needs. The general view of politicians and the public was that these studies favoured the motor vehicle and there was a lack of evaluation and proposals for public transport. Inadequate definition of goals and objectives and a far from comprehensive approach that neglected the role of all transport modes was a major omission. What generally emerged tended to be a highway network plan designed to fit a fixed land use plan with assumptions about the traffic generation of those land uses. Alternatives were not evaluated in most cases and, where they were considered, the concentration was on economic impacts and cost. Despite the relatively sophisticated computer modelling used, the approach of transportation planners was narrow in concept. The approach taken has been seen largely as an engineering exercise to design a physical transport system. The emphasis on the physical at the expense of social, economic, and environmental factors demonstrates its great shortcomings. Today we would condemn this as wholly unsustainable.

The extent to which transport planning and spatial planning are integrated at national and local government levels has been a crucially important issue since the introduction of the 1947 planning system. Peter Hall summarises the position that applied in the 1960s, and continued into the early 1970s, as high mobility in search of increasing choices in employment, education, and leisure.[8]

These opportunities and their consequences were readily apparent to politicians, and Hall concludes that few were minded to deny citizens their rights and freedoms. As a result of this attitude, and the shortcomings of the transportation studies, the predict and provide approach to accommodating private traffic movements continued to be the focal approach.

A White Paper – Public Transport and Traffic – produced by the Labour Government in 1967 suggested that a modified approach might be on the way. However, while reference was made to giving public transport 'a new dynamic role' this statement also referred to 'expanding facilities for private cars'. This was followed by the 1968 Transport Act, which established new Passenger Transport Authorities with the intention that they would have powers covering the coordination of all forms of travel, but this ended up being limited to public transport only. The 1968 Town and Country Planning Act incorporated Buchanan's recommendation that transportation plans should be included as part of the statutory development plan. These new integrated plans, entitled Structure Plans, were mainly, though not exclusively, to be prepared by the large county councils, including the new metropolitan county councils established in the main conurbations by the 1972 Local Government Act. The new West Yorkshire conurbation structure plan preparation therefore integrated transportation and spatial planning at the strategic level while the constituent metropolitan district councils prepared the more detailed local plans incorporating site-specific allocations and development control policies. The position was strengthened for the first half of the 1970s by integrating the Ministry of Transport into the Department of the Environment, only for a separate Department for Transport to be set up within a few years.

* * *

The implementation of city transportation strategies – a planning perspective using the City of Leeds and the former metropolitan county of West Yorkshire as an example

In 1995 I presented a paper on this topic at the First International Conference on Transport and the Environment.[9] I used the Leeds/West Yorkshire example of metropolitan transport planning and town planning, as it provided a good illustration of the initiatives taken, the problems experienced and the outcomes. I looked back to the Buchanan report recommendations for Leeds and to the City Council's City Centre Transport Strategy in the late 1960s, which aimed to take forward the specific proposals contained in Traffic in Towns. In 1977, consultants engaged by West Yorkshire County Council produced a study of the detailed traffic patterns and demands covering the whole conurbation together with specific recommendations for the individual towns and cities.

The Buchanan study had three main objectives:

i) To examine the number and character of vehicular movements that would arise as vehicle ownership and use approach their potential maximum.
ii) To indicate the scale and pattern of alternative distributor road networks that would be required to meet varying proportions of the potential demand for vehicular movement.
iii) To indicate the impact of increased traffic on the town/city centre and the scale of the changes that would be required.

There was a recognition, as stated in paragraph 26 of the introduction, that 'the commuter cannot be forced back onto public transport'. While Buchanan saw the expansion of public transport as making a large contribution to reducing car commuting, he decided that this should not be seen as a solution in its own right and should be treated as one arm of a coordinated policy (paragraph 27).

The main Buchanan proposals for central Leeds were:

i) The creation of an extensive traffic free shopping area.
ii) A new district distributor road to the east of Vicar Lane (this being the approximate eastern boundary of the retail centre).
iii) Redevelopment of 48 per cent of the central area to facilitate the rear servicing of the building blocks which front onto the pedestrianised area.
iv) The regrouping of uses to form environmental areas (a key approach of in the Buchanan Report).
v) A circulation system beyond the central area made up of primary and district distributor roads.

Following the recommendations of the Buchanan Report, the City Council produced their own City Centre Transport Strategy under the direction of the City Engineer and Planning Officer, Geoffrey Thirlwell. This approach, by an integrated professional department under a strong chief officer, enabled the production of a comprehensive strategy. The strategy included the construction of a new inner relief road around the city centre business and shopping districts, four major quadrant multistorey car parks, extensive pedestrianisation of the core with proposed high level pedestrian links from the shopping centre to the central business district to the west. A new bus service was to link the quadrant car parks and take passengers into the heart of the centre. The costs of this strategy proved to be extensive and only parts were implemented. In 1995, some 25 years from the production of the strategy, two phases of the seven-phase inner ring road had yet to be constructed. An extensive pedestrian area was completed by 1995 and has subsequently been extended to provide a welcoming pedestrian experience. Only a small proportion of the upper-level walkways were built and only one of the quadrant car parks was completed by the mid-1990s. The linking bus service was removed following only a couple of years of operation.

These earlier transport/planning strategies had a heavy bias towards the central area of Leeds with some degree of success. These central improvements were not accompanied by any car restraint policies and, in common with many European cities and municipal politics, the aim was to accommodate projected private and commercial vehicle trips on inner relief and ring roads.

The 1977 West Yorkshire Transportation Strategy, produced by WytConsult, identified that the primary problem for Leeds was the accommodation of travel demands in the peak periods, a common conclusion in such contemporary studies. The main proposals were to cater for a modest growth in car usage and to provide as attractive a public transport system as possible, given the limited funds which were recognised as being available. The recommended strategy included commuter parking restraint in the central area. The potential need for cordon pricing and the development of a future light rail system were highlighted. Cordon pricing has not been seriously considered in the succeeding years. At the time of writing this paper in 1995, only the first phase of a light tramway system had been given government approval in principle. This approval was eventually to be withdrawn and two subsequent bids for a three-line tram system for Leeds failed to gain government support, as did a subsequent trolley-bus system. I attended two meetings in Leeds with the transport minister as part of a public/private sector support group where the case was not well presented. The city/city-region remains as the only one of comparable scale within Western Europe, without a light rapid transit (LRT) system. The current government, in 2020, indicated that they would support a conurbation wide mass transit system for the city-region. For many years it has been clear that modern tram or light rail LRT systems providing frequent, clean, and reliable services are the essential component to attract people out of their cars and onto public transport in large metropolitan areas.

In February 1991, the consultancy Steer Davies Gleave developed the Leeds Transport Strategy following full public consultation and a strategic study of future transport. This study predicted that the number of vehicles entering the city centre during the morning peak travel period was likely to rise by 30 per cent in 2010. The study noted the high level of usage of train services by commuters from car owning households. At the same time the study identified the decline in use of bus services, which had been running at 1.7 per cent per annum, as being due to cost, unreliability, and the lack of attractiveness of this transport mode. The scope to attract commuter journeys to public transport in the short to medium term was limited and reliant on major investment on both rail and bus services. Growth in rail commuting to the city centre has increased on several lines, particularly where the services have been upgraded by electrification of lines and the acquisition of new rolling stock. The scope to further increase rail commuting during the peak travel periods was however hampered by crowded trains, the existence of old diesel rail cars on several commuter lines into the city and the lack of platform capacity at the major city centre station.

Surveys for the study carried out in 1987/88 showed that there were some 24,000 parking spaces serving the central area and that 38 per cent of short stay spaces were being occupied by long stay users. Significant commuter car parking pressure existed at the time of these surveys, and this was projected to increase, as a result of new business allocations in the Unitary Development Plan (UDP), leading to a projected city centre employment growth of 15,000 jobs to 2001. During the period 1990 to 2001, Leeds experienced the highest job growth in all UK cities outside London. The City Council's UDP proposed reductions in the amount of long stay car parking provision serving the city centre. The consultant's transport Strategy warned that any such restraint on private vehicular traffic to and from the city centre could only be successful if an attractive alternative package of measures was offered. The Transport Strategy proposed extended pedestrianisation in the city centre together with a public transport 'box' of bus/taxi dominated movements around this area and further outside the 'box' a new one-way gyratory loop road for private vehicles. The whole of this system lay within the confines of the inner ring road. These parts of the proposed system have subsequently been implemented and operate successfully. The Strategy, crucially, also recommended that parking restraints should only be introduced when suitably located, large park and ride site proposals and improvements to public transport had been implemented.

The various city transportation strategies for Leeds from Buchanan in 1963 to Steer Davies Gleave in 1991 did introduce strands of a coordinated spatial planning/land use and transport planning approach with some successful elements. Without exception, the implementation of these strategies has not included the required level of enhancements to public transport in combination with restraint measures encouraging car commuters to make the necessary modal shift in their journey to and from work. The public transport 'box', the inner loop road, and the completion of the final stages of the inner ring road have in combination been relatively successful in maintaining traffic flows around the enhanced retail and business environments of the city centre. However, these highway schemes have also produced a major deficit in terms of substantially reducing the connectivity of inner-city neighbourhoods with the city centre's retail, employment, and leisure activities. In addition, congestion persists on the radial routes approaching the inner ring and loop roads with attendant air pollution concentrations. There is little doubt from the experience in Manchester, Nottingham, Sheffield and other cities in the UK and western Europe that multi-line LRT systems can substantially improve modal shift to public transport. Intermittent bus lanes on radial routes have shaved a few minutes off some bus journeys, but do not begin to meet with the standards and reliability of modern tram systems. It is to be hoped that recent government promises lead to the delivery of a mass-transit network for the Leeds City Region.

Some of the key lessons to be learned from the Leeds experience, and that of other large urban areas in the UK over the period 1960 to 2000, can be summarised as follows:

1. The need for full integration of city spatial planning with transportation planning within local government structures driven by long-term integration at the national level. While transportation strategies for UK cities are usually presented in a comprehensive fashion, the means of implementation and programming of the strategic elements are rarely dealt with on a comprehensive and integrated basis.
2. The funding and bidding mechanisms used by national government have not assisted the delivery of ambitious integrated local transport schemes throughout much of this period.
3. The lack of experience of preparing and delivering transport and planning strategies on an integrated basis within the public and private sectors of the two professions, combined with a lack of political will at the national level.

* * *

2000 to the present day – an improving relationship?

In March 2001 the English government produced the third revision of Planning Policy Guidance 13 – Transport. The original version was issued in 1988 and a significantly revised version followed in 1994. This third edition is also extensively revised and followed a full consultation process and the publication of a White Paper setting out the government's policy for the future of transport.[10] The advice is complemented by a Guide to Better Practice, published jointly by the Departments of Environment and Transport in 1995. The policy aims expressed in this White Paper were to extend choice in transport and to secure mobility in a way that supports sustainable development. PPG13 reiterates this policy context for transport in England, recognising that the quality of life is dependent on transport and easy access to jobs, shopping, leisure facilities and other services and that a safe, efficient, and integrated transport system was necessary to support a strong and prosperous economy. Integration in this new policy context covered the linkages between different types of transport, with environmental policies, with land use planning, and finally with policies for education, health, and wealth creation.

This was the Labour Government with Tony Blair as prime minister, and it had also published a 10-year transport plan, which was based around a partnership between the public and private sectors to provide a modern integrated high quality transport system. John Prescott, on becoming the Secretary of State for a newly combined Department of the Environment, Transport and the Regions in 1997, had declared he would greatly reduce car journeys in line with major improvements to public transport. He also recognised from recent history that transport policy should not be the victim of changes of government or of secretaries of state with policy changes every three to five years. Transport policy and programmes needed to be planned and implemented across a 15-year period to create stability and be able to implement the major changes required. Unfortunately, Tony Blair, in common with his predecessors, was worried about a policy approach that would restrict voter's freedoms of car travel, and Prescott did not get the funding required to fully pursue his transformative policies.

The 2001 version of PPG 13 recognised the key role of land use planning in delivering the government's integrated transport strategy.

'By shaping the pattern of development and influencing the location, scale, density, design and mix of land uses, planning can help to reduce the need to travel, reduce the length of journeys and make it safer and easier for people to access jobs, shopping, leisure facilities and services by public transport, walking and cycling.'

The objectives of this national guidance were to integrate planning and transport at national, regional, strategic, and local levels. Local planning authorities were asked to ensure the complementarity of strategies in the development plan and the local transport plan, and to consider new development allocations together with local transport priorities and investment. This was the beginning of a more sustainable and integrated approach to transport and spatial planning.

Where does the integration of spatial land use planning and transport planning stand today in England? The 1997 Labour Government's attempts at integration did not last very long, as the combined department proved to be too unwieldy and planning guidance, though comparatively progressive on seeking to achieve a shift to sustainable transport, did not 'bite the bullet' of restraining the use of the private car and was not yet a formally adopted national policy. The current version of the NPPF[11] sets the policy approach for local planning authorities on the promotion of sustainable transport. While the current policy requires transport and travel issues to be considered at the earliest stages of plan-making and development schemes and opportunities to promote walking, cycling and public transport use are to be identified, the approach is a weak attempt to achieve a radical shift to the sustainable modes of transport. It provides some encouragement to local planning authorities but does not set a clear and strategic framework that must be followed or provide any programme for implementation.

The weakness of the NPPF in this key area is illustrated by the lack of any direct car restraint policy content. Indeed, one of the few explicit policy statements at paragraph 108 warns local authorities that maximum parking standards for residential and non-residential development 'should only be set where there is a clear and compelling justification that they are necessary for managing the local road network, or for optimising the density of development in city and town centres and other locations that are well served by public transport'. While there is a passing reference to emissions at paragraph 105, there is no specific policy approach to air pollution and to integrated planning and transport actions to reduce emissions and mitigate their impact. A specific policy approach on air quality is also missing from section 14 of the NPPF, which covers climate change. It is surprising that there is no reference at all to integrated working between the two government departments. A search of the Department for Transport's policies and responsibilities reveals a similar gap.

The ways in which the Department for Transport operates and formulates its policies clearly impacts on its relationships with spatial planning policies and whether there are beneficial outcomes for the public. The Institute for Government recently produced some research into the way governments use evidence to make transport policy.[12] This research report opens with reference to the 2019 manifesto of the Conservative Party, which promised an ambitious 'transport revolution' connecting towns and cities by road and rail with significant new investment and supporting electric vehicles, cycling, and walking. The authors comment on the history of previous governments, which demonstrates how they have often failed to deliver on their transport promises, while carbon emissions from transport have remained flat for three decades, so that this remains as the largest emitting sector of the UK economy. This report focusses on the way in which evidence is used and considers that this is crucial to achieving good transport outcomes. How evidence is commissioned, the way it is analysed and then used, are of great importance and transport, as with all government policy making, operates within a political environment. A case for a particular transport policy with a strong evidence base can fail to be introduced if it does not have political support. The authors use the example of the political decision in England to continue freezing fuel duty despite the strong economic and environmental evidence that exists in support of the fuel duty escalator. Similarly, the strength of the case for high investment in public transport enhancements followed by car restraint policies has consistently been made and politically generally avoided, with London being one of the few exceptions.

The Institute of Government report goes on to make the more fundamental criticism of the lack of an integrated transport strategy in England and recommends that the Department for Transport should consider developing a new strategy integrating different modes of transport, rather than past and current working on a mode-by-mode basis. This integrated approach would create beneficial outcomes for travellers and has key implications for the ways in which spatial plans are pursued and new developments and communities are designed. In examining the linkages between the planning and transport systems it is necessary to recognise the significance of the general way in which the Department for Transport funds local government transport responsibilities and activities. Increasingly, via the appointment of city-region mayors and the operation of Combined Authorities transport, powers and funding have been devolved to local government. However, the funding of rail and bus transport is largely via the private sector, though some mayors are seeking to take back control of these services.

What all this means is that the level of integration of planning and transport at the national level is woefully short of what is required to achieve the most successful and valuable outcomes. There are policy vacuums within the two main government departments that underplay or ignore the linkages between transport and planning, and there are types of policy that evidence strongly suggests are essential and are not pursued for political reasons. The relative weakness of national planning policy on the promotion of sustainable transport and the wide extent of the Department for Transport's other responsibilities leaves local government with more scope for interpreting what the local policy approaches should be. This scope is strengthened where powers are delegated to city-region or other groupings of local authorities and a higher level of funding provides further impetus, though transport funding to local government generally falls short of requirements. It is encouraging that the Greater Manchester city-region mayor is the first, outside London, to reinstate a regulated bus system whereby some 47 different operators will be integrated into a new single system with the mayor having powers to set fares and establish routes. The prime minister has supported city-region authorities and given them the powers to take back control of local bus services. The ability to set prices is very significant as journey costs in Greater Manchester are three times those of comparable journeys in London.

A further demonstration of the mixed policy messages emerging from the Department for Transport is the publication of their new road building programme alongside the budget in March 2020.[13] The scale of this programme involves expenditure of £27 billion over the period to 2025. This programme has subsequently subject to a high court challenge by the Transport Action Network (TAN) a crowd-funded action group. The challenge was filed on 27 April 2021 after a judge had given leave for the case to be heard, which is based on the argument that the road building programme failed to take account of the government's statutory pledge to make the UK carbon neutral by 2050. The judge decided that technical evidence would be admissible and further cases by TAN have followed. Others have criticised this strategic policy and programme, arguing that the money could be much better spent on active travel and public transport or on the expansion of superfast broadband. These alternative types of transport and communications expenditure would bring greater benefits to local spatial and land use planning by encouraging households to use their car less and to make different modal choices. The Government's commitment to reach net zero carbon emissions by 2050 under the Paris climate agreement doesn't currently incorporate traffic reduction measures.

There are, however, a few signals that our 'love-affair' with cars is beginning to go into reverse. Statistics released by the Driver and Vehicle Licensing Agency in 2020 show that car ownership has fallen over the period 2010–20 in many neighbourhoods, which accommodate 31.5 per cent of the population of England. There is a counter trend in other areas of the country, which has led to an additional 2.7 million vehicles being acquired over the same period. The main age group relinquishing car use in the first set of areas are 18-

to 34-year-olds. While, not surprisingly, the biggest falls in car ownership were in London, other cities and towns included in the 100 neighbourhoods with the greatest fall in ownership were Newcastle (eight neighbourhoods), Nottingham (four neighbourhoods), Brighton & Hove, Birmingham, and Exeter. The costs of owning and driving a car are a deterrent, particularly for those saving to buy a property. Congestion charging in London and a levy on workplace parking in Nottingham were added costs. The availability of car rental and car-pooling schemes in larger cities provide a cheaper alternative for more occasional car use. The availability of good public transport schemes, with rapid transit systems included, are a further incentive in those city neighbourhoods exhibiting the larger declines in ownership.

Further trends that are manifesting themselves internationally also demonstrate a move away from accommodating the private car in cities and towns. These include actual and planned reductions in road space in cities like Paris and Milan, the latter being encouraged by the experience of travel during the Covid pandemic. Oslo has removed all car parking spaces from the city centre and reduced car journeys to the centre by 90 per cent, while in Greater Manchester a circa 3,000 km cycling and walking network is planned across the city region by 2030. Road space is being removed or redesigned to better accommodate active travel, public transport routes and new public spaces. Although the UK has departed the European Union, the incentive for city and city-region authorities to learn from one another in the field of urban and transport planning remains as innovative schemes are either trialled or fully committed. The many experiences resulting from the Covid pandemic include profound lessons for the future of transport choices and investments in transport and development infrastructure. The Institute for Government research report recommendations on the ways in which the English Department for Transport consider all types of evidence needs to be carefully followed as there is a plethora of evidence to be carefully evaluated by this department and the Ministry of Housing, Communities and Local Government. Only in this way will sound decisions be made on integrated policies and programmes for transport and spatial planning that address the central suite of objectives contained in the NPPF definition of sustainable development and in our climate change commitments.

The importance of establishing a firm and clear relationship between transport and spatial planning, at all levels of government, is further emphasised by the health and wellbeing impacts that arise in this relationship. The ways in which we use land and design our living, working and leisure environments, coupled with how we travel between the different land uses, all contributes in either a positive or negative way to our health and wellbeing. In earlier chapters (particularly Chapters 3 and 5) I have explored a number of the land use and spatial planning relationships, including the ways in which spatial planning can enhance active travel opportunities.

How do the travel opportunities and choices available to us affect the health and wellbeing of individuals and communities? The answers to this question are dependent on certain key travel determinants, which affect individuals and communities in different ways. These include the following:

- The extent to which each individual and their household need to travel taking account of distance, frequency and type of trips and the time absorbed on both a daily and weekly basis.
- The ability to choose where we live and work and the extent to which we possess the socio-economic benefits and freedoms that enable us to determine how we manage the home to work journey and other types of journeys.
- The modal choices available to us from our place of residence for each type of journey and their relative attractiveness.
- The extent to which a multi-modal trip is integrated in terms of time, smoothness of connectivity and cost.

Other factors come into this mix including whether the individual/household live within a city centre, inner city, suburb, or rural area. These relationships are also fluid, and therefore the choices available and the ability to make them can be a complex array. The choices for rural residents are often starker.

General health and wellbeing considerations, which influence the choices of travel mode selected and the route chosen, include prior experience of the mode, previous routes travelled and the experience of others. Safety is a greater consideration for individual travel by car, bicycle and on foot than public transport use. Feeling safe and relaxed in a travel environment avoids stress from travel, but busy roads require a good level of concentration from all travellers thereby removing some of the benefits. Planned segregation and quality facilities for pedestrians and cyclists are key to this relationship and many accidents can be avoided. Spending each working day travelling by car on congested routes is stressful for most travellers and the quality of the journey is further depleted by the ingestion of emissions from other vehicles in the queue. The Covid pandemic created enforced home working for a large part of the employed population for lengthy periods and, as lockdowns ceased, employers and employees were faced with decisions on the extent to which a mix of home and office working was necessary or desirable.

In the late autumn of 2022, Covid-19 continues to have an impact on work and travel patterns, but this is currently much reduced compared with the high average and peak incidence of the pandemic in 2020 and 2021. These patterns are still some way from returning to the norms of the pre-Covid position in 2019. It is too early to reach firm conclusions on the extent to which some days of home working will become the norm for a significant proportion of the UK population and the impact this might have on travel patterns and modal choice.

Air pollution affects all travellers on busy public roads as well as residents living alongside main highways in our urban areas. It is estimated that around 40,000 people die in the UK each year from respiratory disease connected with air pollution from particulates ($PM_{2.5}$ and Nitrogen Dioxide).[14] Traffic-generated air pollution is a major problem that successive governments have failed to tackle and there are still many air pollution 'hotspots' on UK roads where WHO standards for the pollutants are regularly breached. The previous Conservative Government were twice taken to court by the environmental law charity Client Earth, and the EU Court of Justice subsequently ruled that the UK had 'systematically and persistently' exceeded the legal limits for nitrogen dioxide since and had also failed to put plans in place to achieve the required reductions. Rather than tackling the issue head on at national level, the chosen solution was to delegate local government to decide on the necessary actions, including whether to declare charging/clean air zones.

The death of nine-year-old Ella Adoo Kissi Debrah from acute respiratory failure in 2013 hit the press headlines some seven years later, on the 16th of December 2020, when a coroner, following a long legal battle by her parents, finally ruled on the cause of her untimely death. She became the first person in the UK and the world to have the words 'air pollution' listed on her death certificate. The coroner determined that air pollution had both contributed to Ella developing asthma and then had exacerbated its effects due to her exposure to levels of nitrogen dioxide in excess of WHO guidelines. He concluded that 'the principal source of her exposure was traffic emissions'.[15] Ella and her family lived a few metres away from the South Circular ring road in the London Borough of Lewisham.

Action has been taken by the London Mayor, Sadiq Khan with the introduction of an Ultra-Low Emission Zone (ULEZ) covering Central London, with the drivers of polluting vehicles being charged daily to enter. This has resulted in 13,500 fewer cars per day entering the zone with a resulting 36 per cent fall in levels of NO_2. Bath, Brighton, Birmingham, Norwich, and Oxford are some of the first cities to introduce LEZs. Integrated transport plans are required for towns and cities, which combine LEZs with improvements to public transport provision and active travel networks. Where these integrated improvements are phased in alongside reductions in long stay commuter car parking major health and wellbeing gains can be achieved by reducing air pollution, encouraging active travel, and changing the land use of highways that become redundant. England has introduced some initiatives that improve the position, but these fall well short of an integrated travel strategy so that much more needs to be done.

Looking to the future we need to place integrated spatial and transport planning at the forefront of all policies, plans and projects and to demonstrate in guidance at the national and local levels best examples of how this can be achieved in terms of both organisation and design. To achieve this and the sustainability and health and wellbeing gains, to which we should be aspiring, changes in the governance of transport and planning are necessary at the national level and in the ways, we work together as public sector organisations and professionals. This need has achieved some recognition in research papers but is not yet apparent in national policy documents. A recent report by the Government Office for Science, forming part of the Future of Mobility project established by this Office, identified that 'for historical reasons there is no single approach in place across the UK transport system'.[16] This report also concludes that 'reforms of the sub-national governance processes have added further layers of complexity to decision-making. This makes the integration of policy, strategy and funding across different institutions and transport modes even more challenging.' Consequently, governance in the UK remains complex, fragmented, and weakly implemented, particularly in England. This makes it harder to achieve integrated outcomes and to optimise social and other benefits.

An associated report for the Government Office for Science reviewed the evidence for the main report quoted above. This evidence review supported the main conclusions: 'There is a multitude of governance networks rather than a single overarching "governance of the transport system". This makes it difficult to achieve integrated outcomes.'[17] The evidence review report refers specifically to the integration of transport and land use and concludes that this remains a particularly challenging issue and that the balancing of potential conflicts and negative impacts through the planning system should be carried out in partnership with a focus on sustainable transport.

A third report produced by the Chartered Institution of Highways and Transportation, in collaboration with the Transport Planning Society and the RTPI, is the culmination of work by several professionals keen to see the creation of better places by improving the integration of planning and transport. The report, although intended mainly as an advice document to complement new government guidance issued by the DHCLG, has some hard-hitting conclusions and strong recommendations.[18] The report's authors acknowledge an increasing working relationship between the government departments responsible for planning and transport and 'applaud the recognition that professionals have a key role to play in addressing the integration of new development and transport'. The report concludes that limited progress has been made over the last 20 years by government in their attempts to encourage a more sustainable approach to transport within spatial planning. The discredited approach of 'predict and provide' is identified as a continuing problem when planning new developments and three main barriers exist that are preventing the desired integrated and sustainable approach to planning new developments. Firstly, local plans should set out a clear vision and set accessibility and modal share targets; secondly, there are insufficient practical examples that demonstrate how to deliver sustainable transport outcomes and, finally, collaboration between planning and transport bodies is either insufficient or ineffective.

Stronger integrated planning and action between the two national departments is required. I recognise, however, given the large size of both departments, that this is likely to fall short of the full integration trialled in the past. An improved and expanded sustainable transport statement should be incorporated in the NPPF that is the product of joint working and agreement between the two departments, and this is explicitly stated. Similarly, the national guidance in the NPPG should be jointly produced. A formal team comprising civil servants from both departments should be established with the responsibility of producing integrated policies and programmes, and this team should report to a minister who has integration responsibilities in both departments.

Consideration needs to be given to clarifying and, wherever possible, simplifying the transport roles of the main tiers of local government who carry the key transport and planning responsibilities and where greater integration is essential. This mainly lies at the city-region/combined authority level and with the individual authorities preparing local plans and determining planning applications. A transport planner should be involved in all policy/local plan teams and in the determination of all major planning applications (i.e., those residential schemes of 10 or more dwellings). In both plan-making and decision-taking the involvement should be from the beginning of the process and the individual should be integrated into the relevant team rather than acting as a more passive consultee.

New developments clearly present the greatest opportunities to incorporate active travel measures and to introduce new or enhanced public transport connections. In general terms, the larger the development the greater are the opportunities presented for more radical sustainable transport measures, for incorporating mixed uses and thereby reducing many trip lengths as well as attracting people to use sustainable modes. Where the residential base of these new developments is located close to existing communities, the opportunities to provide these new travel benefits to existing residents need to be maximised throughout the planning process from site selection, local plan allocation to master planning, implementation, and future management. Such opportunities exist at the edge of existing urban areas, including in some locations Green Belt releases where overall sustainability can be maximised. Larger urban extension schemes present active travel opportunities to link the existing and new communities to open countryside and to introduce new linear public transport routes as well as enhancing radial routes. These macro designs can also focus on linking new residential areas to existing and planned employment concentrations.

Many of these benefits can also be achieved in city centres and peripheral inner-city communities, particularly where large-scale regeneration schemes are planned and supported by major infrastructure investment. The high-speed rail network planned to link London, Birmingham, Sheffield, Manchester, and Leeds (HS2) and Northern Powerhouse Rail (sometimes referred to as HS3) are major examples of the catalytic impact that can be delivered by such investment. The cities to be linked have prepared regeneration/city centre expansion master plans around the planned new stations and transport interchanges. Via these master plans, new public open spaces in the form of linear and pocket parks, active travel network improvements and reductions in road space occupied by vehicles can be introduced. New travel interchanges with integrated linkages and ticketing systems will make total journeys smoother, more time efficient and less stressful.

In Chapter 9 I briefly summarised the growth plan proposals for the Cambridge–Milton Keynes–Oxford Arc (CAMKOX), which could accommodate up to one million homes and jobs by 2050. The National Infrastructure Commission for England (NIC) ran an ideas competition as part of their government-backed plan and the winning entry, submitted by a professional all-woman team, known as VeloCity, produces a radical new transport plan based around the use of new technologies with the main linking travel mode being cycle routes connecting existing clusters of villages and integrated new homes. The initial proposal concentrates on a cluster of six existing villages around the market town of Winslow, which will have a new railway station on the planned Oxford to Cambridge line. This first of several potential clusters, four of which will be focussed on Winslow, will be located within an 11 km radius of the railway station. Each village within the selected clusters would be linked by walking and cycling networks to the other villages and to the station. The VeloCity proposals also focus on removing vehicles to create space for people, with residential areas being car free and each village having a car club scheme based on electric vehicles.[19] It remains to be seen to what extent these exciting proposals for radically reducing our reliance on the private car will be implemented within the CAMKOX plan, but they demonstrate the great potential of these and related ideas that should be taken into account in developing the future of the spatial planning/transport mix.

Earlier I used examples of Leeds City Council's transport strategies between the 1960s post Buchanan response and 1991, and the extent to which these were implemented, and presented some of the key lessons to be learned from the Leeds experience. In arriving at my conclusions on the way forward for a joined-up approach on transport and spatial planning, I have reviewed the City Council's latest transport strategy which has the strapline 'Connecting Leeds – Transforming Travel'.[20] The Connecting Leeds Transport Strategy was produced as a draft for public consultation in 2018 and the period for responses closed on 11 April 2021. The strategy sets out a vision for Leeds to be a city 'where you don't need a car'. The basis of the overall approach is to give all citizens an affordable zero carbon choice in how they travel with people, neighbourhoods and businesses being connected in the most sustainable ways. The vision statement also includes the aim to create a city centre with efficient well-connected land uses and movement systems that accommodate more people, not more vehicles, with active travel modes, buses and mass transit prioritised.

The strategy has three over-arching objectives: 1. Tackling Climate Change, 2. Delivering Inclusive Growth, 3. Improving Health and Wellbeing. The transport strategy is directly integrated with the individual strategies for these objectives. The City Council declared a Climate Emergency in 2019 and set a target of becoming a carbon neutral city by 2030. The key actions identified under this objective are to reduce the need for travel and the number of car journeys; encouraging people to use active travel and public transport modes; improving the efficiency of the transport network and making better use of road space and leading on and encouraging others to take up zero emission vehicles for freight, public and private transport. The City Council currently has the largest electric vehicle fleet of any local authority in the UK, with over 300 vans and commercial vehicles. Recently the Council have introduced a small fleet of e-cargo bikes, which can be hired by businesses.

The transport strategy components of inclusive growth include assisting the less well-off and deprived communities to access more employment opportunities through the provision of a comprehensive transport network; using investment in transport infrastructure to develop and regenerate communities; improving productivity via investment in a more time and cost-efficient transport system and lowering the cost of mobility by ensuring that sustainable transport is affordable and accessible.

The Council's Health and Wellbeing Strategy sets out 12 priority areas with the aim of making the city a leader in this area, with the aspiration to be 'the most active city in England'. I have examined the Council's stated transport objectives for improving the health and wellbeing of its citizens and compared these with initiatives underway or being considered in other UK cities. The broad actions necessary and resulting from this comparison are summarised in the list below. Integrated spatial and transport planning will be necessary to identify the appropriate mix of actions in different parts of urban areas:

1. Introduce policies, investment priorities and spatial plans that ensure that walking and cycling become the first-choice modes of travel for short journeys within and between neighbouring communities.
2. Remove the negative effects of transport on local communities by achieving planned reductions in all harmful vehicle emissions through the promotion of low emission vehicles, reductions in the number and length of car journeys and enhancements to public transport and active travel networks. Improving the safety of the highway network within and bordering communities via the segregation of active travel routes from the main vehicular highways and providing more space and enhanced visibility for pedestrians through reductions in road space and other measures.
3. Reducing central parking areas to the minimum that can be justified under the new 'transforming travel approach', while enhancing the scale, location, pricing and general attractiveness of park and ride schemes and their linking public transport networks.
4. Devise neighbourhood plans that achieve permanent solutions for reclaiming parts of the street network for active travel, greening and public recreation. These need to be achieved via full public engagement and learning the lessons of the temporary closures undertaken during the Covid pandemic.
5. In new housing developments and residential regeneration schemes, introduce ideas and designs that remove private cars from the core of the residential neighbourhoods to well-designed indoor parking areas on the periphery. Lessons can be learned from Freiburg in Germany and recent residential developments in Norwich, Cambridge, and Leeds.

Our travel routines and preferences will have to change if we are to achieve the climate change targets set by the Paris Protocol and to enhance our health and wellbeing by becoming more active and freed from debilitating ill health caused or exacerbated by air pollution. Changes to transport policy, investment decisions and pricing policy for sustainable travel modes are essential, together with new approaches to spatial planning and the association and integration of land uses. Failure to make the brave or radical decisions in the past means that we need to catch up on leading towns and cities as quickly as possible. Freiburg, in southern Germany, has for some time led the way on implementing several of the necessary policy changes. The city of 230,000 people made early strides on sustainable urban development from the 1970s and has used a highly integrated approach to introduce sustainable energy supply, active and sustainable travel, contained and safe urban neighbourhoods where spatial planning has been used to reduce distances travelled, and a distribution of accessible services. By 1999 some 50 per cent of all trips were on foot or by bike. After a one-third cut in the price of the unified public transport pass there was a 23 per cent increase in patronage in the first year, rising to 100 per cent in subsequent years, when public transport use had more than doubled.

The Freiburg story is both impressive and inspiring, but it might be argued by some that it is easier to achieve these changes in a smaller city where politics and individual environmental choices were strongly aligned in the early years. While there is some truth in this there can be little doubt that there are major lessons to be learned from Freiburg and other pioneering cities such as Oslo, Amsterdam, Vancouver, and Copenhagen.

Changes to the existing ways in which we carry out spatial and transport planning, including greater integration of policymaking, decision-taking, and implementation, need to form a core part of a renewed town and country planning system.

Chapter 11

Homes for Everyone?

The issues of housing need, demand and supply have often been central themes in the UK development and planning system. However, over the last three decades the problems in and surrounding the housing market have become more apparent, of some greater complexity and of more pressing political concern. National politicians have increasingly realised that not enough homes are being built to accommodate the housing needs of a growing and changing UK population. My experience of working in the housing sector stretches back over some 45 years, from Leeds City Council into the subsequent consultancy part of my career, but it is the last 25 years of practical experience, research for specific projects and general observations of market trends that have had most influence on the views expressed in this chapter.

The issue of providing sufficient housing had previously reached peak importance in the early years following the war. In the post-war Labour Government of Clement Attlee, it was Aneurin Bevan, the passionate force behind the creation of the NHS, who also, surprisingly, was given responsibility for housing. His overt policy to meet the pressing housing needs of the reconstruction of war-damaged areas, and the repatriation of the armed forces, was to restrict private housebuilding and to concentrate resources on the provision of new local authority housing.

The incoming Conservative Government of October 1951 was led by an ageing Winston Churchill. At the party conference held the previous Autumn in Blackpool, an ambitious target of building 300,000 homes a year was adopted as part of the manifesto. Housing at this time ranked as only second to defence as a political issue. In an article written for *Conservative Home* in 2013, Andrew Gimson recounts how Harold Macmillan, selected by Churchill as housing minister, was able to achieve and indeed surpass this target.[1] Macmillan records in his diary his feelings after being given the job at a meeting with Churchill: 'He asked me to "build the houses for the people". What an assignment! I know nothing whatever about these matters, having spent 6 years now either on defence or foreign affairs. Churchill says it is a gamble – make or mar my political career. But every humble home will bless my name if I succeed. On the whole it seems impossible to refuse – but, oh dear, it is not my cup of tea – I really haven't a clue how to set about the job.'[2] Under Bevan and a concentration on public house building, Labour achieved an average of 200,000 homes per annum between 1945–51. Macmillan was, however, a shrewd businessman who had been involved with the family publishing company. He was also very good at recognising talent amongst others and determined that the Ministry would utilise lessons learned from wartime business and procurement. The target was achieved for three successive years via a combination of council and private sector housing. However, this success had its drawbacks as Macmillan and his team paid little attention to the quality of the product and to aesthetics.

While previous conservative governments had facilitated the sale of council houses to tenants, the numbers sold were limited. The 1979 incoming Conservative Government led by Margaret Thatcher introduced a statutory right to buy via the 1980 Housing Act, which was opposed by many local authorities including several led by Conservative majorities. The number of sales to former tenants increased from 568 in 1980 to 196,430 in 1982, with the total sold between 1980 and 1991 reaching 1.46 million. These sales were paralleled by a decline in new council house completions and a generally insufficient level of investment in maintenance of the retained stock of council houses. During the Conservative Governments of Margaret Thatcher and John Major spanning the period 1979 to 1997, the level of new house building fell back to around 200,000 per annum. The contribution of council housing had dwindled to a low contribution and the indication that it would be replaced by that of housing associations failed to materialise to a sufficient extent.

These short reviews of past actions and performance by national government at various stages help to demonstrate some of the impacts that national politics and consequent policies can have on the performance of the housing market and in particular the supply of new housing. Aneurin Bevan all but excluded the private sector from his plans and concentrated on the delivery of public sector housing and the qualities of the product in terms of the space available and some of the design issues. He recognised the links with the health and wellbeing of the occupants but was not able to meet the full level of housing need and demand. While the build-up from the end of the war to a peak of around 230,000 homes in 1947/48 was quite an achievement, given the post-war problems such as the supply of materials and trained labour, the average annual build-out was 200,000 homes. Macmillan's approach was wide-reaching and embraced the public and private sectors. It was successful in quantitative terms due to the full engagement of councils and private housebuilders, the well organised business approach, and the talented team that he assembled. I can only describe the approach of Margaret Thatcher as destructive and singularly unsuccessful in terms of meeting housing need and demand. The only apparent benefit was to extend the level of house ownership, but this created at the same time a widening gulf between advantaged and disadvantaged households based on whether they remained in a council house or had acquired a property.

This brief review provides an insight into one set of factors influencing the productivity of the housing market, and it has provided some lessons for successor governments on how to organise and deliver on their promise of meeting in full the housing needs of the population. My main aims, as in previous chapters, are to examine the role and performance of the planning profession and the planning system in the quantitative and qualitative aspects of housing market delivery and the impact of adequately satisfying housing needs on the health and wellbeing of the population. Planners and the planning system have been much criticised by politicians, developers, landowners, and some members of the general public for their contributions to the failures to meet national housing targets and deliver the kind of homes that the public need or desire. Many reports have been commissioned and written that examine the under-performance of the housing market. While I have read several of these it is only possible to refer to a few, which I have selected based on their coverage, influence, and the regard in which they are held by a wide audience.

When I set out to try and provide a blend of my professional experience with that of others who have investigated these issues, I initially neglected to go to the very heart of the question posed in the chapter heading. On first examination, housing everyone appears to be an unrealistic goal, which is so difficult to achieve that many governments worldwide will either not commit to or will fail in their attempts to achieve the goal. The Universal Declaration of Human Rights was agreed by the General Assembly of the United Nations on 10 December 1948. Article 25 of the Declaration recognises the right to housing as part of the right to an adequate standard of living. 'The Right to Adequate Housing' is the title of UN Fact Sheet No. 21 on Human Rights produced by UN Habitat. This states that the right to adequate housing should not be interpreted narrowly and should be seen as 'the right to live somewhere in security, peace and dignity'. This fact sheet goes on to make it clear that four walls and a roof do not meet the standard of adequacy. The definition adopted by the UN is far-reaching and states that the needs of marginalised and disadvantaged groups need to be taken into consideration. Habitability includes the provision of physical safety, adequate space, and protection from threats to health and structural hazards.

The unwritten constitution of the UK does not contain a 'right to housing' and calls for Parliament to introduce such a right are not new. These were repeated following the Grenfell Tower fire disaster in central London in 2017.[3] Recent government policy statements in the form of two planning-related White Papers and the current version of the NPPF do not directly mention or allude to the right of all households in England to be safely and adequately housed. The 2017 White Paper, via its stark title, accepts that there are major problems.[4] These policy proposals were issued for consultation and set out the Government's plans to reform the housing market and boost the supply of new homes. In August 2020, a second White Paper sought views on a further package of proposals for reforming the planning system. The provision of more homes is a central aim of these proposed reforms. In his foreword, the prime minister, Boris Johnson, seeks to primarily lay the blame for the insufficient supply of homes at the inadequacies of an outdated and ineffective planning system. 'Thanks to our planning system, we have nowhere near enough homes in the right places. People cannot afford to move to where their talents can be matched with opportunity.'[5] While I do not seek to totally exonerate planners and the planning system from any blame, I will demonstrate that there are many other more significant contributions to this ongoing core problem and, as is apparent from the historic comparisons I have already summarised, national governments carry a large responsibility. The current version of the NPPF concentrates on boosting the supply of homes and enhancing delivery. None of these policy documents refer to the full make up of housing need and the social justice aspects of a right of all households to a home.

It is left to Shelter, the country's main housing charity, and a few other charitable and research bodies, particularly the Joseph Rowntree Foundation, to fully set out the extent and nature of housing need. Shelter states on its home web page that it 'exists to defend the right to a safe home'. In a report on its website, Shelter assesses that some 17.5 million people in England are impacted by what they term is a 'housing emergency'.[6] A total of around 250,000 people are regarded as homeless because they are either sleeping on the streets or are in temporary accommodation, while councils and other organisations try to get them back into a permanent home. The high cost of housing, both purchase and rental, is at the heart of the housing emergency, alongside the successive failure to deliver the level of housing required to meet fully assessed need. There has been a progressive decline in the safety net of social housing provision since 1970 and this is a further contributor to the problem of a deficient supply of new housing. Shelter arrive at their figure of 17.5 million people by including the homeless; people in overcrowded dwellings where basic facilities and available space and privacy are wholly inadequate; those living in properties that are regarded as unsafe for one or more reasons; households who lack security of tenure in private rented accommodation and finally households locked into accommodation by mortgage or rental agreements that they can no longer afford due to income not keeping pace with housing costs.

Within these groups there are clear links between the housing problems faced and the physical and mental health and wellbeing of the occupants. In overcrowded damp or unsanitary conditions, people are more prone to contract illnesses. Cramped and inadequate spaces and worry about meeting housing costs have a clear adverse impact on the mental health of many individuals. The post-Grenfell discovery that there are an estimated 700,000 people living in flats wrapped in several kinds of flammable materials is a stark example of modern, often recently built, apartment blocks, which have been declared unsafe and, until remedial works are carried out, unsaleable. While the government has set up a significant fund to assist in progressing recladding and related work, there is a huge backlog, and many owners are faced with large bills. The final major report of the Grenfell Inquiry is awaited, and the Government will need to act on many key issues the inquiry has considered, including the responsibilities/liabilities of the developers and construction companies involved in erecting these unsafe buildings. Far more people are impacted because they cannot prove that the flats they occupy are safe.

The problems summarised in Shelter's report comprise a large part of the issues that have faced English governments over the last 30 years or more, and which now comprise major components of what they refer to as a 'housing crisis'. The Coalition Government (2010–2015) and the successor Conservative Governments, while not ignoring some of the key issues covered in the Shelter report, have focussed their attention over the last ten years on the underperformance of the housing market in terms of delivering the required levels of new supply and improving access to the market especially for younger people. While the planning profession are involved in those parts of the housing market associated with home improvements via extensions of existing homes and conversions to residential from other uses, their major involvement lies in the identification of suitable sites for new housing development within local plans and comprehensively assessing the suitability of specific residential development proposals via planning applications. It is in the performance of these specific roles that the planning system and local authority planning officers and planning committees have been frequently singled out as the main source of delays and under-delivery of new homes. Before responding to this recurring criticism, it is important to first understand the problems that the market for providing new homes is seeking to address, then to examine the roles and performance of the other main players in this sub-market and finally to identify what are the main contributing factors to under-delivery in both quantitative and qualitative terms.

This brief overview of the problems facing those parts of the housing market concerned with new supply relate primarily to the position in England, though certain figures quoted refer to the UK. The problems in the devolved nations, while significant, are generally not as great. The first and major problem is the rate at which house prices have increased in the UK. In the thirty-year period to 2004 the trend rate of real house price growth was 2.4 percent, which is shown to be considerably higher than the European average of 1.1 per cent. These figures are taken from the landmark report prepared by the economist Kate Barker in 2004.[7] Between 1996 and 2019, 160 per cent real terms increase in house prices in England presents the scale of the problem.[8] This situation is further emphasised by the increase in average UK house prices of 7.5 per cent over the single year ending in January 2021. At first sight this is somewhat remarkable given the impacts of the Covid pandemic, but the main reasons for this will become apparent. At this point the average house price in England was £267,000. This rise in house prices, interrupted only by the 2008 to 2012 financial recession, presents a major affordability problem, with the average age of first-time buyers in the UK now standing at 34 years. This represents a rise in this average of 6 years since 2007 and 8 years since 1997. The third main problem is the under-delivery of new housing supply that comprises the number of new dwellings added to the existing housing stock via new build and the conversion of properties from other uses.

Having outlined these primary problems, it is necessary to explore the nature of housing need and demand, the differences between them, and how their quantification has been a key focus for national government for so many years. There is no specific definition of housing need, or to put this another way there is no clear consensus amongst the experts involved in housing need and market assessments. In an excellent briefing paper from the House of Commons Library entitled 'Tackling the under-supply of housing in England'[9] the authors state that, while there is no strict definition of housing need, 'it can be understood as the amount of housing required for all households to live in accommodation that meets a certain normative standard'. The paper goes on to refer to the projected growth in the number of households being used as a proxy for housing need, but importantly recognises that this doesn't present the whole picture as household projections don't attempt to accurately forecast future changes and allowance needs to be made for an existing backlog of need, which includes households living in unsuitable or overcrowded accommodation.

The terms housing need and housing demand are often used interchangeably, but it is important to recognise the clear difference between them. Housing demand is the amount of housing in terms of both quantity and quality that households will choose based on their preferences and ability to pay. While the majority of households able to enter the housing market are looking to purchase a home, private rental has increasingly become a selected option. Planners and some demographers use the term housing requirement to refer to the combination of need and demand, especially where market and affordable housing provision are under consideration.

In preparing development plans, policy planners in local authority planning departments will, to varying degrees, become involved with demographers and other expert consultants in the preparation of Strategic Housing Market Assessments (SHMA). These either include, or are supported by, the assessment of housing need and components of that need, including the amount and type of affordable housing required over the period that the development plan will cover. The strategic market area covered in the SHMA may be the same or very similar to the boundary of the local planning authority however, increasingly the extent of this area has spread across more than one planning authority area and the SHMA is produced on a joint basis. The individual housing need assessments are based on nationally provided population and household projections, which are broken down into separate figures for each local planning authority area and are prepared by the Office for National Statistics (ONS). The population projections include natural change (the difference between births and deaths) and the level of net international migration. The projection of how many new households will be formed across a particular period are assessed at national level based largely on past trends in the rate of new household formation and the conditions that influenced this and the extent to which these and other conditions might apply in future.

In Chapter 9 I summarise the rise and fall of Regional Spatial Strategies (RSS) and how these were prepared by Regional Planning Bodies (RPBs), following national planning policies, and using national demographics and trend analyses, derived the housing requirements for the constituent district planning authorities. This approach to deriving and handing down the housing requirements for individual local planning authorities at the regional level in the statutory RSS led to the major negative reaction from the incoming Conservative/Liberal Coalition Government in 2010. This in turn produced the abolition of regionalism and a major move towards localism, with the individual local planning authorities given responsibility for identifying the long-term housing requirements for their areas.

It is therefore surprising in political terms that the Conservative Government in 2018 decided that it was necessary to introduce a new standard method for assessing housing need for all the English local planning authorities. The new method of assessing baseline need for each authority was in the form of a nationally derived algorithm, and changes to the NPPF made it a requirement to use this method, which was set out in the related National Planning Practice Guidance (NPPG). Any departure from using this method had to be justified by exceptional circumstances. Two sets of changes were made to this approach as the government received adverse reactions from many local authorities. Put simply there was a north-south divide in the responses as many authorities in the south were asked to provide significantly greater amounts of housing than those contained in their existing local plans, while several authorities in the north had a significantly reduced housing need compared with previous assessments. The southern authorities argued that, due to environmental and other constraints, it was either not possible or would be very harmful if they were to agree to pursue the enhanced requirements. Several northern authorities had significant growth proposals and argued that reductions of a third or more in their housing requirements would have a wholly negative effect on their plans. It is now possible for individual local planning authorities to apply uplifts to the baseline need assessment if these can be justified based on economic growth, affordability, or regeneration requirements.

It can be seen from this short review that the issues surrounding the assessment of housing need are in themselves quite complex, and individual planning authorities and the residential developers operating in their marketplace can arrive at different positions on what the end-housing requirement should be for the next 20 to 30 years. Many factors can affect the rate at which the formation of new households will take place in the future. These include the following influences:

1. The ways in which the reliance on past trend information is applied. For example, in England MHCLG's projections were informed by trends from 1971–2011, but when the ONS subsequently took over the responsibility for making the projections they used different assumptions about the rate of new household formation using more recent short-term trends.
2. The effect of marked economic downturns, such as the recession that followed the financial crash of 2007/8 tend to suppress the rate of new household formation leading to younger potential new households remaining with parents or sharing a household with other young people.
3. The length of the period used in deriving projections is a simple but key factor. Alan Holmans of the Cambridge Centre for Housing and Planning

Research has argued that it is necessary to select projection periods which cover 20 years or more when estimating household formation rates for various planning purposes including town planning and the planning of future infrastructure. This enables short term socio-economic trends to be 'smoothed out' in order to derive a more realistic long-term trend.[10]

4. Variations in the amount and type of net international immigration into the country will impact on future household formation rates depending on a range of factors relating to the socio-economic condition of the immigrants and whether, for example, they are joining an existing household made up of family members.
5. The rate at which dwellings are built in the future and the mix of housing types and tenures can encourage or discourage new households to form. If many fewer dwellings are built, shortages of accommodation and worsening affordability can force household formation downwards.[11]
6. National government can use various financial incentives that encourage household formation, recent examples deployed by the UK government include the 'Help to Buy' scheme, incentives targeted at first-time buyers and reductions in stamp duty on the purchase of a dwelling.

The current Government's standard method for assessing housing need uses the population and household projections as a baseline before adjusting for affordability and then allowing local planning authorities to make changes for other factors such as growth strategies, major infrastructure investment and regeneration plans. The government have retained their target of building 300,000 homes per year despite lower household projections. They have argued that new household formation is indeed constrained by housing supply and that this is part of the reason for the fall in projections, that historic under-delivery of housing needs to be included and that low supply has led to declining affordability.[12]

As we move on to examine the involvement of the planning system in the supply side of the housing market it is important to recognise how the preparation of local plans are now dependent on the assessment of housing need by other parties at the national level and by consultants asked to advise at the local level. This responsibility has shifted between national, regional, and local government levels. This has not created a helpful, clear, and stable working context for the planning system and criticisms based on the alleged poor performance of that system in the under-supply of housing must first appreciate the many problems associated with assessing actual housing need with any high degree of accuracy.

I consider Dame Kate Barker's 2004 report on housing supply[13] to be the most knowledgeable, balanced, and considerate of the many papers and reports on this focal topic produced in the last 30 years. Her wide understanding of the subject and its many relationships is demonstrated in the coverage and clarity of her writing and by her subsequent appearances as an expert contributor to later reviews. The Report was commissioned in 2003 by the Treasury and the Office of the Deputy Prime Minister (who carried responsibility for housing and planning). The terms of reference were to 'conduct a review of the issues underlying the lack of supply and responsiveness of housing in the UK'. In particular, she was asked to consider the roles of competition, capacity, technology, and the finance of the housebuilding industry, and to examine how these factors interact with the planning system and the Government's sustainable development objectives.

Her foreword to the Report is in the form of a letter to the Chancellor and the deputy prime minister and it opens with the following sentence: 'Housing is a basic human need, which is fundamental to our economic and social wellbeing.' While this directly stops a little way short of the UN Declaration that all people have a right to adequate housing, the strong sense of social justice for all households is present, as is the alignment with the definition of sustainable development, as now contained in the NPPF. The foreword goes on to summarise some of the problems caused by a weak supply of housing, including macroeconomic instability, reducing labour market flexibility, and constraining economic growth as a result. The issue of affordability is recognised as increasingly problematic with its potential for creating an ever-widening social and economic divide. Equally she gives recognition to the dilemmas facing politicians and decision-makers at all levels of government, including public concerns about the environment, the loss of open spaces and pressures on support infrastructure. She concludes that the problems of increasing homelessness, affordability and social division are too great not to take a series of positive actions to increase supply.

The Barker Review puts forward a series of policy recommendations to address both the lack of supply and the responsiveness of those acting as providers, who need to both up their game quantitatively and to substantially improve design, innovation, and sustainability of provision. A main suggestion in the recommendations is the need to bring economic considerations, including market information, into the planning system. Several other recommendations are made to improve the working of the planning system. These include the central recommendation of allocating more land for housing development while at the same time giving more attention to the preservation of the most valuable land in terms of environmental and other qualities. While the Review is now some 18 years old and was completed under a Labour government committed to regional planning, a great deal of its content remains valid to the housing market position today as the problems of supply, affordability, and the social divide in access to the market have all increased.

The second report about weak housing supply in England, which I have found to be most helpful and insightful, was produced in a yearlong collaboration by the leading accountancy firm KPMG and Shelter.[14] The report, produced in 2014, contains a proposed programme of action for the incoming 2015 government to carry out over the following five years. The authors urge the government 'to show real leadership and embark on a programme of reform and investment' that tackles all four of the interrelated problems that they consider are, in combination, limiting house building. It is notable that they do not single out the planning system for criticism in this context, and also that the annual target their programme relates to is 250,000 dwellings.

The four main actions identified are: i) to fix the broken land market; ii) invest in new affordable housing; iii) increase the diversity of housebuilders and iv) help cities and towns to take a lead through enhanced powers and finance. The process by which housebuilders acquire land, particularly in the form of competitive bidding between the volume builders in a complex and opaque market, means that the value of the land today has to be assessed on the expected value of the dwellings over the period of completions. The more that the successful developer pays for the land, the more adverse are the impacts on quality, affordability, and prices in general. The report recommends the establishment of a new National Housing investment bank and new financial incentives to increase the provision of affordable housing with a stronger direct provision role for local authorities. This report, along with others, recognises that, while the main volume housebuilders have increased their outputs, there is an obvious need to widen the diversity of providers. Over the short period of the financial recession, 2008 to 2012, around one third of small housebuilders went out of business and KPMG/Shelter proposed that the Government should guarantee loans to small new building companies to help them break into the market and propose that the Government also encourage increased output from housing associations, the self-build sector and industrial system building.

The KPMG/Shelter programme of actions rightly implies that the lead responsibility for addressing the under-performance of the housing market lies with the national government and recognises that they must also carry a significant part of the responsibility for the generally weakening position in recent years. This conclusion is supported to varying degrees from within government by the House of Commons – Housing Communities and Local

Government Committee, the Select Committees of the House of Lords on i) Economic Affairs (2016) and ii) Intergenerational Fairness (2019), and in the Conservative Home article by Andrew Gimson quoted earlier.

What has become clearer in my mind during my more recent professional practice and the research for this chapter is that there are several parties and factors responsible for contributions to the overall sequence of under-delivery in the English housing market. At this point I turn to examine the extent to which the planning system and professional planners must carry some of this responsibility. At the outset I must emphasise that this level of responsibility falls well short of the primary blame all too frequently implied by leading figures in national government and some in the housebuilding industry. Earlier in this chapter I quote a stark example of this level of blame from none other than the former prime minister, Boris Johnson.

An examination of the stages in which the planning system impacts on housing delivery starts with the preparation of local plans. The time taken to reach adoption of the final plan by the local planning authority, the amount of land, which is allocated in the plan, when measured against the assessment of housing need over the plan period, comprise possible sources of under-provision. The assessment of baseline need in the currently modified version of the government algorithm may fall some way short of the actual locally assessed need in the SHMA. If the local planning authority fail to apply any of the potential uplifts to the baseline need figure, then the selected housing requirement in the plan is likely to be insufficient to meet the full level of need over the plan period. There are also situations where authorities do not provide sufficient land allocations in the plan to meet the estimated housing requirement. This can, however, happen in a number of different ways where the responsibility for the outcomes does not lie solely with the planning officers. The following examples illustrate this.

In the first example, the officers select a range of sites for allocation and assume certain housing numbers will be achieved on each of these either on a gross or net density per hectare, which, following site master planning by a development team, proves to be unachievable. Cumulatively this can lead to under-delivery against the requirement in the local plan. The independent government inspector may have the opportunity to correct some of this shortfall if he or she recognises the potential problem, or it is drawn to their attention by other participants at the Examination in Public (EIP).

A second example relates to the inability of a local planning authority to find sufficient land to allocate within their boundaries without having an on-balance, unacceptable adverse impact on primary and secondary environmental constraints such as landscape, nature, heritage, and flood risk designations. In these situations, which should really become apparent at the early stage of plan preparation, local planning authorities are encouraged to work together to jointly resolve where residential allocations can best meet their combined housing requirements. The example of the West of England grouping of four authorities in the greater Bristol/Bath housing market area, covered in the chapter on regional planning, demonstrates where such cooperation and hard work by planners can still break down or fail to meet the housing targets as the coverage of primary constraints is so extensive.

A third example often occurs where a high proportion of a local planning authority's area is designated as Green Belt and the planning officers and the committee responsible for overseeing the preparation of the local plan give very great weight to the national policy in the NPPF of seeking to minimise the amount of land that needs to be released from the Green Belt to meet the fully assessed housing requirement. In these circumstances, where the geographical spread and amount of brownfield land is limited, the choice of Green Belt sites for allocation can and should form part of the sustainable development mix. However, in my experience this leads to protracted and confrontational plan-making with Inspectors at EIPs faced with making the difficult decisions. It is not unknown for a local MP to ask the Secretary of State to 'call-in' the plan for review, which in my local area of Bradford Metropolitan District added a further 6-month delay to the process of achieving an adopted plan.

The next stage in the planning process is assessing individual planning applications, either following the allocation of a site in the local plan or dealing with what is referred to as a windfall proposal (an unexpected application for a development that is not subject to an allocation in the development plan). If the developer/applicant chooses to make an outline planning application followed by a subsequent reserved matters detailed application, this clearly lengthens the decision-making process. It has long been argued that refusals and delays at the planning application stage has been responsible for under-delivery. This has been refuted by the Local Government Association (LGA) and the RTPI amongst others. The LGA, using national government data (MHCLG), identified that over the period 2010/11 to 2020/21 a total of 2,782,00 homes were granted planning permission, but only 1,622,730 were built over that same period. Councils are approving 9 out of every 10 planning applications and very high proportions, exceeding 80 per cent, are granted within the 8- and 13-week deadline periods, which have been in operation for several years. In 2017/18 permission was given for 382,997 dwellings, well in excess of the government's target of building 300,000 homes per year.

The issue of time lags between the grant of permissions and starting and completing dwellings on site is subject to several contributing factors, one or more of which can be in operation to create delays in development. The grant of planning permission is always subject to conditions and certain of these, known as conditions precedent, must be resolved prior to the commencement of development. Conditions precedent on large or complex residential developments can include matters such as design details, final approval of materials and approval of long-term management arrangements for areas of public open space. The Home Builders Federation (HBF) point out that while it is encouraging to see more permissions being granted, these are recorded once one of the conditions attached to them is discharged. This can typically leave several other conditions precedent to be discharged before development can be legally commenced on site. The extent of delays in resolving these conditions is dependent on both the time taken by the developer and the planning officers in respectively providing the necessary information and assessing its adequacy.

Persistent public opposition to residential development proposals, as well as delaying the grant of planning permission, can continue afterwards in various forms. I acted as planning consultant for a major housebuilder on a 175-dwelling development adjacent to the Wharfedale village of Menston in Bradford District. This was one of two sites that were allocated in the development plan in 2004, with my co-director acting on the other site proposals for 130 dwellings. The principle of development was established, and the local planning authority had commissioned and approved a development and design brief covering both sites. Full planning permission was granted for the first site following detailed scheme design negotiations. The village action group raised further funds and applied for the site to be designated as a village green, arguing that some of the fields had been used for public recreation. The evidence was very slender and, following a public inquiry, the application was refused by an independent inspector. This was followed by a high court challenge on drainage and flood risk grounds. After several years of delay the development of both sites could proceed. However, due to the long delays and additional expense for the national housebuilders on both sites, they withdrew. New residential developers entered into development agreements with the landowners of both sites and development is now progressing. While these are somewhat extreme examples there have been too many cases where public action groups have been able to delay or frustrate development that has otherwise been supported by the local planning authority. The production of fully binding neighbourhood plans, including residential allocations, should help to prevent many such delaying actions by groups, who often do not represent the majority of residents.

The difference between the dwelling capacity contained in the planning permissions approved annually and the number of housing completions achieved each year has been a recent focus for the national government as the investigation of the reasons for under-delivery of supply has widened to include this specific gap. The first of a small number of reports was the 'The Lyons Housing Review', produced in 2014 for the shadow Labour government prior to the national election of 2015. In the foreword, Sir Michael Lyons acknowledges and endorses the major work carried out by Kate Barker and her prime conclusion that the single biggest constraint on building more houses is the shortage of land made available for building. He then goes on to state that: 'We have tried to get behind the sometimes over-simplified, public discussion about the constraints imposed by land use planning to understand more fully why communities are so often resistant to new development and see it as imposing costs which greatly outweigh any benefits. So, we are clear, the problem is not about those who sit on planning committees, or even those who advise them, it is about addressing the public's concern that houses are often built in the wrong place, for the wrong people and without adequate attention to the pressures created for existing infrastructure.'[15] Chapter 4 of this comprehensive review deals with the issue of ensuring that land with planning permission is built out and proposes disincentives for landowners or developers holding onto a planning permission in addition to measures that would incentivise earlier delivery of land allocated in a development plan. These included reducing the lifetime of a planning permission to 2 years, strengthened compulsory purchase powers for local planning authorities to make it easier to acquire land where it is not being brought forward, and giving councils the ability to charge taxes on the landowner that are not completed in a given time period.

The investigations for the Lyons Review found that a significant number of the planning permissions for developing housing were not controlled by housebuilders but by investment companies, landowners, and other developers and that this was particularly the case in London and some of the larger cities in England.

At the time of the 2017 Autumn Budget, the Chancellor of the Exchequer commissioned an independent review of build out rates to explain why the gap existed between the number of planning permissions being granted against the numbers being built in areas of high demand. Sir Oliver Letwin MP was appointed to chair this review.[16] The main conclusions of this review were targeted at future large sites with a capacity of more than 1,500 dwellings in areas of high housing demand and sought to introduce a much greater diversity of market housing in terms of design, layout and housing types and tenures.

Sir Oliver had identified the lower-than-expected market absorption rates on large sites as a key determinant of build out rates due to the similarities of the housing type and design offer of the large housebuilders involved. The current NPPF allows local planning authorities to consider the application of a planning condition imposing a shorter timescale for the commencement of development than the current default position of 3 years and requires closer monitoring of progress in building out sites. The 'Planning for the Future' White Paper (August 2020) proposed, via further revisions to the NPPF, a requirement for master plans and design codes for substantial sites to seek the inclusion of a variety of development types from different builders, which would allow more phases to come forward together.

This overview of several of the factors contributing to the persistent under-delivery of new housing supply demonstrates that there are several key participants involved in contributing to the overall position. The national government is chief amongst these not only because it carries the lead responsibility for putting policies and actions in place but because historic national policy approaches have contributed to the current problem. The most notable of these contributions has been the great decline in council house building and sales of existing local authority housing stock, most in vogue during the 10 years of Margaret Thatcher's leadership. In 1969/70 council house building starts in England amounted to 175,550. Ten years on in 1979/80 the number had dramatically declined to 67,450 and by 1989/90 only 16,110 were commenced. The nadir was reached in 1999/2000 when only 390 starts were achieved. In recent years there has been a slow recovery, but the figures achieved are far too low to have a significantly beneficial impact on bridging the large gaps in overall supply and affordable provision.

Other contributors to the problem include the larger national housebuilders, some landowners, and the organised NIMBY (Not in my back yard) action groups in many settlements, which are all too often led by retired individuals looking for a cause to follow and a means of protecting their living environment, with limited or no involvement from the younger age groups, several of whom may be seeking market or affordable housing. However, the issues for those who become deeply involved in the planning and housing market processes are far from black and white and require carefully balanced considerations. My own experience and investigations align with several of the reviews referred to here when assessing the contribution made by the planning system. There are undoubtedly improvements which can be made to the production of local development plans both in terms of speed and the quality and clarity of outputs. However, as is now increasingly recognised, this cannot be achieved without substantial re-resourcing of local authority planning departments following many years of cuts to services. The government commit, in 'Planning for the Future' (August 2020), to 'developing a comprehensive resources and skills strategy for the planning sector' with several action areas. An investment of £500 million over the next 5 years was proposed. The main planning-related issue identified at the stage of local plan production is the failure to allocate sufficient land, but this, in itself, is the product of many interactions, rather than simply the result of the direct actions and approach of the professionals producing the plans.

At the planning application stage of the residential development pipeline the evidence is clear, the decision-making part of the planning system has in recent years been providing the overall approved dwelling capacity required to achieve the Government's annual target of 300,000 dwellings. Issues with the rates of housebuilding at the post-permission stage need to be addressed by the government following reviews by the House of Lords Economic Affairs Committee (2016), Lyons and Letwin.

There is also increasing recognition and action regarding the largest potential solution to the under-delivery of the required supply, which is to increase the outputs from other contributors as well as local authorities. These include helping more small builders to enter the market; continuing to widen the role of housing associations and the intervention activities of Housing England; encouraging more self-build and industrialised methods of producing complete dwelling units. The Government are actively promoting measures to achieve these contributions and new companies, both large and small are entering the market.

Earlier in this chapter I reviewed the background to the UN 1948 Declaration of Human Rights, certain of the health and wellbeing implications of being denied adequate or safe housing and how we define and assess what constitutes housing need. There are those who have examined the factors affecting housing need, demand and supply who have questioned the extent to which increasing supply can lower or even stabilise house prices and directly improve affordability. The House of Commons Briefing Paper on dealing with the problems of under-supply of housing in England[17] refers to a report prepared by Oxford Economics for the Redfearn Review into the decline of home ownership.[18] The baseline forecast in the Oxford model assumes the steady addition of 210,000 dwellings per annum. The modelling results found that to put downward pressure on prices new supply would need to outstrip underlying household formation and that boosting annual supply by a further 100,000 dwellings would help to keep prices in check. However, by the end of the forecast period prices would only be about 5 per cent below the baseline forecast. Their conclusion was that, even at historically high rates of building at 310,000 dwellings per annum, this would have to be sustained each year over a long period of time if it is to have a substantial impact on prices.

The report prepared by Ian Mulheirn,[19] formerly the Director of Consulting at Oxford Economics, goes much further than the conclusion reached by his former employers. He calls for a fundamental rethink based on his research and analysis around the impact of additional supply on house prices, stating that the growth in house prices over the last 23 years does not appear to have been the result of under-supply and then goes on to examine whether greater supply could nonetheless be the solution. His report accepts that meeting the Government's target of 300,000 houses per year would put more downward pressure on prices and rents but having reviewed the available academic evidence states 'that no plausible rate of supply would significantly reverse the price growth of the last two decades'. His review of a range of modelling exercises in the UK and elsewhere finds that a 1 per cent increase in the stock of houses generally leads to a decline in rents and prices of between 1.5 per cent and 2 per cent. This leads to his conclusion that building 300,000 homes per year in England would only reduce house prices by around 10 per cent over the course of a 20-year period. When we consider some of the recent large annual increases in house prices and the overall rise over the last two decades, his conclusion that the answers to creating more affordable houses to buy and rent lie elsewhere appears credible.

Mulheirn's report provides us with a relatively up to date review of the operations of the housing market, but while its coverage is quite wide it cannot be described as comprehensive. In the alternative policy approaches suggested he does include the lack of sufficient affordable housing and past erosion of the social housing stock. I do not find his arguments, relating to the growth in concealed households and almost eliminating new supply from the solutions he suggests, at all convincing. The strong aspirations of those young people living with their parents and young people living in immigrant households as well as those sharing with their own age group are part of the underlying housing need. Mulheirn concentrates on other financial solutions for these young, concealed households and, while acknowledging the need for more affordable housing, neglects the ways in which this is provided via market housing delivery. More than half (52 per cent) of all affordable homes delivered in England in the year April 2019 to March 2020 were funded through section 106 planning agreements with private sector providers and this rate was higher than in previous years. A total of 92 per cent of all affordable homes built in England in this period were new build. The proportion delivered by local councils was 11 per cent of the total, the highest since 1991/92.[20] The private sector were the highest direct providers linking with a registered social provider (usually a housing association). Therefore, until the Government make radical policy changes, the private housebuilders will continue to provide the majority of affordable houses as part of their market housing schemes under policies contained in local plans.

The above reasoning, combined with the health and wellbeing and social justice arguments for the widest possible access to housing, lead me to support the arguments for an enhanced overall housing supply incorporating allowances for meeting the backlog of provision and further increasing the supply of affordable housing. A report was prepared by Professor Glen Bramley of Heriot Watt University for Crisis (a lead national charity for homeless people) and the National Housing Federation (NHF is the national organisation representing housing associations in England). Crisis help people directly out of homelessness and campaign for changes that will come up with a full resolution of the problem. The NHF's vision is to achieve 'a country where everyone can live in a good quality home they can afford'.[21]

Glen Bramley concludes from his research that 'there is an urgent need for more housing that provides people on low incomes with security, decent living conditions and affordable rents'. In the foreword to his report, he states the failure over many years to build enough homes for people in the greatest need. This research is differentiated from previous studies by its specific focus on low-income households and those experiencing homelessness. The research also adds to the existing evidence on housing need by assessing the backlog of unmet need and via the new methodology introduced by the author for assessing housing requirements. He assesses the backlog of housing need in England at 4 million households, of which 3.15 million are made up of concealed and overcrowded households, those with serious affordability or physical health problems and people living in unsuitable accommodation. The total of 4 million is completed by 240,000 homeless, 250,000 older households with unsuitable accommodation and 410,000 households whose housing costs are unaffordable. From this large backlog of need analysis, he projects the overall levels of supply required on a 15-year time frame. The required supply in England is 340,000 dwellings per annum with the component of social housebuilding in England needing to rise to 90,000 per annum, shared ownership of 25,000 and intermediate rent 30,000. He uses three partially distinct methodologies in his assessment, two of which are based on a traditional demographic framework, which he then enhances to reflect affordability. The third is based on a sub-regional housing market model including the range of housing components relating to affordability, poverty, housing need and homelessness. Affordability emerges as the key criterion for adapting housing targets away from those adopted in previous plans or demographic projections.

Bramley concludes: 'If the goal is to make a significant and proportional response to housing need, particularly the most acute needs such as those experiencing core homelessness, quite strongly differentiated housing targets are appropriate.' He identifies the need for further analysis to cover questions relating to resource requirements including the proportion of costs that can be borne by new developments and the extent of the investment required from the national government. His report and its findings are important evidence in relation to enhancing housing provision for reasons related to the overall health and wellbeing of the population. The provision of good standards of housing for those in the areas of greatest need is proven to be beneficial to mental and physical health with consequent savings created in other areas of government expenditure.

The recent granting of permitted development rights enabling private sector developers and property owners to convert commercial property to housing has resulted in some conversions of low space and design quality with commentators referring to 'the creation of the slums of tomorrow'. It is important that greater control is provided through the planning application system to avoid the all too obvious problems being created. Along with many of my colleague professionals I consider this 'quick fix' approach to achieving enhanced dwelling numbers as a wholly retrograde step.

The Covid pandemic has heightened the general importance of the design of home interiors and exteriors to provide for working from home, adequate space for children to do their homework and the great desirability of incorporating sufficient garden space for recreation and relaxation of individual members of each household. Communal open areas including spaces for play, social interaction, landscaping, and biodiversity are also essential provisions for all aspects of our health and wellbeing and become an absolute necessity for those living in high density housing schemes lacking individual garden space.

In Chapter 3, Health and Wellbeing, I briefly described two recent award-winning housing schemes in Norwich and Cambridge that demonstrate that very attractive and more sustainable neighbourhoods can be achieved without adding significantly to overall build costs. The word attractive is used in the widest sense to encompass design, living environment and social conditions, all of which contribute to wellbeing. I am adding a third multi-award-winning example (including the Sunday Times British Home Awards Sustainable Development of the year) conceived and developed by the Citu Group in the City of Leeds, where more than half of my professional career has been based.

The Climate Innovation District, located on both banks of the River Aire in the expanding city centre, is an inspiring example, which is delivering a residential development of 516 town houses and apartments, offices (including the company's new headquarters), leisure facilities and the manufacturing unit where Citu fabricate the majority of all the dwelling components and currently has the capacity to produce 750 low carbon homes per year. The former industrial land is being totally regenerated and currently ranks as the largest sustainable housing development in the UK. All the dwellings are being constructed with an airtight thermal envelope, triple glazing, and a mechanical ventilation heat recovery system. Active solar panels, photovoltaic modules and passive solar gain combine with the main construction features to create close to Passivhaus standards.

While a supply of undercroft parking spaces is available to purchase for £15,000 per space, resident households are being encouraged to further their sustainable lifestyle by opting to use the segregated riverside footpath and cycleway connections into the heart of the city. The whole development is being raised to a height above the river, which will achieve very high flood resilience. A system of greenspaces and high-quality landscaping are designed to provide all residents with accessible open space in addition to the gardens provided for the townhouses. The external living environment is further enhanced by the total absence of motorised traffic.

On completion the greenspaces, infrastructure and renewable energy systems will be transferred to a Community Interest Company (CIC) controlled by the residents. The CIC will be able to decide on what new renewable energy sources to invest in and how they want the revenues generated by feed in tariffs from the solar panels to be spent. All such monies generated must remain within the development.

When I visited the project recently on a warm sunny afternoon, I was highly impressed by the emerging development, its design qualities, and the impromptu greetings I received from clearly happy residents in their gardens fronting the river.

These three examples illustrate what can be achieved and how the residents of all three schemes are clearly benefitting from experiencing their internal and external living environments. These are the standards we must achieve in all new developments to maximise public health and wellbeing gains and ensure developments have a very small carbon footprint and high resilience to climate change impacts.

In England and the devolved nations of the UK, providing enhanced housing supply can over time contribute to stabilising and marginally reducing market house prices. In England there is clear evidence, especially, but by no means exclusively, from Glen Bramley's work, that the large backlog of affordable housing need must be met in a prolonged programme of action over some ten years or so, and to do this requires market housing supply to be boosted to continue and potentially enhance its contribution to affordable housing.

There is one further means available to reduce house prices, in the mix of market economics and government controls and influences that are at play, and that is to find a successful way of reducing the amount of development value gained by landowners and the land prices paid by developers. As mentioned in other chapters, many ways of 'taxing' or otherwise reducing the level of gain have been attempted, and of those actually introduced none have had great success or longevity. In their Planning for the Future White Paper the Government put forward the option of abolishing section 106 agreements and reforming the Community Infrastructure Levy system, so that it is charged as a fixed proportion of the development value of land with planning permission above a threshold with a mandatory nationally set rate or rates. The alternative of aiming to capture more land value to better meet necessary infrastructure delivery, with the cost of the increased CIL levy being capitalised into land values would, the Government argue, ensure that landowners make a greater contribution from the increases in land value created by the development partners' negotiated planning permission. I consider that this approach would need some mandated government framework that provides a basis for resetting the land value negotiation between the landowner and prospective development partner. Many factors come into this equation and all need to be weighed in the balance before coming up with a workable and sufficiently attractive system, which can operate in different subregions/housing market areas.

In the Raynsford Review of Planning in England, a set of four recommendations are made to tackle betterment values (i.e. the increase in value to a landowner by virtue of a planning permission), which concentrate on strengthening the powers in local plans, the greater use of development corporations with land acquisition, including compulsory purchase, powers to assemble large sites at values closer to existing use value and the redistribution of national land tax revenues.

Much has been written about land value capture (LVC) over the decades and it is not my wish here to go into any detail, but it is helpful to simply highlight the broad results from several past attempts at some form of development land taxes or betterment levies. The early work of Ebenezer Howard has been summarised in Chapter 1, including his pioneering efforts in devising a system of retaining increasing land values within the specific 'municipal corporation' established to build and run the new garden city. The MHCLG parliamentary select committee provide a useful recent review of LVC having taken evidence on the subject from a wide range of witnesses.[22] Previous attempts at LVC from the Development Charge introduced as part of the 1947 Act planning system to the Planning Gain Supplement recommended in Kate Barker's Review of Housing Supply were political in nature, all being introduced, though not necessarily implemented, by Labour governments. A common result was for landowners to hold back their land hoping that a change of government would repeal any statute or otherwise cast the idea aside. Having been involved in the implementation of the 1976 Development Land Tax System while at Leeds City Council I can confirm its complex nature and certain of the adverse impacts.

The fact that LVC is still actively being explored confirms the central virtue of seeking to extract a sizeable, or sufficient, proportion of enhanced value resulting from the grant of planning permission for residential development in order to pay for the essential infrastructure required to support it. The complexities and sometimes the scale of previous attempts at LVC have led to opposition from the development industry, but it is fair to say that housebuilders are not generally opposed to the principle if a reasonably clear, simple, and equitable system can be derived as advocated in a survey of participants in the housebuilding industry carried out by the Royal Institution of Chartered Surveyors (RICS).[23]

A key lesson, accepted in the MHCLG review, is that to be successful, any selected methodology will need cross-party political support. The RICS survey recommends specific skills training and support would be needed for local planning authorities to implement a new LVC mechanism. This survey also recommends that further research is required to identify which components of CIL, section 106 agreements and any wider taxation system are captured from the uplift of land values from the granting of planning permission as well as the extent to which they are captured from other sources such as the housebuilders profit or house prices.

Planning authorities, and officers, are already involved with housebuilders in examining the extent of financial returns for landowners and developers where issues of the affordability of physical and social (affordable housing) infrastructure are at issue. This usually involves the developer in commissioning a viability assessment incorporating what are regarded as reasonable returns for that company and the landowner alongside the costs of infrastructure required as a result of the specific proposals in a planning application. It is necessary for the planning case officer to have some understanding of the basic methodology of the assessment and then usually involves a chartered valuation surveyor being engaged either from the Council's estates department or an outside consultancy. A recent academic paper on this sub-topic of LVC provides a helpful evaluation of the detailed approaches to considering what constitutes a 'competitive return' to the landowner when carrying out a viability assessment.[24] The key issue arising in this consideration is what level of value should the landowner be assumed to receive as a priority over any land value capture in the form of whatever planning obligations should be applied based on planning policies in the local plan and how they are used in a particular project planning application and its location. This is generally known as the Benchmark Land Value (BLV). The Government have required that in formulating and implementing planning policies the planning authority should use viability assessment, primarily but not exclusively at the plan-making stage to ensure that the policy requirements they develop for all physical and social infrastructure contributions are deliverable. This important and demanding task is to ensure that the plan policies are realistic and that the total cumulative cost of all policies will not undermine deliverability of the plan. The National Planning Practice Guidance directly encourages housing developers to engage in the plan-making process and to take account of costs, including their own profit expectations. The Government warn developers that the price that they pay for land is not a relevant justification for failing to accord with the relevant policies in the local plan (NPPG Paragraph: 002 Ref ID: 10-002-20190509).[25]

The NPPG goes on to provide the Government's recommended approach to viability assessment for planning and the standardised inputs required together with essential engagement with developers, landowners, and infrastructure providers. In plan making and decision taking viability assessment 'helps to strike a balance between the aspirations of developers and landowners in terms of returns against risk, and the aims of the planning system to secure maximum benefits in the public interest through the granting of planning permission' (Paragraph: 010 Ref ID: 10-010-20180724). The guidance advises the use of BLV in defining land value in the viability assessment and that the main components that should be used in determining such value are the existing use value, the provision of a premium for landowners and a consideration of abnormal costs, site specific infrastructure costs and professional fees (Paragraph: 014 Ref ID: 10-014-20190509).

The NPPG approach to viability assessments for testing the delivery of plan policies and determining planning applications appears relatively straightforward, but to arrive at the required inputs requires a great deal of expertise, negotiation, and analysis. Arriving at developers profit margins is relatively simple within the 15–20 per cent range advised in the NPPG. The derivation of BLV is more complex and a significant amount of evidence is required to provide the level of landowner premium that will work firstly in the local plan housing market area and secondly with respect to negotiations associated with a particular residential planning application project. The

Government have not yet put forward any specific proposals as to how the derivation of BLV could be used to reduce the net land value payable to landowners, after the deductions of infrastructure costs and professional fees, to enhance the housebuilder's ability to reduce actual house prices. This needs further early exploration as such an approach has the potential to reduce prices more quickly than supply increases though the latter and is an essential requirement to meet full housing needs including provision for the backlog.

In this chapter and in Chapter 3, my introductory chapter on planning and health and wellbeing, I have referred to those situations where the standards and quality of housing provision can have beneficial or detrimental impacts on the health and wellbeing of households and communities. In Chapter 1 the poor housing and living conditions of the nineteenth and early twentieth century urban areas, and the health and wellbeing consequences, were identified as the major influence on the 'movement for planning'. The need for greater cooperation between planning and health authorities has been accepted in recent years and planning and health policies and guidance have turned their attention to tackling the sources of poor mental and physical health as well as health inequalities. Housing need and inadequate provision, particularly in low-income households, are key determinants of poor health. A recent report by the Health Foundation (December 2020) emphasises these relationships and reinforces the need arguments put forward in Glen Bramley's review of housing need.[26]

The Health Foundation's report identifies that, at the beginning of the Covid-19 pandemic, some 32 per cent or 7.6 million households had at least one major housing problem relating to overcrowding, affordability, or poor-quality housing. Problems like these can affect health outcomes, including physical health directly from poor quality homes and mental health from affordability or insecure housing. The Foundation assess that one million households in England experience more than one of these problems and, consequently, having such multiple problems is associated with even worse health. While the number of homes classed as non-decent have declined over the last 10 years, overcrowding and affordability problems have increased. The Covid-19 pandemic has further highlighted the health implications of housing as poor housing conditions such as overcrowding and high density have been associated with greater spread of the virus. Many people have had to spend more time in these inadequate housing conditions during lockdowns, working from home or with a sheer lack of access to the outside environment due to disability, accessibility to open spaces and other factors. In 2022 the energy price crisis, experienced in the UK and across Europe, has further highlighted the problem of poorly insulated homes and the energy cost implications facing many households. This crisis has required financial support from the Government. Despite this and other impacts we still await a national programme of retrofitting cold and damp homes.

The report concludes that several interventions are needed in the housing market to improve health and remove health risks and that these will create opportunities to address interrelated environmental and economic challenges. Research from this charity has shown that investing £1 in housing support for vulnerable households delivers £2 of savings to public services, including health and social care. The challenge for cooperative working is to ensure that housing makes a positive contribution to the health of more and more people. Greater provision of housing, which is affordable, of high quality and secure will offer great benefits for the health and wellbeing of individuals and communities. Highly sustainable and attractive neighbourhoods incorporating active travel at the expense of car access with landscape and biodiversity led designs have the capacity to transform people's health and wellbeing.

My experience and an assessment of much of the best research leads me to the conclusion that there are many factors and parties contributing to the under-supply of housing, continuing rises in market prices and increasing affordability problems impacting on a growing number of households. The operation of the planning system and its main components of preparing development plans and determining planning applications is not the prime reason for any of these problems. There are however aspects of plan-making that contribute to delays in housing allocations entering the 'development pipeline'. The overall lack of sufficient allocations being provided in local plans is influenced by national government policies, and Green Belt policy is a prime example. Public opposition, also influenced by Green Belt and other national policies, is all too often led by NIMBY elements within local communities. Improvements to performance in determining planning applications lead to the clear conclusion that this part of the system is not generally delaying housing starts on site, though post-permission determination of section 106 agreements and conditions precedent do contribute. A key positive contribution needs to be driven by the national government and that is the requirement to expand the number of organisations operating in the market. There are some positive trends in this direction in terms of more small builders becoming engaged. The direct involvement of local councils needs to be fully activated.

From this analysis it can be seen that the planning system and the professionals operating it have a very important and multi-faceted set of roles to play and that generally these have over recent years become more complex and testing as the resourcing of local authority planning departments has declined. These roles and skill sets include assessing housing requirements in local market areas; identifying and assessing suitably sustainable housing site allocations; providing plan policies that are clear, positive, proactive, and capable of delivering viable solutions; determining planning applications in a timely and balanced way and ensuring quality outcomes in terms of design, beauty, and sustainability. These are extremely challenging roles requiring considerable experience and skills. I consider it is remarkable that so many quality schemes have been delivered through the planning system in recent challenging times.

Chapter 12

Climate Change – The Role of Town and Country Planning

In the days leading up to writing a first draft of this chapter I was pleased to read three positive articles in a national daily newspaper and a countryside magazine I subscribe to. *The Guardian* newspaper of Friday 2 July 2021 carried a report in its business section covering Nissan's decision to produce a new electric car model at its plant in Sunderland. The Japanese company's partner Envision will build a new electric car battery plant alongside a new battery recycling facility. Initial plans also include up to 10 solar farms to power the project. The whole integrated project will create a total of 6,000 new jobs in direct manufacture and the supply chain. In the August 2021 edition of *Country Living*, the founder of the Eden Project in Cornwall (the world's largest indoor rain forest), Tim Smit, announced the completion of a new geothermal plant at the Project, which has been ten years in planning and involved drilling 4.7 kms into the continental crust to release energy from the rock and creating enough hot water to heat all the greenhouses and generate electricity. The same magazine reported the community creation of the 2,185 hectares (5,400 acres) Tarras Valley Nature Reserve in southern Scotland. The £3.8 million required to purchase the land was raised by a crowdfunding scheme established by the Langholm Initiative with support from other organisations including the Woodland Trust. The land improvements to date have already created a haven for threatened bird species including hen harriers, merlins and short eared owls.

Hardly a week goes by now without similar good news stories of projects inspired and carried out by individual entrepreneurs, landowners, communities, and local and national government bodies. However, at the same time, there are shocking reports of the devastating impact that climate change is already having across the globe. The Guardian of 2 July 2021 devoted its front page, and an internal double-page spread, to the intense heat being experienced by communities in the western areas of Canada and the United States. The formation of a massive 'heat dome' resulted in record temperatures, which at one point reached 49.6 degrees Celsius in part of British Columbia. The resulting forest fires caused rapid evacuations of certain communities and recorded higher rates of sudden deaths, which medical experts attributed to the record high temperatures. The physical impacts of these high temperatures also cause great damage to infrastructure, agriculture, and ecosystems. Other adverse impacts are experienced from the increased incidence and magnitude of flood events, mud slides and rapid soil erosion, sea level rises caused by melting glaciers, ice sheets and rising ocean temperatures. Exceptional rainfall affected parts of Western Europe over a two-day period (15–16 July 2021) when parts of the catchments of the river Rhine and its tributaries received the equivalent of two months' rainfall leading to devastating floods, landslides, loss of life and extensive infrastructure damage. Over this short period dramatic shifts in the pattern of the jet stream were identified as a potential factor in both the North American heat dome and the devastating floods in Western Europe. While there is now a very high level of consensus amongst the world's climate scientists that human activities in the form of various industries (primary extractive, agriculture, and manufacturing), consumption and travel are collectively the main contributors to global warming, further investigative and modelling work is still required to determine the nature, rapidity, and inter-linkages between the dramatic changes we are experiencing.

In *The Guardian* articles of 2 July, Professor Peter Stott, from the UK Meteorological Office, is quoted as saying: 'The risk of heatwaves is increasing across the globe sufficiently rapidly that it is now bringing unprecedented weather and conditions to people and societies that have not seen it before. Climate change is taking us out of the envelope that societies have long experienced.' Professor Simon Lewis of University College London described the situation as 'scary' and warned that extreme heat events could have huge impacts on many aspects of life including food prices and power supplies. He went on to say that: 'Everywhere is going to have to think about how to deal with these new conditions and the extremes that come along with the new climate that we are creating. That means everyone needs plans.' There is a view emerging amongst climate scientists that such extreme events are moving ahead of previously modelled outcomes and, in common with Professor Lewis, they conclude that governments and policymakers need to pay great attention to these warnings and to dramatically increase plans to halt fossil fuel emissions.

A year later from my first draft, revisions are necessary as record daily and nocturnal temperatures are recorded in the UK, with a peak daytime temperature of 40.3 degrees Celsius recorded on 19 July 2022 in Lincolnshire. The heat caused random grass wildfires around London and other towns, resulting in the destruction of nearby houses. The heat buckled old steel railway lines and overhead electricity lines bringing large sections of the intercity rail network to a halt. The new peak heat event is a rise of 1.6 degrees on the previous record. The level of this rise, the incidence of several hottest years within the last two decades and the damage to key infrastructure bring home the lessons we must learn on the rate of warming and the extent and urgency of the mitigation and adaptation actions required.

As part of my preparation for this chapter I took down from my bookshelves two yellowing copies of landmark paperbacks, both of which were first published in 1972. The first of these *The Limits to Growth* is a condensation by four authors of a report for the Club of Rome's 'project on the predicament of mankind'. In July 1970 an international research team at the Massachusetts Institute of Technology began a study into the effects and limits of continued worldwide growth in global population, agriculture, resource use, industry and pollution and demonstrated how these factors interact with one another. Their conclusion was that, even with the advances in technology proceeding on the most optimistic upward curve, the world could not support the then current rates of economic and population growth for more than a few decades. The use of a large-scale computer model, as part of the study demonstrated that only by a concerted attack on all these major problems at once could 'man achieve the state of equilibrium necessary to his survival'. The second book, *Only One Earth – The Care and Maintenance of a Small Planet*, was commissioned by the Secretary-General of the United Nations Conference on the Human Environment to provide expert background information for the UN Conference held in Stockholm in June 1972. The report has been described as 'a Domesday Book of the kingdom of man'. Key content in the report includes the planet's unity, the presence of only one shared biosphere for humanity, the price of pollution, including air pollution, humankind's use and abuse of the land and the need to work together on the science and the solutions to ensure survival. These books served as early warnings that we needed to collectively change our lives and attitude to, and care for, the one-world environment. They demonstrated the close relationships between human actions on economic growth, the relentless search for and use of natural resources to grow our economies, feed our expanding populations and fuel our energy demands, leading to the exploitation and depletion of the environment and its life-giving qualities.

* * *

Some recommended background reading

In more recent years my reading has included just a very small number of the hundreds, if not thousands, of books written on and around the subject of climate change, its causes and effects, the interactions, the solutions, and the rate of progress being made towards a carbon-neutral existence for humanity. I have listed this small number of books in my bibliography as a sample of potential reading for all those involved, directly and indirectly, in the central role that the town and country planning system needs to play in helping to deliver the essential processes of adaptation, mitigation and resilience to climate change impacts. Many more books are being, and will be, written and I hope that these both enable and inspire the next generation of planners to select their own sample of background reading to provide them with the basic knowledge and understanding they will need in making their important future contributions.

James Lovelock, regarded as one of the world's leading scientists, and a renowned environmentalist and futurist, conceived the Gaia Hypothesis back in 1972. Since then, he has written several books on this subject based on the concept that the earth is a complete living organism made up of its relatively thin spherical shell of land and water and the atmosphere that surrounds it, and the whole being is occupied by an interacting system of living organisms. As Lovelock himself says, it is a metaphor for the living earth, which is one major global ecosystem. Support for the Gaia Hypothesis came from a meeting of scientists from the four great international global research programmes in 2001, which agreed the following statement:

> 'The Earth system behaves as a single, self-regulating system, comprised of physical, chemical, biological and human components. The interactions and feedbacks between the component parts are complex and exhibit multi scale temporal and spatial variability.'[1]

Lovelock examines climate change and human contributions to this alongside destruction of biodiversity, all kinds of pollution, the depletion of resources, including the degradation of land and soils and population increase in the form of damaging impacts on Gaia. I have selected three books from his wide list of publications and scientific papers.

My next choice is two books written by the former Vice-President of the USA in the Clinton administration, Al Gore. These are chosen because they are clearly and passionately written by a man who, having reached high office and then in 2000 ran unsuccessfully as the Democratic candidate against George W Bush for president, subsequently dedicated his professional career to campaigning on the climate crisis. A second reason for choosing them is that they have had a very wide public audience, extended by the first book, *An Inconvenient Truth*, being made into a film which went on world-wide distribution. The second book presents a 'plan to solve the climate crisis' based in part on some 30 'solutions summits' organised by Gore's team.

Stewart Brand's book entitled *Whole Earth Discipline* is also a clear and passionate response to the three profound transformations which are underway on earth: climate change, urbanisation, and biotechnology, with the former of this trio pushing us towards recognising the need to manage the planet as an entity. Brand, who originally trained as an ecologist, argues in his book that the environmental movement must change some of its long-held opinions and embrace certain of the tools it has traditionally distrusted, including science and engineering. Only then, he believes, can it slow down and halt the cataclysmic deterioration of the earth's resources. He comes up with several key observations that run counter to commonly held views, a key example of direct interest to planners and their wider public is that cities are greener than the countryside.

Nicholas Stern (Lord Stern of Brentford) is a leading economist and academic with a highly distinguished career. He previously held posts as Chief Economist, first at the European Bank for Reconstruction and Development and then, from 2000 to 2003, at the World Bank. He is now Professor of Economics and Government at the London School of Economics and Chair of the Grantham Research Institute on Climate Change and the Environment at Imperial College London. The Grantham Institute is one of the world-leading research and innovation centres on matters related to climate change.

In July 2005 he was appointed by the Chancellor of the Exchequer, Gordon Brown, to conduct a review into the economics of climate change. The Stern Review, published in October 2006, concluded that there was still time to avoid the worst impacts of climate change if strong action was taken; that the impacts on growth and development could be very serious and that the costs of stabilising the climate are significant but manageable and delay would be dangerous and much more costly. The Review also concluded that a range of options existed to cut the harmful greenhouse gas emissions, but strong and deliberate policy action is required to motivate their take-up. Future international frameworks were recommended including economic measures such as universal carbon pricing and emissions trading schemes, as well as high levels of technological cooperation and using economic tools to understand the risks and uncertainties and the relative costs of action and inaction. The plaudits for the Review flowed from a number of Nobel laureates in economics, the World Bank and several international institutions and the press, but there was some unease amongst leading politicians that the document was too doom-laden.

His subsequent book on the subject, published in 2009, has a positive title and positive solutions, while continuing to integrate into the narrative the warnings contained in his earlier Review. Both the Review and the book include some salutary examples associated with land use, spatial and infrastructure planning, and the cost appraisal of development projects, taking into account climate change impacts.

Three naturalists/environmentalists form an important part of my bibliographical sample. Tony Juniper and Sir Jonathan Porritt both had leading roles at Friends of the Earth and Sir David Attenborough, the renowned naturalist, has, in his tenth decade, become a world leading campaigner and broadcaster on climate change and declining biodiversity. Tony Juniper is now chair of Natural England the government body responsible for habitat protection and enhancement. His book, *Saving Planet Earth*, was published in 2007 as a companion volume to the BBC television series and with great clarity presents the relationships between earth's biodiversity, global warming and our use and depletion of planetary resources.

Jonathan Porritt followed his period at Friends of the Earth by becoming a founder director of the Forum for the Future, the UK's leading sustainable development charity, which now has 70 staff and some 100 partner organisations, including world leading companies. He has also served as chair of the UK Sustainable Development Commission advising government on all aspects of what is now central to national planning policy. Jonathan's book, *The World We Made*, is, in my experience, a unique and very positive approach to delivering the message of how we can combat the trio of interrelated threats of climate change, declining biodiversity and depletion of the earth's resources. It is written in the form of a history and a memoir produced by a fictional history teacher, Alex McKay, looking back from 2050 to 2022 when he started his teaching career. The book charts, with the help of his senior students as research assistants, how we arrived at a world that had achieved sufficient advances to avoid the worst impacts of the trio of threats and could look forward to a more balanced, sustainable, and fairer existence.

Sir David Attenborough needs little introduction and is now recognised worldwide for his messaging on climate change and declining biodiversity. He can communicate effectively with leading politicians, global organisations, and people of all ages. He became widely known for his long history of work as a naturalist, writer, and broadcaster with the BBC's Natural History Unit, eventually producing nine landmark series from 1979 to 2008 which together form the *Life Collection*. His latest book is entitled *A Life on Our Planet, My Witness Statement and a Vision for the Future*. In the first part of the book, the Witness Statement, Sir David provides an autobiographical review of his life with each period, from 1937 (when he was 11 years old) onwards to 2020, charted against the backdrop of three whole-earth statistics: world population, the amount of carbon in the atmosphere and the extent of the remaining wilderness. In 1937 world population stood at 2.3 billion, the carbon in the atmosphere was 280 parts per million and the wilderness area was estimated at 66 per cent. By 2020 the comparable figures were 7.8 billion, 415 parts per million and 35 per cent.

In his 'Vision for the Future' David Attenborough draws on all his personal environmental knowledge and the work of other scientists and one economist. Reference is made to two Earth System scientists, the work of the Intergovernmental Panel on Climate Change (the IPCC) and the Intergovernmental Platform on Biodiversity and Ecosystem Services (the IPBES) and that of Kate Raworth on the Doughnut Model of Economics (see Chapter 3 of this book). The Earth System scientists are Johan Rockstrom and Will Steffen, who developed a system of nine planetary boundaries to define a safe operating space for humanity. The Doughnut Model introduces an inner ring to the Planetary Boundaries Model to identify the minimum requirements to achieve human wellbeing. To achieve a future equilibrium, all people on the planet should be able to access their full range of baseline needs for a sustainable life while at the same time collectively ensuring that the threshold of stability or ecological ceiling of all nine planetary boundaries are either not exceeded, or where the boundaries have been crossed everything possible is done to redress the balance. It is both fascinating and encouraging how David Attenborough and others have achieved agreement on the need to follow the Doughnut Model approach to achieving sustainable development on a global scale. He concludes that the challenge set by this integrated model is both simple and formidable.

We must see climate change as one of the nine inter-related planetary boundary crises that are underway. These include loss of biodiversity, air pollution, change of land use from natural ecosystems (forest, moorlands, marsh, and natural grasslands) to agriculture and urban development, pollution of the land and water bodies by the overuse of fertilisers that disrupt the nitrogen and phosphorus cycles, depletion of freshwater resources, ocean acidification, chemical pollution, and depletion of the ozone layer. It is clearly established by scientific research that a number of these crises for the earth's ecosystem contribute to loss of biodiversity; that climate change is a major contribution to the acidification of the oceans and that air pollution from the burning of fossil fuels, agriculture and other sources are major contributors to climate change. David Attenborough suggests that the restoration of the earth's biodiversity is a major part of the solution to these crises and that this should include the rewilding of the land and the oceans, while recognising that this must be achieved in a balanced way and that existing and new cities and towns can make a positive contribution.

This is, I believe, the basis of the strategic approach we must follow, and it leads me to the conclusion that the balancing acts required, in the integrated delivery approach we need to follow, are a key factor that should lead to the fullest possible engagement of the town and country planning profession. Before moving on to examining the current and future role of the planning profession there is one final book I have added to my sample, which is wholly focussed on the solutions and actions we must implement.

Drawdown is a truly dramatic book, with a bold subtitle: *The Most Comprehensive Plan Ever Proposed to Reverse Global Warming*. A coalition of researchers and scientists worked under the leadership of Paul Hawken. These Drawdown research fellows comprise a total of seventy individuals from 22 countries, a truly international collaboration. The book includes a total of 100 solutions comprising 80 extant solutions and 20 'coming attractions' that are known to be deliverable. The 80 tried and tested solutions are given a ranking based on i) their contribution to the reduction in greenhouse gas emissions in gigatons of CO_2 equivalents; ii) their net cost and net savings in dollars, all of which are assessed across the period 2020 to 2050. The 'coming attractions' are known to be deliverable and include carbon capture from the air, living buildings that are to meet 20 performance imperatives including food production, the production of net positive waste and the generation of more energy from renewables than they utilise. Each planning department and private consultancy should have this book in its library.

* * *

Current National Policy and Guidance

The NPPF section 14[2] and NPPG Climate change[3] provide the policy and guidance framework on climate change for the work of town and country planners. They provide for mitigation and adaptation to climate change via development plan policies and individual planning application decisions, where climate change and related issues (e.g., loss of biodiversity) should figure strongly in the balance of material planning considerations. Both the national policy and guidance documents refer planners to the significance of the 2008 Climate Change Act, which established a legally binding target to reduce the UK's greenhouse gas emissions by at least 80 per cent in 2050 from 1990 levels. Since this enactment the UK became one of 196 signatories to the

Paris Agreement, signed at COP21 in December 2015, which came into force in November 2016. The Agreement is to limit global warming to well below 2 degrees Celsius and preferably to 1.5 degrees of pre-industrial levels. The UK and international policy are to achieve carbon neutrality by 2050.

The latest version of the NPPF now places specific emphasis on mitigating climate change and adapting to its impacts when preparing plans. This represents a change of emphasis from the earlier February 2019 version. The revision is encapsulated within the important presumption in favour of sustainable development stated in section 2 of the document.

The current version of the NPPF sees the planning system as supporting the essential transition to a low carbon economy via placemaking that contributes to radical reductions in greenhouse gas emissions, minimises vulnerability and improves resilience to adverse climate impacts such as flooding and overheating. When preparing development plans, planners are expected to take a proactive approach to mitigating and adapting to climate change, taking account of the long-term implications for flood risk, coastal change, water supply, biodiversity, and landscapes. Policies in development plans are to support 'appropriate measures' to ensure the future resilience of communities and infrastructure to climate change impacts. The only examples given in the NPPF are the provision of space for physical protection measures or making provision for the possible future relocation of vulnerable development and infrastructure.

Plan policies need to include a positive strategy for the provision of energy from renewable sources and should consider the identification of suitable areas for renewable and low carbon energy sources together with supporting infrastructure. In addition, plans should identify opportunities for development to draw its energy supply from renewable or low carbon energy supply systems and to consider the identification of sites for the co-location of energy suppliers and customers. When considering development applications, local planning authorities need to consider factors in the location, siting and design of development that will minimise energy consumption. These factors include topography, layout, building orientation, massing, and landscaping, though the NPPF currently neglects the inclusion of other criteria including materials, other aspects of building design such as ventilation, passive solar systems, and accessibility to sustainable transport.

The NPPF policy statement remains rather generalised and vague in parts. However, it is necessary to examine all sections of the policy document as a number include policy content of direct relevance to planning for climate change, transport being a prime example. The NPPG makes up for this to some extent by providing advice on how to identify suitable mitigation and adaptation measures in the planning process to address the impacts of climate change. This advice is presented in the form of a schedule of key questions and answers and the guidance is updated on a relatively frequent basis. It is clear from an examination of the NPPG advice that policies incorporated into local plans and the application decisions based on these policies will need to be far-sighted, and planners will need to integrate the advice of climate scientists, engineers, macro designers and a range of environment professionals. National policy and guidance require locally developed policies to provide for the future resilience of communities and infrastructure to climate change impacts, and the NPPG places a great deal of emphasis on the lifetime of developments when considering adaptation and mitigation measures.

The NPPF for England incorporates mitigating and adapting to climate change and moving to a low carbon economy into the environmental objective of sustainable development. The NPPG provides further emphasis by stating that responding to climate change is central to all three dimensions of sustainable development. This, together with the legal position in the 2004 Planning Act, places action on climate change at the very heart of the planning system. This should therefore be given much greater emphasis in the required changes to planning legislation, the NPPF, and in the enhanced resourcing requirements of the whole system. The national statutory framework, policy and guidance leave considerable scope for local planning authorities and individual communities to take specific actions to ensure that existing and new developments are resilient to climate change. This includes the ability to introduce and deploy integrated mitigation and adaptation measures in local and neighbourhood plans and individual planning application decisions.

Prior to leaving the topic of national policy and guidance it is important to include a footnote on the Government's reform proposals for the English planning system. It is extremely surprising that the Planning for the Future White Paper, published in August 2020, does not include actions on climate change amongst the most important national challenges, which are central to the future work of planners and planning authorities. Indeed, it is the greatest challenge in which the planning system must play a central role as advocated by the RTPI and the TCPA.[4] Several organisations, including the Housing, Communities and Local Government Committee of the House of Commons, criticised the omission of climate change and the delivery of sustainable development from the August 2021 Planning White Paper and sought further consultation on these and other omissions before the drafting of new planning legislation.[5] Criticism also came from the government's advisors, the Committee for Climate Change (CCC). The planning bill, which was drafted to implement changes recommended in the August 2020

White Paper, was withdrawn and a mixture of revised planning reforms were subsequently incorporated in a new Levelling-Up and Regeneration Bill, which was progressing through the parliamentary process prior to various changes of national government in the summer/autumn of 2022. This new Bill and the Levelling-Up White Paper, which preceded it, are considered in my concluding chapter. At this stage limited reference is made to the planning systems role in achieving the required actions on climate change in the UK.

* * *

The work of the CCC, RTPI and TCPA on Climate Change and its relevance to the future role of the planning system

The CCC is an independent statutory body made up of a wide range of experts, which was established under the 2008 Climate Change Act. The in-depth and very well researched work of the CCC provides all departments of the English Government, and those of the devolved nations, with regular progress reports on meeting the net zero goal by 2050, and adaptation and other requirements, in accordance with the monitoring system required by the Paris Agreement. The most up-to-date reports from the CCC demonstrate how the English Government's actions on policy formulation, the achievement of the Paris Agreement carbon reduction targets and the incorporation of adaptation into current plans and policies are currently falling well behind what is required to achieve the intermediate 2030 goals and net zero by 2050. The CCC's assessments and recommendations contained in their 2020 and 2021 reports add urgency and degrees of difficulty to the realisation of town and country planning's central role. There are other professions and public and private sector agencies who will share this central ground in the UK's drive for improved performance and a high degree of collaboration will need to be established and maintained.

The CCC and the RTPI agree that a failure to act on many fronts now will defer costs to future generations with consequent additional expense and impact. Town and country planning's existing and potential future role extends across several sectors including buildings, land use, transport, energy and the environment, with placemaking and spatial planning forming an overarching and coordinating responsibility.[6] The RTPI's response to this 2020 CCC report to the English Parliament also includes a proposal that would see local authority planners playing a key role in the early delivery of a national strategy for retrofitting buildings and established communities with new sustainable energy, green and transport infrastructure. The proposal envisages local retrofit strategies being included in local plans and being integrated with the need to regenerate deprived areas. While I agree with the sound logic of this additional proposal, the scale of the challenge must be rapidly recognised and acted upon at all levels of government. Again, the re-resourcing of local authority planning departments becomes crucial to successful outcomes.

The 2020 and 2021 progress reports spell out the gap between government aspirations (the UK was the first major economy to set the net zero goal for 2050) and achievements to date. The first of these reports is the Sixth Carbon Budget (December 2020).[7] Each of the Carbon Budgets look 12 years ahead so that the Sixth Budget is for the five-year period 2033–2037. These budgets place a five-year projected cap on greenhouse gas emissions, which should not be exceeded if the emission reduction commitments of the Paris Agreement are to be met. The CCC's Sixth Budget recommended pathway requires a 78 per cent reduction in UK territorial emissions between 1990 and 2035. The nature of this considerably enhanced target is demonstrated by the fact that it brings forward the UK's previous 80 per cent overall target by nearly 15 years. The foreword to the Report by the Chairman of the CCC, Lord Deben, describes the 2020s as the decisive decade for action and refers to the pathway meeting the stipulation of the Paris Agreement of 'highest possible ambition'. While this is regarded as challenging, he also describes it as 'hugely advantageous' in terms of the industrial opportunities created and the 'wider gains for the nation's health and for nature'. The Government have adopted this revised target, which accepts that emissions will have to fall more quickly than those projected in earlier carbon budgets.

The Sixth Carbon Budget Report identifies four key areas for action to meet the targets. These are as follows: reducing demand for carbon intensive activities, which includes reductions in travel demand, slower growth in flights and in waste; the take-up of low carbon solutions, which involves all new cars and vans and all boiler replacements in homes and other buildings to be low carbon by the early 2030s; the expansion of low carbon energy supplies in the form of electricity production and use of hydrogen, and finally a transformation in the UK's use of land.

The second CCC report produced in June 2021 is the UK's third Climate Change Risk Assessment.[8] This third independent assessment of the UK's climate risks, as required by the Climate Change Act, is based on extensive new evidence in a detailed technical report. A total of 61 risks and opportunities are identified, which are described as 'fundamental to every aspect of life in the UK: our natural environment, our health, our homes, the infrastructure on which we rely, the economy'. The report describes the gap between the level of risk faced and the level of adaptation that is underway as alarming as 'adaptation action has failed to keep pace with the worsening reality of climate risk'. In her foreword to the Report, the Chair of the Adaptation Committee of the CCC, Baroness Brown, describes adaptation as the Cinderella of climate change as it is 'under-resourced, underfunded and often ignored'. This combines to make action on adaptation more urgent, for without it we will simply enhance future costs, miss opportunities to avoid the 'lock-in' of climate change impacts and make all the societal goals we have set harder to deliver. The planning system must urgently engage in and perform its adaptation role in formulating key policies in local plans, ensuring that development decisions incorporate adaptation measures into proposals and provide for policies and implementation of retrofitting existing communities with new low energy heating systems, enhanced active travel opportunities and urban greening and cooling systems.

The CCC's second report assessing UK Climate risks and adaptation requirements identifies eight areas of risk requiring urgent action over the next two years. While planners need an awareness of all these urgent risk areas two are of particular significance to their work. The first comprises risks to the viability and diversity of terrestrial and freshwater habitats and species from multiple hazards. The comprehensive recasting of environmental policy, which is underway following the UK's exit from the EU, provides a time- limited opportunity to ensure that adaptation is explicitly integrated into policies. While the drafting of environmental policy falls predominantly under the Department for the Environment, Food and Rural Affairs (DEFRA), this comprises an increasingly important part of the planning system particularly at the sub-national level. More effective integrated working between DEFRA and the new Ministry for Levelling-Up, Housing and Communities is essential for future success in this area, together with absolute clarity on responsibilities for policy formulation and implementation. The devolved Welsh Government has taken a strong lead by setting up a new 'super-ministry' for Climate Change with a portfolio spanning planning, housing, the environment, transport, and climate change. While such a combined ministry would probably prove to be too unwieldy and too taxing for an English secretary of state it emphasises the great need for early and more effective collaboration on all climate and environmental policy.

The seventh of the eight risk areas is a key action for the planning system. This relates to risks to human health, wellbeing, and productivity from increased exposure to heat in homes and other buildings. The CCC Report identifies a review of the NPPF as one of the main opportunities for integrating adaptation on this risk area into major policy documents. Local planning authorities must also take up this urgent challenge in formulating local plans and related policies and design guidance. Spatial planning policies in local plans for new developments and communities, plus the retrofitting of existing communities, provide the opportunities to integrate adaptation measures into the macro-design of these developments. They also provide the scope to combine mitigation and adaptation measures in plans and projects. The CCC Report strongly advocates such integration:

> 'The best way to address climate change and to avoid unintended consequences is to ensure adaptation and mitigation are considered together in those areas where there are major interactions: especially across policies for infrastructure, buildings and the natural environment.'

The NPPG also advises that planners give particular attention to integrating adaptation and mitigation approaches 'and looking for "win-win" solutions that will support sustainable development.'[9] Examples include the provision of district heating networks that include tri-generation (combined cooling, heat and power) and the provision of multi-functional green infrastructure, which can reduce urban heat islands, manage flooding, assist species to adapt and contribute to a pleasant environment encouraging active travel.

Finally, the CCC Report provides ten principles for good adaptation that the Committee consider will improve understanding of risk and enable effective adaptation to climate change. These principles include integrating adaptation into policies, especially together with mitigation measures, avoiding the 'lock-in' of irreversible changes and addressing inequalities. The practice of building new homes without designing in adaptations to counter future extreme heat events is quoted as an example of 'lock-in' where expensive retrofitting is then required, the retrofitting of new windows and shutters being assessed as four times more expensive than designing them into all schemes now. The identification of addressing inequalities when seeking to introduce adaptation measures is of considerable importance to planners as the planning system is already responsible for taking account of the relative deprivation of communities across local authority areas in relation to the integrated approach to creating sustainable developments. The CCC conclude that climate change is likely to widen existing inequalities as it has disproportionate effects on socially and economically disadvantaged groups. These matters will need to be addressed in local development plans.

The two leading planning bodies in the UK, the TCPA and the RTPI, have produced an excellent guide for local authorities on all aspects of planning for climate change.[4] The production of the guide had strong funding support from a number of bodies including EPICURO, the European Partnership for Urban Resilience and the European Union Civil Protection and wide endorsement from consultancy, academia, local authorities and the Centres for Sustainable Energy and Sustainable Healthcare. The guide provides a positive 'can do' approach recognising that we possess the tools to respond positively to climate change impacts, but we must deploy these quickly and effectively. The guide was first produced in 2018 because the author organisations recognised that local planning authorities were in many cases not producing effective and up-to-date policies or delivering the adaptation actions necessary to secure the long-term resilience of communities and the required radical reductions in greenhouse gas emissions. The aim is to provide local authorities with the confidence to act, and advice on the ways in which to frame and implement appropriate local policies; choose sites for development by integrating all relevant climate change factors into the site selection process and determining applications for development by ensuring climate change factors receive significant weight in the decision-making process.

The guide concentrates on mitigation (particularly in relation to energy use and generation), adaptation and resilience. It does not contain detailed advice on matters such as green infrastructure, biodiversity, food security and flood risk, but does provide important cross references to other guidance on these impact topics. Several local authorities are more advanced than others in adopting policies on carbon reduction, and appropriate mitigation and adaptation measures and exemplar policies are included in the main text and several case studies are provided in an annex to the guide.

In addition to providing this guide for local authorities, the RTPI has produced several research papers about climate change, which provide the planning profession and national and local government with insightful conclusions at a policy level, on actions which need to be taken, the existing and future role of the planning system and practical examples of what can be achieved. I regard these documents as a very important contribution to the national and international research base for action on climate change and further evidence for the key central role which planners must play in achieving the goals established in the Paris Agreement and augmented at COP26 in Glasgow.

A research paper produced in June 2020 examining how planning can contribute to a sustainable, resilient and inclusive recovery from the Covid-19 pandemic, and the associated economic crisis, provides major parallels with the planning required to meet the challenges of climate change.[10] This paper makes specific references to the climate and ecological crisis and to the wider responsibilities for improving health and wellbeing, many of which have been underscored by the Covid pandemic.

In a Position Paper on Climate Justice the RTPI produce a strong argument for building the social dimensions of disadvantaged communities into the spatial planning response to climate change.[11] As the climate crisis deepens these communities will be the most impacted due to a complex array of factors that combine to make them the most vulnerable to climate change impacts, including high average ages, levels of disability and existing ill health, low incomes, and areas that have suffered most from the cuts to local government resourcing. Climate justice presents a challenge to government and decision makers to provide for well-resourced spatial planning that provides for equitable mitigation and adaptation. Strong public engagement will need to be at the centre of planning for climate change to understand the key vulnerabilities of a place and to be able to best explain the interventions necessary to mitigate and adapt to the changes likely to impact a particular community. Climate justice will provide a focus for considering the wider social benefits and costs of adaptation and mitigation measures.

The consideration of climate justice and the actions needing to be taken in different communities introduces important questions about the governance and resourcing of those areas and the institutional capacity to carry out the planning for and delivery of the actions deemed to be necessary if an equitable and fair approach is to be taken. The RTPI paper and associated research, which was reviewed in its production, provide specific examination of the differing abilities of local government to respond to individual community's levels of vulnerability to climate hazards. England is selected as a particular example of the under-resourcing and deregulation of local government, a national policy focus on housing and economic growth and uncertainty around national planning policy as factors combining to affect the ability of local authorities to respond to climate change through planning. The impact of these factors varies greatly across the UK. Planners are seen as having a key role in communicating 'the story' of what is required by way of change and adaptation in individual communities. Getting across the significance of risk and other climate change related concepts are regarded as very difficult matters to communicate, but there is a good body of research covering how this can be done by relating the technical issues to local daily needs, community identity, care for the future and equity in delivery of solutions.

In a 2020 report the RTPI advised the Liverpool City Region Combined Authority on how they should provide for climate adaptation across several local government sectors and in collaboration with adjoining authorities via their emerging Spatial Development Strategy. This uses the expanded concept of climate resilience, which covers the ability of large urban areas, as ecological, social and economic systems, to resist, recover from and continue to develop despite the experience of adverse climate change impacts.[12] A subsequent RTPI research paper in 2021 focussed on the importance of a 'place based' approach to climate change and examines how collaboration between in-house specialists in local authorities can integrate climate action within planning processes, as well as informing and contributing to other departmental responses. This research identifies significant opportunities for joint working between planning officers and specialists in climate change, sustainability, engineering, and the environment. However, there is a skills and knowledge gap within the planning profession that needs to be urgently addressed to facilitate the central role of planners in the place-based response to the climate and ecological emergency.[13] The report draws on a review of case studies of plans produced by individual authorities including, for example, climate emergency plans prepared by Glasgow City Council and Cornwall County Council. The Amsterdam 'Doughnut' developed by Kate Raworth and colleagues is summarised as an international example and it is notable that Cornwall County Council's Climate Emergency Development Plan has tested proposed policies against their own decision-making wheel which is derived from Raworth's Doughnut Economics.

The RTPI's place based research paper incorporates recommendations to local authorities from the CCC.[14] The CCC and the RTPI agree that a place-based approach is key to achieving their recommendations and the contribution of local authorities to achieving net zero. This place-based approach must be based on partnership working, both externally with public and private bodies and internally across local government departments. A further paper produced by the TCPA and the Centre for Sustainable Energy as a response to the Planning White Paper consultation focusses on planning's key role in responding to Climate Change.[15] This paper argues that the planning system and its reform must have climate change at its heart. If this does not occur, it is perceived that the operation of a system with economic and housing growth at its core will risk adding to the climate crisis. The TCPA/CSE paper aligns with a view I have expressed earlier that the NPPF is far too vague on the issues of climate change and fails to give a strong and specific lead to local authorities.

I have summarised how the RTPI, the TCPA, the CCC and local authorities themselves have sought to plug this considerable national policy void. These research papers provide us with a clear framework for the many contributions the planning system can make to help in achieving a country's climate zero targets. I would like to stress, as the RTPI do, that while we see the planner's role as crucial, they cannot deliver climate action in isolation from other council departments and from public and private bodies who also have considerable responsibilities for the necessary actions. Planners also have an important role in public engagement to explain planning policies, development decisions and certain groups of the required actions as part of a wider attempt to persuade the public to play their part in achieving our climate goals. Here again we must recognise that planners are only one part of this wider public engagement process. The personality and approach of planners has always been important to carrying out their role in a professional and effective way. More than ever before they will need to approach their work in an honest, well-informed, and clear way without any degree of arrogance and at least a touch of humility. They will need to recognise the increasing importance and challenge of their role without expressing any self-importance to those who they communicate and negotiate with.

The key areas of contribution that the planning system can and must make, now and in the future, arise from a relatively weak national policy framework in England, a reasonable standard of guidance in the NPPG and considerable research by the RTPI, TCPA and other bodies, particularly the CCC. Initiatives taken by local authorities in the UK and in other countries and cities add to a large resource of learning and expertise that can be shared. International collaboration on many fronts is necessary, as we will see from the latest report of the United Nations International Panel on Climate Change (IPCC).

Spatial planning is the first of these key areas and a place-based approach has been strongly emphasised. Planners and related professionals in design and the environment, transport and infrastructure planning have developed expertise over many decades and further expansion of this expertise is required for this area of planning to play its full central role in climate mitigation, adaptation and ensuring the overall resilience of communities of all sizes and locations. Spatial planning for climate change needs to be carried out at four levels if we are to meet the net zero target by 2050. The first level is regional/city-regional, and in Chapter 9 I advocated a return to some form of organised regional planning following the abolition of the statutory regional component of planning in England in 2010. Subsequently an ad hoc set of arrangements have arisen as the result of devolution of powers and finance to city-region mayors, the establishment of Combined Authorities and local arrangements between groups of planning authorities. The English Government will need to quickly decide whether these current arrangements can deliver macro spatial plans that integrate comprehensive solutions for delivering net zero by 2050. My conclusion is that there is insufficient time and a lack of political will to establish a replacement system of regional governance, and it is therefore necessary, given the urgency of the required actions, to rapidly organise collaborative working by groups of local planning authorities so that the whole of the country is covered. This needs to be achieved as a policy imperative under an expanded NPPF with additional guidance provided in the NPPG. Making this a statutory requirement could be achieved by small amendments to existing planning and climate change legislation.

These regional spatial plans will provide a strategic tier of planning to bring together public and private investment and enable many cross boundary/large catchment issues and development decisions to be effectively addressed, as individual authority plans and the current loose duty to cooperate have not proved adequate for these purposes. These regional spatial plans can and must be produced in a short timescale compared with the former Regional Spatial Strategies (RSS). Several lessons can be learned from these former plans as well as international experience. The content of these plans should be based on natural and socio-economic catchments, including river catchments and water gathering grounds, main landscape and habitat types, major wildlife corridors and large recreation areas and landscape designations. National parks will produce their own plans, though some of the smaller parks will need to work with adjacent local planning authorities. The planning of macro natural solutions to climate change in the form of both mitigation and adaptation will emerge from the regional analyses, with most of this work being covered by digital/GIS mapping of physical, environmental and land use data. National plans for re-afforestation, peatland restoration, water resources and other initiatives will be incorporated at this regional level to test the suitability of proposals in terms of sustainability appraisal, environmental impact, and competition between land uses.

Specific large-scale mitigation and adaptation measures can be identified, and a 'best-fit' approach taken to their integration with other environmental and land-use proposals. In common with spatial planning at local, neighbourhood and location/site-specific levels, planners, the professionals they work with, and the general public must look for 'win-win' solutions across the spectrum of planning for climate change, the enhancement of the natural environment and the health and wellbeing of communities. Identifying socio-economic catchments such as housing market areas and journey to work zones together with main areas of deprivation and regeneration in rural and urban contexts should form another layer of these regional spatial plans. This will add to the information required to plan for transport, energy supply and other key infrastructure, including that necessary for climate adaptation.

Regional and local development spatial plans will need to identify locations and main site opportunities for off-grid sustainable energy generation schemes as well as components for the national supply network. Planners will need to work closely with the energy companies and their engineers and surveyors to blend their expertise in site finding to ensure the optimum locations for efficiency and best environmental fit.

Local development plans produced by individual authorities will need to incorporate a strong spatial planning dimension in order to integrate a new central stream of planning for climate change via mitigation and adaptation solutions. As with regional spatial planning this can be facilitated by mapping several layers of information and analysis. Where local plans have recently been adopted, the district spatial plan could be incorporated as an addendum to the adopted local plan. Existing broad policies on planning for climate change will need to be adapted and expanded to include more specific actions and solutions, including clarity on energy efficiency standards for residential and commercial buildings to be met by the private sector.

At the neighbourhood planning level, local authority planners will need to advise those drafting/redrafting plans on the appropriate policies and scheme proposals that might be incorporated in planning for climate change at the community level. This advice will be informed by government policy and guidance, the content of emerging regional and local plans and best practice examples. Full and early community engagement will be essential and local authority planners, subject to further training, will be best placed to organise and deliver the engagement process.

The final layer of involvement for planners is at the location/site-specific level for new development proposals and the regeneration of existing communities, including the retrofitting of infrastructure for climate change mitigation and adaptation. The use and appropriate expansion of existing housing and development design guides can facilitate the layering of climate change mitigation and adaptation measures into the range of environmental, physical, and other constraints and opportunities, which must be considered in the planning balance when evaluating development proposals. The layout and design of buildings, open spaces, landscape, and sustainable travel connections present several opportunities for reducing carbon emissions while at the same time providing benefits for the health and wellbeing of occupants. The role of planners in the public and private sectors is to maximise those combined benefits.

Looked at in combination, these expansions to planner's roles and responsibilities seem daunting at first. I believe they are achievable, and that the planning profession is best placed to deliver them but, as is becoming widely recognised, substantial re-resourcing of local government planning departments is necessary in the immediate future. This will include training, reskilling, and new appointments to expand the capabilities of individual departments. There is also scope within this programme to establish sharing of specific skills and expertise across neighbouring authorities.

There is now considerable worldwide evidence and supporting research relating to the adverse impacts of climate change on the health and wellbeing of individuals, and national, regional, and local populations. Climate change events including heatwaves, floods, landslides, and forest fires cause death and severe or significant physical injuries, and there are increasing numbers of examples of these impacts over the last two years. These and other events, or the risk, known or otherwise, of such events, cause anxiety and distress leading to, or exacerbating existing, mental health conditions. Such events can cause temporary or permanent loss of homes and property and, in extreme cases, lead to climate migrants who are unable to return to their previous settlement, with consequent distress. There is evidence that young children worry about the impacts on the world which they will inhabit and the potential consequences for their future livelihoods.

The Covid pandemic of 2020/21 has provided some clear parallels that strongly illustrate the higher impact of disease and other risks, such as those brought by climate change, on people and communities with existing health and deprivation problems. Deprivation in this context includes lack of access to open spaces and nature, poor active travel and public transport connections to essential services and poor housing in terms of insulation, costly heating, and poor air quality.

Planners need to work with health professionals and researchers to gain a better understanding of health inequalities and health and wellbeing impacts, together with the public health plans to combat these. There are considerable synergies with town and country planning. It is important to recognise the number of co-benefits arising for the health and wellbeing of communities and individuals from policies and actions to mitigate and adapt to climate change. Examples of such co-benefits include improving provision for active travel, access to recreation opportunities, improvements to biodiversity and connections with nature, which improve physical health and reduce stress and anxiety. Reductions in air pollution via switches away from fossil fuels, gas fired domestic boilers and improvements to the energy efficiency of dwellings and the provision of local sustainable energy generation systems can reduce fuel poverty and improve mental and physical health. Engaging in community action on climate change helps individuals with mental health problems by building social connections, alleviating eco-anxiety, and providing a greater sense of purpose and control over climate change.[16]

As I was formulating my thoughts on how to conclude this chapter, the IPCC published the first of three major working group reports on 9 August

2021: 'The Physical Science Basis of Climate Change'.[17] A total of 234 scientists from 66 countries contributed to this first report and it has been built on more than 14,000 scientific papers to produce the final 3,949 pages of the document. Prior to publication, the report received the approval of 195 governments, while the Summary for Policymakers document, drafted by scientists, was agreed to on a line-by-line basis by all governments. The report contains the starkest warning yet produced by the IPCC of major inevitable and irreversible climate changes and represents the strongest scientific consensus to date. Global warming by 1.5 degrees Celsius is now expected to be reached by as early as 2040 unless global emissions are drastically cut and the whole world reaches net zero by 2050 with 50 per cent of the cuts occurring by 2030. The irreversible changes include the continued melting of land and sea ice with consequent sea level rises dependent on the rapidity of the warming and melt rate. The report does contain some positive conclusions including the future ability to reduce warming from the 1.5-degree level if net zero is fully met before 2050.

The publication *Drawdown* provides a strong and confident basis for achieving net zero by deploying a series of tried and tested and new technologies across the following sectors of national economies: energy, agriculture and food production, buildings and cities, land use based around natural systems management (e.g., afforestation and forestry protection, peatlands, land management systems), transport and materials. Urban and rural planning will need to be at the heart of delivering these technologies.

There can now be little doubt that there is a new urgency to the introduction of appropriate policies, plans and firm action programmes. The UK governments need to provide clear strategic and financial frameworks for public and private sector contributions. The role for the planning system needs to be rapidly endorsed by the UK and devolved nation governments. The re-resourcing of planning departments and university courses must be put in place very quickly to ensure that the contribution of the system can be fully delivered across the remainder of this decade, and beyond to 2050.

Chapter 13

Planning and the Environment – Achieving Enhancement and the Mitigation of Harm

There is a general perception that exists within the UK, but which is predominantly held by the English, that the land area is highly urbanised. This view is not altogether surprising given that 83 per cent of the population of the UK lives in urban areas and there are many households who do not, for one or more reasons, get to appreciate the life-enhancing qualities of the open countryside. The position based on detailed land classification data from 2012 and recorded in mapping produced in 2017[1] shows that only 6.28 per cent of the UK was classified as urban development. This can be compared with 56 per cent identified as agricultural (land used for animal grazing and crop production), 9 per cent peatland, 7.5 per cent moor and heathland and 8.4 per cent woodland. England is the most urbanised of the home nations and urban development was recorded at some 9 per cent of the land area. By way of comparison, and as recorded in Chapter 7, this proportion is significantly exceeded by the area of the English countryside designated as Green Belt, at 12.4 per cent. When examining the extent of urbanisation in the UK, and in England in particular, and comparing this with the extent of land that is environmentally protected, it is vitally important to use datasets that are as accurate as possible and use very clear definitions of what constitutes 'protected land' and the nature of that protection in terms of legal and other requirements.

Before returning to this most important comparison, in the context of our enhanced need for accessible and conserved natural environments, it is important to stress that Green Belt policy, as contained in the original legislation and guidance and the current NPPF, is not based on an assessment of environmental qualities. The prime purpose of Green Belts is to contain development. The second misconception is that areas designated as Green Belt are permanently protected. They do not have such permanent protection and, due to urban development and other needs such as essential infrastructure, they can, in what are described as exceptional circumstances, be subject to boundary reviews. Green Belt designation does not take account of relative environmental quality, and areas can range from wholly degraded environments to locally protected areas, some of which may enjoy permanent protection. As I have argued earlier, the Green Belts around our urban areas need to be repurposed with more sophisticated local policies, which ensure environmental and recreation gains alongside areas of enabling development that satisfy the full definition of sustainability. I include this argument at the outset because of the large land area covered by this designation, the misunderstandings surrounding the nature of the Green Belt and its considerable environmental enhancement potential close to where high proportions of our population live. Recent government statistics include the Green Belt, together with national parks, AONBs and other national environmental designations, as part of the total area of protected land in England assessed as 40 per cent of the total land cover of 13,046,000 hectares.[2]

If we look at the same 2012 land classification data for Bradford, one of the 20 largest cities in England, a more urbanised pattern emerges with a total of 30 per cent of the land area made up of what is classed as urban fabric, broken down into three separate categories of continuous urban fabric (0.3 per cent), discontinuous urban fabric (26.53 per cent) and industrial or commercial units (2.86 per cent). The extremely low proportion of continuous urban fabric is accounted for by the number of areas of parkland, woodland, open space/sport and leisure facilities and water bodies. Pasture, woodland, natural grasslands, moors and heathland and peat bogs make up 65.5 per cent of the total land area, with peat bogs comprising 11 per cent of that total. Virtually the whole of this open land (97.5 per cent) is designated as Green Belt in the local development plan. The urban areas comprise the main urban area of the city, three separate principal towns and several villages. Cities such as Manchester and Birmingham are more urbanised as a result of constricting local authority boundaries.

The Land Cover Atlas of the UK, compiled from Corine data, presented the extent of urbanisation in terms of land occupied when surveyed in 2012. The MHCLG first produced land use statistics for England in 2017 and land use change statistics for the period 2017/18. In 2018, 8.3 per cent of England's land was assessed as developed and 91.5 per cent non-developed and 0.2 per cent vacant.[3] Following consultation with user groups these statistics are to be produced annually. A third source, which assesses the spatial extent of urbanisation, is the UK Centre for Ecology and Hydrology, which uses satellite data to produce Land Cover Maps. They assessed changes over the period 1990 to 2015 for the UK as a whole and calculated that there was a net increase in urban areas of 3,376 km^2 during this period and an increase of 5,236 km^2 of woodland, primarily from the change of use of open pastures. A further measure of changing urbanisation is the split of population between urban and rural areas. Over the period 2010 to 2020 the percentage of the UK population in urban areas increased slowly from 81.3 per cent to 83.9 per cent. The impact of the Covid pandemic and working from home has established what may be a short-term trend favouring rural locations.

Since the year 2000 there has been a significant increase in the land area protected by national environmental designations. New national parks designated in England include the New Forest in 2005 and the South Downs in 2010. The Lake District and the Yorkshire Dales National Parks were extended in 2018 so that they are now almost joined together along the line of the M6 motorway. Natural England announced a landmark new programme for protected landscapes following the Glover Review in 2021.[4] This review made a series of recommendations to government including the need to do more to support the recovery of nature in the national parks and Areas of Outstanding Natural Beauty (AONBs), that the status of AONBs should be strengthened, that the management and strategic oversight of these protected landscapes should be more closely integrated and that more funding should be provided to support public access. The proposals for two new AONBs, the Yorkshire Wolds and the Cheshire Sandstone Ridge, together with extensions to the Surrey Hills and Chilterns AONBs, are now being formally progressed.

The Secretary of State for DEFRA has also asked Natural England to consider the designation of new National Nature Reserves. Natural England is also pursuing the biggest initiative to restore nature ever launched in England, known as the Nature Recovery Network (NRN) Delivery Partnership.[5] This is a product of the Government's 25-year Environment Plan and, working with conservationists, businesses, farmers and landowners and local environmental organisations, aims to establish an England-wide network for nature. The objectives include the restoration of habitats within protected sites and landscapes together with the establishment of new wildlife-rich habitats extending to at least 500,000 hectares. This initiative is a key part of the programme to address the triple challenges of biodiversity loss, climate change and people's isolation from the natural world. Part of the programme is to explore and create large scale nature recovery areas, recognising that this can create new and expanded wildlife corridors, support and enhance ecosystem services and bring nature closer to people. Natural England have recently demonstrated how the nature recovery system can work across the country by establishing the first 'super' National Nature Reserve (NNR) at Purbeck Heaths in Dorset. Working with seven lead partner organisations, this project is integrating 11 types of priority habitat.

The mapping and detailed recording of land with environmental qualities, which are protected under national and international designations in the UK and elsewhere in the world, is rapidly improving as is the assessment of the condition of these areas in terms of their biodiversity. In the UK, in addition to the Land Cover Atlas based on Corine data, the Joint Nature Conservation Committee (JNCC), which was established in 1991, advises the UK Government and those of the devolved nations on the development and implementation of domestic and international policies for the protection of natural resources, as an integral part of sustainable development. Amongst the many JNCC publications providing evidence and advice is one that reviews UK Biodiversity Indicators covering the extent and condition of protected areas in the UK and its four constituent nations. The latest version was published in 2020 and covers the indicators dealing with the extent and condition of the protected areas.[6]

The protected areas information covered by the JNCC includes the following designations on land and at sea: Areas and Sites of Special Scientific Interest, National Nature Reserves, Marine Conservation Zones, Nature Conservation Marine Protected Areas, Ramsar Sites (sites designated under the Convention on Wetlands of International Importance), Special Areas of Conservation and Special Protection Areas (both of which were designated under EU Habitat Regulations), national parks and Areas of Outstanding Natural Beauty. Though AONBs, national parks and Ramsar sites were only added in 2014, and National Nature Reserves in 2015, the series were recalculated to the start so that those new site types appeared in all the years from which the designation first existed. By July 2020 a total of 28 per cent of the land area of the UK was covered by these designations and 26.5 per cent of the land area of England. The extent of these protected areas is continuing to increase, and the UK government have set a target of 30 per cent of the total land area being included within these designations by 2030. The addition of the proposed new and expanded AONBs and new National Nature Reserves would ensure that some 50 per cent of this target is achieved in the short term.

The condition of those protected areas relating to nature conservation are monitored annually in the JNCC publication. Various targets have been set and are assessed by the constituent national nature conservation bodies in the UK to establish whether the biodiversity of these protected areas is in a favourable condition (all targets are met) or should be classed as unfavourable and recovering.

What constitutes 'protected land' in the UK and internationally has varied according to various factors including the approach of national governments, the advice from their scientific and professional advisors and the influence of the public in pressing for certain designations (e.g., national parks in the UK). The influence of international bodies concerned with nature conservation and declining biodiversity has led to pressure for greater conformity of designations and higher standards of management and governance to create nature-based solutions. Chief amongst these is the International Union for the Conservation of Nature (IUCN), which was established in 1948 as the first Global Environmental Union. IUCN is active in at least 160 countries and more than 15,000 experts are organised in six Commissions. Through its World Commission on Protected Areas, the IUCN has set global standards for them.

The IUCN National Committee for the UK prepared a handbook to assist in the identification of protected areas within this country and to assign the IUCN management categories and governance types to them.[7] Essentially, the National Committee found that application of the IUCN guidelines revealed that national parks and AONBs did not meet the nature conservation guidelines and that further improvements could be made in those protected areas that were more directly concerned with nature conservation.

A further up-to-date study considers the extent to which positive nature conservation measures are being pursued in protected areas in the UK and are contributing to halting the decline in biodiversity both globally and nationally. It concludes that this is not yet being achieved.[8] Current programmes being promoted by Natural England and the nature conservation bodies of the devolved nations are aimed at reversing these trends. Large Scale Conservation Initiatives are seen as one of the ways in which nature recovery can be successfully achieved. The Purbeck Heaths 'super' National Nature Reserve is one example of this approach. A paper published in 2017 looks at the value of bringing together or expanding smaller nature reserves and Sites of Special Scientific Interest (SSSI) on a landscape scale.[9] The extent to which the planning system needs to be involved in the spatial and land use planning of such initiatives is a key matter for consideration.

Other areas of protected land exist in the UK, which planning authorities need to take account of when determining development applications, and in many areas the relevant development plan contains policies that seek to conserve the environmental qualities of these sites. Some have local designations as nature reserves and country parks, others are protected as areas of public open space for recreation or wildlife purposes and benefit from legal protection. Ancient and other biodiverse woodland may have a protective designation that is carefully guarded by local environmental groups. The Woodland Trust is a leading example of a charitable ownership approach to protecting existing woodland and planting new mixed native species through ownership or agreements with other landowners. Common land is protected and, dependent on the grazing regime and management applied, can contribute effectively to nature recovery. Increasingly, local river catchments and inland water bodies are being protected and enhanced to support their habitat potential. Within towns and cities existing parks, public gardens and open spaces are already protected.

A series of factors including the decline in biodiversity, over-intensive farming of land to the detriment of soils, interference with natural ecological processes via pollution and development and the degradation of specific habitat types by human interventions, such as the draining of peat bogs, have led to a growing movement for 'rewilding'. This concept has been around for several years and there is now a broad spectrum of opinion which supports some form of nature recovery system, but the term and what it means and involves is a subject of dispute. 'Rewilding' has been misrepresented in the popular press as a predominating wish, by certain large landowners and conservationists, to reintroduce animals, once prevalent in the UK, including wolves, lynx and bears into suitably modified habitats. While such approaches do exist as isolated examples, they are not reflective of the main vision held by leading proponents of the concept.

Foremost amongst the advocacy organisations is Rewilding Britain an organisation formed by experts in conservation and community engagement in 2015. It is a charitable trust and is the first and only countrywide organisation in the UK focussing on rewilding and advocating the major benefits it can bring in achieving our climate change goals, helping to reverse our biodiversity loss and improving people's connectivity with nature to the benefit of their physical and mental health. In my opinion their definition of rewilding is the clearest and most universal in its application:

> 'The large-scale restoration of ecosystems where nature can take care of itself. It seeks to reinstate natural processes and, where appropriate, missing species – allowing them to shape the landscape and the habitats within.'[10]

Rewilding Britain have a specific vision, which partly aligns with that of the UK government for achieving their target of 30 per cent of the land area being made up of nationally designated protected areas. The great difference is that their vision includes the creation of a large-scale mosaic of species-rich habitats restored and connected across the whole of the UK. This vision of upscaling and improving habitat quality and connectivity is shared by Natural England and several organisations including the RSPB and the Woodland Trust.

The scale of activity required in terms of land use and the competition from other potential land uses demands the involvement of the planning system at national and local levels of organisation. This approach will make planning authorities and planning professionals cofacilitators of the Nature Recovery Network (NRN), as proposed by Natural England, and being pursued in similar forms in the devolved nations, rather than the inhibitors that some perceive them to be. This work should be progressed together with the role outlined in the previous chapter on climate change. Such an approach is part of the shared vision of Rewilding Britain and like-minded organisations who see collaboration and coordination as essential ingredients for success across the three inter-related fields of planning for climate change, biodiversity recovery and improving the health and wellbeing of the inhabitants of these islands.

I believe it is important for all those involved in delivering the benefits of the planning system to appreciate what is already being achieved on larger scale nature recovery projects in the UK so that the commitment and achievements of several landowners, communities and organisations can be celebrated. From my reading on this and the related topics of climate change and health and wellbeing I am aware that the national governments, some local authorities, and the national nature conservation bodies are already conscious of the great potential for collaboration that exists and needs to be harnessed, though this does not yet appear in firm plans and action programmes. These are urgently needed as is full engagement with the public. A brief review of further examples of landscape-level nature recovery projects is appropriate before turning to an examination of the legislative framework, which is emerging in the UK, which has the capability to drive the expansion of this vital work, but currently lacks some of the necessary content.

The trustees of Rewilding Britain include Charlie Burrell who, together with his wife, the author Isabella Tree, began to turn his former arable and dairy farm, the 1,400 hectares Knepp Castle Estate in West Sussex into the Knepp Wildland, which has become a classic demonstration project of rewilding leading to nature recovery. The story of this journey from 2001 is best told by Isabella in her book *Wilding: The Return of Nature to a British Farm* and in several articles she has written. Two of these articles are included in excellent source books. The first is in a review of examples of projects in the UK and Ireland, plus related short topic papers, entitled *Rewilding*, edited by the conservationist and photographer David Woodfall.[11] The second article appears in an anthology of British and Irish nature writing entitled the *Wild Isles* and edited by Patrick Barkham.[12]

Further examples presented by individual authors in David Woodfall's collection demonstrate the variety of people, communities, charitable trusts, and organisations involved. Keilder Wildwood in Northumberland is a woodland restoration and creation project, but it is also part of a much larger nature restoration plan, covering land up to and across the Scottish border and the large Kielder Forest and Water Park, working with major landowners including the Forestry Commission. In the introduction to Chapter 12 I referred to the large Tarras Valley Nature Reserve established by a crowdfunding scheme with assistance from the Scottish Land Fund and Buccleugh Estates, a large local landowner. Other schemes in Scotland include the establishment of the Carrifran Wildwood in the Southern Uplands, where a group of locals and others came together to form the Borders Forest Trust, and the Dundreggan project to restore part of the wild woodland of the Caledonian Forest on a 4,000 hectares estate in Glen Moriston, near to Loch Ness. There are many more rewilding projects underway in the UK including fenland and peat bog restorations and schemes to improve and expand the habitats of species such as red squirrels and rare butterflies. However, as Rewilding Britain discovered in their ambitious Summit to Sea project in mid-Wales, strong opposition from farmers can derail schemes where major changes to habitats and farming practices are proposed.

* * *

The changing statutory and policy framework relating to the environment of the UK

When I first examined the proposed statutory framework in 2020, three new statutes were to establish the legal and policy framework for the future protection, enhancement, and management of our urban and rural environments, though their inter-relationship was, and remains, unclear. No integrated approach has been established. The first of these statutes, the Agriculture Act, was given royal assent on 11 November 2020. A key part of this Act covers the Environmental Land Management Scheme (ELMS), which will replace the Common Agricultural Policy (CAP) of the EU following the UK's exit from the Union. The CAP payments system for UK farmers is to be phased out over a 7-year period and progressively replaced with the new ELMS farm payments and land management system. Farmers will be paid to provide public goods including environmental, soil and animal welfare improvements and the production of food in an environmentally sustainable way.

The second, the Environment Act, became law on 10 November 2021 during COP26, following three years of debate and re-drafting since the original bill was introduced in Parliament. The Act makes provision for targets, plans and policies for improving the natural environment; for statements and reports regarding environmental protection; the creation of an Office for Environmental Protection; provision for waste handling and resource efficiency; air quality requirements; water, nature, and biodiversity; conservation covenants and the regulation of the use of chemicals. The Act introduces the requirement for a 10 per cent biodiversity net gain for new development. A new system of local nature recovery strategies is introduced.

A new Planning Bill was announced in the Queen's Speech at the reopening of Parliament in May 2021, which was based on the general proposals in the August 2020 Planning White Paper. While this Bill was intended to apply to the whole of the UK, most of the proposed changes were expected to apply to England only. There was a dominant focus on housing provision in terms of both quantity and quality. The headline statement on the purpose of the Bill stated that it was intended to introduce 'laws to modernise the planning system, so that more homes can be built.' Considerable opposition to the Bill followed from all parties in the House of Commons and the Johnson Government suffered a large defeat in the Chesham and Amersham (Hertfordshire) parliamentary by-election, with the main proposals in the Bill, particularly regarding the lack of provision for public consultation on housing proposals, constituting one of the key reasons for this defeat. Many Conservative MPs subsequently expressed strong opposition to the proposed contents of the Bill.

The Parliamentary Select Committee for MHCLG argued for pre-legislative scrutiny of the Bill proposals as they were not persuaded that the broad three-part zoning system in new local plans would produce a quicker, cheaper and democratic planning system.[13] They identified a number of key omissions with particular emphasis on environmental considerations, including the designation of protected areas and species, proposals for net gain in biodiversity and reforms to the process of environmental impact assessment. The Committee considered that climate change was also a major omission. In September 2021 the prime minister, Boris Johnson, announced that the Ministry of Housing, Communities and Local Government (MHCLG) would be expanded and renamed the Department for Levelling-Up Housing and Communities (DLUHC) and that Michael Gove would be its new Secretary of State. The earlier unpopular planning reform proposals were replaced by a new set of reforms, which were outlined in the Levelling-Up White Paper, published in February 2022, and included in the Levelling-Up and Regeneration Bill. Significantly, during this summer period of legislative and political change Michael Gove, the architect of the 25-year Environment Plan and Environment Act, when Secretary of State for the Environment was removed from his post by Boris Johnson. What followed in the late summer/Autumn of 2022 can only be described as one of the strangest and most destructive periods in UK politics. During this period, we had three Conservative prime ministers and numerous changes at Cabinet level. Boris Johnson resigned from office on 6 September, to be succeeded by Liz Truss who was only in post for 45 days, being replaced on 25 October by Rishi Sunak. Michael Gove is now back in his former job as Secretary of State for Levelling-Up and Regeneration.

The planning reforms proposed within the new Levelling-Up Bill, though very significant, comprise an eclectic mix within much broader and lengthy legislation that focusses on the levelling-up of the UK. Chapter 14 provides a summary review of these reforms, but it is important to focus on proposed changes affecting environmental law and policy. In his introduction to the Bill, Michael Gove stated that those provisions which would become environmental law would not, if enacted, 'have the effect of reducing the level of environmental protection provided for by any existing environmental law'. The Bill proposes a new system of Environmental Outcomes Reports (EOR), which are to be prepared for relevant planning consents and plans and follows with the consequential repeal of the current system of environmental impact assessment (based on European Law previously adopted via statutory regulations in the UK). Before making any new regulations for EORs the Secretary of State 'must have regard to the current environmental improvement plan (within the meaning of Part 1 of the Environment Act 2021)'. This proposed legislative link is the only cross reference available and further exposes the lack of a strategy for integrated environmental legislation.

While the form and content of the Agriculture and Environment Acts are relatively clear, the result of subsuming planning reforms within the Levelling-Up and Regeneration Bill and making some consequential amendments to clauses in the existing planning acts is a retrograde step. It is essential to have clarity and an integrated approach, within and across these three key legislative areas, which are of fundamental importance to the future conservation, management, and enhancement of our environment. Such an approach is demanded by the need to achieve truly sustainable development, climate change mitigation and adaptation, biodiversity enhancement, balanced land-use and the future health and wellbeing of the population.

The current national planning policy for England relating to the environment is set out in section 15 of the NPPF.[14] All planning policies and decisions are to contribute to and enhance the natural environment and six headline approaches set out how this should be achieved. Valued landscapes and sites of biodiversity value are to be protected and enhanced in a way which is commensurate with their statutory status or their identified quality in the development plan. The intrinsic character and beauty of the countryside should be recognised together with the wider benefits arising from natural capital and ecosystem services. The character of the undeveloped coast is to be maintained, net gains for biodiversity are to be provided and strong ecological networks established which are resilient to current and future pressures. Local environmental improvements are to be provided as part of developments, including enhancement of air and water quality and the avoidance of contributions to all forms of pollution. The list concludes with the remediation and mitigation of 'despoiled, degraded, derelict, contaminated and unstable land where appropriate'.

As with other topics covered in the NPPF, these laudable policy objectives leave much room for interpretation on responsibilities and delivery. There can be little doubt, however, that the environmental issues to be covered in local planning policies and in decision making are particularly wide reaching. The policy content covers new concepts including natural capital, ecosystem services, biodiversity net gain and the wider environmental net gain. The NPPG provides some of the answers in a summary format and identifies further sources of more detailed guidance from Natural England and other bodies.[15]

A brief explanation of each of these new concepts and their inter-relationships is helpful for planning professionals and public understanding. Environmental net gain is wider than biodiversity net gain and relies on the delivery of benefits to ecosystem services and natural capital assets. Ecosystem Services are services provided by the natural environment that benefit people and include agriculture, forestry, grassland, wetlands, the sea, and urban environments. The benefits are grouped in four broad categories: Provisioning (food and water); Regulating (control of climate and disease); Supporting (nutrient cycles and pollination); Cultural (spiritual and recreational benefits – can be translated as health and wellbeing).[16]

The term Natural Capital can be misconstrued by some professionals, in particular economists, as I have seen a definition that includes hydrocarbons, a resource encased within the earth, which is a key contributor to climate change and also to environmental degradation. The Government's 25-year Environment Plan provides a definition of Natural Capital and a helpful description of why this is to be the central policy tool in all future environmental planning including all aspects of future development planning.[17] The definition used by the Government is based on advice from the Natural Capital Committee (NCC), which was established following the publication of its 2011 White Paper, 'The Natural Choice', based on the key objective of being 'the first generation to leave the natural environment of England in a better state than it inherited'.[18] The NCC was set up to advise on how to deliver this key objective and recommended that government should produce a comprehensive plan and establish a number of 'Pioneer Projects' in order to explore the challenges and opportunities raised in pursuing a natural capital approach in practice. The 25- year Environment Plan responds to this advice and includes the following definition:

> 'Natural Capital is the sum of our ecosystems, species, freshwater, land, soils, minerals, our air, and our seas. These are all elements of nature that either directly or indirectly bring value to people and the country at large. They do this in many ways but chiefly by providing us with food, clean air and water, wildlife, energy, wood, recreation, and protection from hazards.'

These values are not captured in traditional accounting methods and have been too often ignored in management and policy decisions. Using the natural capital approach will facilitate better and more efficient decisions that support environmental enhancement and deliver benefits such as reduced flood risk, increases in wildlife and improvements to health and wellbeing. Ecosystem services are therefore those that are delivered by the cumulative value of the main types of Natural Capital. The ONS have now derived some headline values for these main habitat types. Using the example of England's forests and woodlands the value of the services they deliver has been estimated at £2.3 billion and only an estimated 10 per cent is made up of the value of timber. The remainder comprises benefits provided to society in the form of recreation, other health and wellbeing benefits, carbon sequestration and flood risk mitigation.

The concept of Environmental Net Gain is incorporated in the 25-year Environment Plan following recommendations from the NCC.[19] The NPPG provides a helpful answer to the question of what this comprises and how it can be achieved.[20] The aim of environmental net gain is to reduce the pressure on, and achieve overall improvements to, natural capital, ecosystem services and the benefits that they deliver. The general example of habitat enhancements is outlined, which can provide improvements to soil, water and air quality, flood risk management and opportunities for recreation. The guidance advises that strategic planning for the wider environmental net gain, which can be expressed as the enhancement of natural capital, will be able to use evidence held by Natural England and other nature organisations. As part of the development of this evidence base several metrics are in production to measure and monitor aspects of wider environmental gain. The NCC expressed their disappointment that the February 2019 version of the NPPF, while briefly introducing net environmental gain and a reference to natural capital enhancement, did not go further by stating a specific policy requirement for achieving such gains via development proposals. The July 2021 version retains the same wording and does not advance the policy position.

Biodiversity encompasses the variety of all life on earth and includes all living things and the places in which they live. Biodiversity net gain (BNG) is the delivery of measurable improvements for biodiversity by creating or enhancing habitats in association with development. This can be achieved on-site, off-site or via a combination of on- and off-site measures. At the present time such gains are encouraged by national policy in the NPPF, and the government have signalled that this will become mandatory for all planning developments with the general proposal contained in the Environment Act and subsequent details being incorporated via amendments to the planning acts. It is not expected that this will be achieved until late 2023 and in the meantime the policy aim is to encourage this and achieve it on a voluntary basis.

Based on work initially commissioned by DEFRA in 2008, a biodiversity metric has been developed by Natural England working in collaboration with a wide range of other bodies, consultancies and individuals including DEFRA, the Environment Agency (EA) and the Wildlife Trusts. The latest version, Biodiversity Metric 3.0, was published in July 2021.[21] The 25-year Environment Plan set out the wish to see BNG mainstreamed within the planning system with a move towards an approach that integrates natural capital benefits. The sooner such an integrated approach is achieved the better as otherwise the planning system is faced with different but overlapping assessment systems for the various forms of enhancement being sought. For the foreseeable future the Biodiversity Metric 3.0 User Guide and associated documents are designed to provide ecologists, planners, developers, and other interested parties with a means of assessing changes in biodiversity value, both gains and losses, as a consequence of a particular development proposal or change in land management. The metric uses a habitat-based approach to determine a proxy biodiversity value.

The Environment Act provides for a minimum 10 per cent net gain from any project and the approval of a net gain plan with the enhanced habitat being secured for at least 30 years via planning obligations or a form of conservation covenant.

From this initial assessment of the four concepts for delivering environmental enhancements, considering their inter-relationships, and including measurable net gains, there emerges a clear case for a rationalisation or careful consolidation. This will directly influence the future formulation of planning policy and the assessment and measuring processes used in terms of quantitative and qualitative criteria, which ecologists, landscape and other environmental professionals will produce, and planners will assess. It is also necessary to establish how these concepts are incorporated into existing planning assessment processes, including environmental impact assessments, sustainability appraisals, site selection and the planning balance, when evaluating planning applications.

* * *

Consequences for the future work of town and country planners and the role of the planning system

The preceding review of the rapidly changing context for the protection, conservation and enhancement of the UK's natural environments presents further challenges for planners and the planning system, including the roles of national and local government. Collectively, the changes being proposed represent the biggest, and potentially the most positive, ambitious, and essential programme, for the environment and future land use planning, attempted by a national government since the introduction of the suite of legislation comprising the 1947 planning system. They are driven by the loss of biodiversity, the lack of adequate ongoing protection for the environment, the need for enhanced public access to open spaces to enjoy the health and wellbeing benefits of nature and recreation, and to fully address the parallel pressures of the challenges of climate change.

At the national levels there is good evidence of increased cooperation between the main agencies involved. In England, Natural England is enhancing its joint working with National Parks England and the National Association for AONBs through a new delivery agreement, which is aimed at achieving a step change in the provision of multiple and integrated benefits for people, nature, and the climate. The new approaches by Natural England to driving the NRN Delivery Partnership, which brings together organisations with coordination at national, regional, and local levels to provide the restoration of protected sites and landscapes, is already producing some benefits in exemplar landscape scale project initiatives. In Wales the establishment of a single ministry across all the key areas involved in contributing to climate change and environmental enhancements signals an integrated approached comparable with that taken for health and wellbeing in the Wellbeing of Future Generations (Wales) Act of 2015.

The position in England gives rise to two major concerns. Firstly, the three statutes which are key to the required enhancements of the environment and its future management are not proceeding in concert and there is no clear umbrella strategy defining roles, cross departmental working, coordination, and delivery. While the 25- year Environment Plan goes some way towards integrating approaches on policy areas across the responsibilities of DEFRA, there is little content in this Plan on the nature, extent and importance of town and country planning's role. No other government paper has yet emerged to address this. The second concern is strongly related. The previously proposed Planning Bill, and now the Levelling-Up and Regeneration Bill, make scant reference to environmental improvement as a key component of either an entirely new and comprehensive planning act, or as appears more likely the inclusion of planning reforms in a composite Levelling-up and Regeneration Act. The result is that the much needed and urgent changes to the planning system, covering legislation, policy, and guidance, on the lead topic of environmental enhancement, are being delayed as are those concerned with climate change.

The positive initiatives and projects underway at national, regional, and local levels to create a range of environmental improvements, while very welcome and, overall, highly beneficial, are being progressed on a disaggregated basis. This disaggregation arises from the lack of national and regional spatial plans for the environment and a current isolation from other land use planning requirements. A partial exception is the proposed programme of Local Nature Recovery Strategies (LNRS) being pursued by DEFRA and Natural England. Six pilot projects were established in English local authorities in August 2020 and concluded in May 2021. The aim of LNRS is to create new and enhanced habitats based around the mapping of existing valuable habitats and the identification of sites and locations where nature can be restored. The Strategies are being designed as tools to drive more coordinated, practical, and focussed action. The Environment Act requires all English local authorities to produce an LNRS. This will become part of the evidence base for making planning policies and taking decisions on planning applications. It would be beneficial if the production of these strategies were integrated at local planning authority level into the production of the wider mapping and assessment of all environmental constraints and opportunities. This can be achieved either on an individual local authority basis or more desirably on a collaborative basis with adjoining authorities. Such groupings already exist in some areas and benefit from a shared ecological data bank and advisory service.

The Purbeck Heaths NNR in Dorset is an example of a Nature Recovery Network led by a national agency. Other examples are being pursued by the regional Wildlife Trusts, environmental charities, major landowners, and farmers already anticipating the requirements of the ELMS. The rewilding projects referred to in the introduction to this chapter have been pursued by a variety of landowners and pioneering communities and individuals. A further example involving the Yorkshire Wildlife Trust is located in the western part of the Yorkshire Dales National Park. This regional Wildlife Trust has partnered with Natural England's NNR team, the University of Leeds, the United Bank of Carbon, and WWF UK to create a nature restoration project which will return the large limestone pavement and adjacent habitats to their biodiversity rich state. A total of 1,200 hectares (3,000 acres) are already in restoration and it is the intention to acquire more land. The Wildlife Trusts work with local wildlife and environment groups and many of these also have their own localised programmes and projects. A good example of this is my home village environment group, part of our Civic Society, with more than 200 volunteer members engaged in monitoring and surveying wildlife, planting new woodland and wildflower meadows, restoring streams, and managing village open spaces.

All this work is gradually creating a new 'backcloth' or more appropriately a 'patchwork quilt' that planners will have to consider in all facets of their work, as these areas become more treasured by communities and the extent of protected area designations increases. This activity places the planning system and planners somewhat 'behind the curve' in the core activity of land use planning and accommodating competing or complimentary uses.

At a policy planning level, a typical local development plan produced by a UK based local planning authority will include approximately 10 strategic and topic-based policies for the various forms of environmental enhancement and protection, climate change, active travel (walking and cycling) and the delivery of sustainable development. The new environmental protection and enhancement processes summarised in this chapter will give rise to a need to amend and expand this policy coverage in development plans. The evidence base production requirements will expand, though wherever possible this should be assisted by advances in mapping and digital recording processes. The criteria used in plan, policy, and site selection processes, including sustainability appraisals and environmental impact assessments, will similarly need to be adapted.

Spatial planning of environmental plans and programmes will be required and as identified in the preceding chapter on climate change, will need to take place at four levels: regional, local, or district-wide, neighbourhood or community and finally location, site and project specific. The case for a return to some organised and preferably statutory form of regional planning has been advocated in Chapter 9 and exemplified in Chapter 12 – Climate Change. The regional spatial plans would be more useful as a framework for the lower tier spatial plans if they were produced on a simplified, nimble, and regularly updated basis with a strong mapping and graphic content. Spatial planning at the regional and local levels is a relatively urgent requirement given the immediacy of the actions which need to be taken on the twin challenges of climate change and the loss of biodiversity.

The case for greatly enhanced site and project specific spatial master plans is well established and generally encouraged, though not yet mandated. This case arises from the need to deliver developments that are fully sustainable and provide benefits against all three of the objectives that must be met to achieve a truly sustainable development. The August 2020 planning White Paper encouraged the use of design guides and the best of these already incorporate many of the requirements to provide for environmental enhancements. In the next two to three years the requirement to achieve the wider environmental net gain on developments is most likely to become a legal requirement. The urgency of the position should be used now to encourage developers to prepare high quality master plans for their site-specific proposals. The wide range of criteria and considerations involved require the project coordination and assessment skills of planners to be deployed in a leading role, with the involvement of ecologists and landscape architects at the very outset of the project. Developers, when preparing master plan schemes, should no longer proceed based on development 'capacity-testing' versions followed by attempts to retrofit the environmental space and quality standards required. Invariably, such an approach has led to poor quality outcomes.

It is now essential to develop an approach to master planning where environmental and climate change measures are given the fullest consideration from the first stages of the process through to more detailed design versions. This should result in net development cells or footprints of buildings and connecting highways emerging as a later product from a preceding multi-layering process of overlapping plans, each based on sub-groups of gathered local and site-specific information, including identified constraints and opportunities. In this layering process environmental criteria and policy considerations will need to play a major introductory role, rather than being left as an afterthought or end process where the development form and design has reached a fixed state in the mind of the developer and their architect. These requirements are recognised by some architects and a few enlightened developers and are encouraged in a small number of high-quality design guides and in the recommendations of the Building Better Building Beautiful Commission Report.

Prior to commencing the production of a development project master plan, the planning consultant should prepare a brief for the professional team and client, which includes a review of all relevant local and national policy background ensuring that a distillation of environmental policies and guidance has prominence. This brief should also include preliminary mapping of all environmental designations within and close to the site boundaries. Some sites being progressed by developer applicants will already be subject to a specific allocation in a draft or adopted development plan and information supporting the allocation from plan evidence base documents and any site-specific policy should also be incorporated into the brief. A good brief discussed at a first team meeting will facilitate the production of the master plan and help to achieve the necessary team integration and understanding. The preparation of the master plan can then begin based around a series of overlapping layers of information and interpretation.

The extent, physical orientation, features and topography of a site, together with its neighbouring land uses and their characteristics, will form the initial base layer. The physical features mapped would include landforms, water bodies, woodland, hedgerows, and other distinctive natural features. A second layer would map identified physical and environmental constraints at the locality/site level as distinct from the higher order constraints, including protected environmental designations already identified in the planning brief. Secondary environmental constraints such as ancient and locally protected woodland and locally designated nature reserves would typically be included in this layer. A third layer would comprise the results of on-site surveys of habitat and landscape features together with their condition and relative quality. The early engagement of professional ecologists and landscape architects will be an essential prerequisite of this process. A fourth layer would examine potential linkages to existing ecological, greenspace and landscape networks in the wider locality of the site, plus the potential to create linkages to existing active travel routes. The layering process should not be considered on a rigid basis but rather as a flexible and interactive approach dependent on the nature and location of the site.

As the early-stage development proposals are integrated into the emerging framework master plan, opportunities for environmental enhancements, biodiversity gains and climate change mitigation and adaptation measures should be considered and incorporated into the plan. The development team and the local authority planners should meet to carry out a review of an early draft plan with a view to agreeing potential environmental and other enhancements, including the ability to link an existing community into the active travel network, to give them access to the habitat, landscape and open space improvements being planned.

This approach is applicable to all types and locations of development but is particularly suited to residential developments. Adaptation of the approach is likely to be necessary for most development sites within urban areas. In order to enhance living and environmental conditions within towns and cities and provide much improved public access to natural sites and networks of greenspace and habitats, spatial planning is required at four levels: city-region, city/town-wide, localities/neighbourhoods and site/project specific. Databases covering city-regions provide evidence of the current extent and quality of habitats, open spaces, and landscapes, together with their connectivity in the form of corridors intermingled with urban communities.

From these records the constituent local planning authorities can prepare policies and plans that seek to enhance provision of all these environmental resources in terms of both quantity, quality, and accessibility. Future enhancements will principally be provided in two ways. Firstly, national government will provide grants and local authority finance to support direct investment in all types of green infrastructure, based on locally prepared plans that provide the sound evidence for financing improvement programmes. This evidence base will include detailed city/district wide spatial plans identifying areas requiring expanded provision and qualitative improvements to existing sites and networks. The second essential contribution to environmental enhancements is via the private sector, as landowners and developers, principally through planning applications for development. Local planning authorities need to be encouraged to initiate negotiations with landowners on sites they have identified as potential allocations in a draft development plan. Where developers already have an option on the future development of draft plan allocations they should be included in these negotiations. In my experience the private sector parties to such negotiations are often willing to make wider environmental enhancement contributions beyond those being sought in local plan policies.

Developers, when preparing planning applications and their constituent master plans, should be required to consider existing deficiencies of habitat and open space provision in the wider urban community, via equitable and carefully considered development plan policies and design guidance. In this way it will be possible to positively address existing deficiencies of provision and the level of environmental improvements required to address climate change adaptation and mitigation, biodiversity enhancement and the achievement of a range of required benefits for the health and wellbeing of many urban communities.

The scale of deprivation in many cities and large towns in the UK is often compounded by a lack of local provision of all types of greenspace and sustainable travel access to alternative provision. The Covid-19 pandemic has added immediacy and emphasis to the need to solve the problems of access to greenspace and natural habitats. The need to retrofit the provision of all forms of greenspace has not generally been a matter for consideration when preparing urban development and master plans, or too little emphasis has been given to this vitally important consideration. To ensure that poorer communities benefit from the same levels of greenspace and ecosystem services provision, as those who enjoy ease of access to domestic gardens, parks, open spaces, and natural habitats, requires a high level of retrofitting in large parts of some of our major urban areas. The basic human privileges of enjoying cleaner air, the experience of nature and access to recreation opportunities should be available to all. Local planning authorities responsible for communities with inadequate greenspace provision, in quantitative and qualitative terms, must rework earlier urban master plans to devise ways of addressing these deficiencies. At the same time, they should make provision for climate change measures and positive planning solutions for the health and wellbeing of individuals and communities. Collectively these will, in the medium and longer term bring major financial benefits in the form of reduced ill health, enhanced wellbeing and consequent savings on healthcare services. These changes will also make important contributions to efficient and cost saving ways of providing for climate change mitigation, adaptation, and resilience planning. The values of existing and new developments will be increased by multi-purpose greenspace provision. When the natural capital and ecosystem service benefits are weighed against the costs of full greenspace provision on new developments and retrofitting older communities, the values gained can be readily appreciated and should motivate action.

A third way exists to close the gap in urban greenspace provision where the community, through the activism and leadership of individuals, decide to act themselves. There are now many urban examples in the UK of poorer communities being allowed and helped to reclaim derelict, vacant and underutilised land. These initiatives should be welcomed by local planning authorities and planned for by identifying future opportunities in local development plans and civic forums. Some authorities have taken bold actions, in some cases motivated by the experiences of the Covid-19 pandemic, to reclaim and rationalise areas from existing uses. This is particularly the case regarding reductions in the scale of certain urban highways, with Paris, Milan, and Oslo notable amongst those seeking to provide more greenspace and active travel opportunities by removing highway lanes and rerouting and discouraging private vehicles.

London recently became the first city in the world to declare itself a National Park City (NPC) on 22 July 2019. On the previous day, the National Park City Universal Charter was launched, with its main aim being to achieve enhanced connections between people and nature. The London NPC plan is to make life better for city dwellers, wildlife, and nature, with human wellbeing at its heart. The London Mayor and Assembly have committed to increasing London's greenspace from 30 per cent of the Greater London administrative area to 50 per cent by 2050, which also aligns with the national net zero target. By comparison the city of Paris has 10 per cent of its area devoted to greenspace and New York 27 per cent. The proposals include greening small spaces, such as front gardens that have been hard surfaced, creating green roofs on domestic and commercial buildings, planting more trees and larger expansions of greenspace. The idea was developed by Dan Raven-Ellison, an ex- geography teacher, following a tour of all 15 national parks in the UK. Like many great ideas it rapidly gathered support, including that of councillors and the Mayor of London, Sadiq Khan. The NPC will be a model for other major urban areas across the world and all large towns and cities should follow the lead of London by setting out action plans and programmes, as well as raising public awareness and the multi-use potential of available and proposed greenspaces. The creator of the NPC concept has suggested that a combination of small efforts by individual households turning one square metre of hard surface into greenspace would be a major step towards surpassing the 2050 goal.

Given the lack of greenspace in many large urban areas of the UK, retrofitting plus expansion by new provision are particularly pressing needs and appropriate policies should be incorporated into development plans to engage the public, private sector, and community/charitable bodies. Specific policies, action plans and programmes are also needed, and such is the urgency of this approach that these should be prepared simultaneously. The RTPI have recently proposed the production of Local Environment Improvement Plans (LEIPs) which would guide on the ground actions and set the context for involving the public and private sectors and individuals. The RTPI's proposals also include the establishment of local Green Growth Boards to oversee the detailed planning and implementation of environmental improvements.[22] These proposals were made as part of the response to the August 2020 Planning White Paper. They are centred on establishing collaborative strategies that bring together local authorities and key stakeholders with a focus on the challenges of climate change, environmental recovery, the need to deliver levelling-up and the provision of new housing. This approach requires action at regional, local and neighbourhood levels. These plans should incorporate all types of environmental improvement proposals and cover all aspects of natural capital. This will assist the public, landowners, and developers in positively engaging in environmental enhancement strategies.

Due to higher population densities in central and inner urban areas, the presence of deprived communities, and an overall lack of greenspace in many of these, the planning and action programmes for them should be progressed as a top priority. This need not be at the expense of the outer suburbs and the urban periphery and its adjacent countryside, much of which lies within Green Belts. The agencies involved in delivery of environmental enhancements in the outer areas will differ, at least in part, from those involved in the inner urban areas with private landowners and developers having a lead involvement in the outer areas. Private developers also need to play a leading role in the inner areas working closely with local authorities. The reality now facing government, landowners and the development industry is that actions on environmental improvements are required across the board to fully address the triple major impacts of climate change, biodiversity loss and improvements to health and wellbeing. Most of the improvements required to reverse the loss of biodiversity and achieve enhanced environments will result in cumulative benefits. They will contribute to climate change mitigation and adaptation and create health and wellbeing benefits.

The linkages between improvements to the natural environment and the availability and accessibility to natural and active recreation areas has been well researched and documented. Health, planning, and other professionals are agreed on the many benefits which access to, and enjoyment of the natural environment can provide. A 2016 report by the Institute for European Environmental Policy researches the case and provides guidance and recommendations for policies and actions to further improve the social and healthcare benefits provided by nature.[23] A key proposal is the identification and integration of health, social and nature issues into strategies, plans and implementation methodologies across all relevant sectors and at all levels of governance. Such integration will also offer opportunities to use common tools, such as the mapping of physical characteristics with socio-demographic data, to improve town and healthcare planning. An example is the mapping of available natural spaces and access routes in association with the incidence of vulnerable populations (the elderly, disabled, ill or those with mental ill-health).

There is a great deal of work to be done at the four levels of planning intervention identified. This will ensure that appropriate policies, spatial plans, local action plans and programmes for environmental enhancement and access to nature, are achieved as soon as possible. This wide range of work needs to be carried out in ways which are integral to policy formulation, spatial planning and implementation for climate change and health and wellbeing. This is the most efficient and effective way of achieving the urgent progress required on this interlocking agenda. The scale of the challenge is considerable and will require skilled working within local authority planning departments, on an interdepartmental basis and with other public bodies. Planning is other key role in managing and controlling development will also broaden as material planning considerations in these areas expand in both nature and importance.

Chapter 14

Conclusions – Learning from the Past and Meeting Future Challenges

My review of the UK planning system traces its origins back to the Victorian slums of the mid-nineteenth century and then to the major social and economic impacts brought about by the Second World War. As demonstrated in my opening chapter, these two major influences are quite different in terms of the time taken to produce any tangible benefits, the extent and nature of those benefits and the people and organisations involved in producing them. Concern over poor living conditions and the consequences for individual health and life expectancy was slow to develop amongst the urban masses in the last half of the Victorian era. Industrial philanthropists and a few leading medical practitioners and researchers were the first to realise that radical actions were necessary to improve the life of the urban poor. Salt, Cadbury, Lever, and Rowntree were notable leaders in developing new model towns and villages for their workers with greatly improved infrastructure and amenities. Towards the end of the nineteenth century, Ebenezer Howard almost single-handedly came up with a comprehensive approach to urban planning centred around the Garden City and the decanting of people from the overcrowded conditions of the major towns and cities. The establishment of the Garden City Association in 1899, which later became the Town and Country Planning Association, was in reality the origin of a movement for town planning. The Town Planning Institute followed in 1913.

The first Housing and Town Planning Act of 1909 was a product of public health and housing concerns and enabled local authorities to produce town planning schemes with the aim of improving health and living conditions. However, its influence was limited as it only applied to land that was, or was likely to be, developed. The impact of the First World War had little influence on the case for a town planning system. Despite the efforts of Howard and other proponents of town planning who followed him, notably Raymond Unwin, Patrick Geddes and Frederick Osborn, the subsequent inter-war years did not achieve a comprehensive and binding system. The planning acts of 1919 and 1932 had little positive impact. The only significant common thread was the expression of a wish to improve the health and wellbeing of the population.

This slow progress towards a comprehensive system of planning, with the exceptions of the lasting influence of Howard's work and to some extent the new settlements produced by the Victorian industrial philanthropists, adds great emphasis to what was achieved in the seven years during and immediately following the Second World War. The concern for the health and wellbeing of the UK population was eloquently expressed in the Beveridge Report of 1942 and in the 1944 Planning White Paper. The resulting messages were fully embraced by the incoming Attlee Government and a series of statutes were produced which covered most aspects of an integrated welfare state including the Town and Country Planning Act of 1947 and the other acts, which are collectively described as the 1947 planning system. Looking back the collective commitment to action by Attlee's cabinet shines through, as does their concern to improve the lives of their contemporaries. Their achievements in such a short space of time are truly outstanding and inspirational. They were undoubtedly helped by the clarity of purpose arising from the wartime reports, the central objectives that had to be urgently pursued and the high level of support from the general public.

The three main challenges and associated issues and impacts we now face in the third decade of the second millennium are even greater than those which confronted the immediate post-war government in 1945. The urgency for rapid statutory and policy review, and above all action on many fronts, could not be greater and is emphasised by inadequate levels of progress towards our climate goals of net zero by 2050 and the necessary intermediate targets the UK Government has set for 2030. The recent reports from the UK's advisory Committee on Climate Change (CCC) produced in 2020 and 2021, and from the UN's International Panel on Climate Change (IPCC) published in August 2021, demonstrate starkly the lack of sufficient progress to achieve net zero by 2050 and a limitation of global warming to no more than 1.5 degrees Celsius.

The loss of biodiversity and degradation of habitats and landscapes due to humanity's adverse impacts on planet earth are now well understood and demand inter-related policies, programmes, and actions. The environmental and climate change emergencies, aided by the recent Covid pandemic, have brought into stark focus the adverse impacts on our health and wellbeing arising from a lack of access to natural habitats and recreational opportunities. While this has been a temporary situation for some during Covid lockdown periods, for many, especially those living in our urban areas, the disconnect is permanent and needs to be addressed. The positive potential of nature and recreation to contribute to our health and wellbeing are now well recognised, as are the preventative benefits which can remove the need to access our healthcare services with consequent financial savings. The realisation of these benefits has brought a new movement for a wider 'right to roam' in the vast areas of the countryside that remain in private ownership.

* * *

A Renewed, Rationalised and Reformed Planning System

All the main components of a renewed, rationalised, and reformed planning system need to be fully engaged in the comprehensive plans and programmes required to meet the three interlocking challenges. This is both a daunting and an urgent task. There has been a cumulative loss of precious time due to a series of worldwide socio-economic shocks, including the financial recession of 2007–2012, the Covid-19 pandemic of 2020–2022, the war in Ukraine and associated commodity and consequent cost of living price rises leading to rising inflation and squeezed household incomes. These impacts have diverted the UK Government's attention and funding away from the required progress on the three interlocking challenges and a holistic and positive reform and renewal of the planning system.

It is not necessary to carry out a root and branch reform of the existing planning system and frankly it is important that time is not lost and that the best components of existing legislation, national policy and guidance are retained in a new holistic system. The main components of the planning system that need to be fully energised and engaged include policy formulation and implementation; the production of development and action area plans; the extension and updating of planning guidance; the production of spatial plans and the determination of planning applications via a streamlined development management approach. The project coordination and team building skills of planning professionals need to be fully deployed within a revitalised, comprehensive, and understandable planning system.

There are some key lessons that can be learned from the establishment and progression of the 1947 system which remain relevant today. These include:

- Retaining and wherever possible enhancing the centrality of the health and wellbeing of all citizens.
- Ensuring that a holistic system is created and maintained.
- Fully acknowledging that the renewed planning system is a force for public good.

- Gaining public support for a revitalised planning system including enhanced measures for public engagement throughout the process.
- A strong, committed, and lasting approach from national government.

For several years local planning authorities have been denuded of resources including the loss of skilled and well-trained staff who have not been replaced. This earlier trend was exacerbated by the impacts of the financial crash of 2007/8 and the ensuing years of austerity, which hit local government financing particularly hard. Planning departments have for too long been the 'Cinderella' of local government financing. While income from fees for planning applications has helped to stem the losses to some extent, there remains a pressing need for refinancing and restaffing local government planning departments. The Housing, Communities and Local Government Committee of the House of Commons, in their First Report on the Future of the Planning System in England, published in June 2021, recognised the need for re-resourcing in terms of both skills and finance with a proposal that some £500 million needed to be spent over the next four years on local planning departments. This case was fully endorsed by the RTPI. New practical training programmes are urgently needed both within the planning departments, consultancy and in further education.

The 'Planning for the Future' White Paper published for consultation in August 2020 contained a set of proposals for reforming the English planning system. Proposal 23 in the White Paper recognised the general need for re-resourcing and stated: 'As we develop our final proposals for this new planning system, we will develop a comprehensive resources and skills strategy for the planning sector to support the implementation of our reforms.'

While this was a welcome and overdue recognition of the resourcing problem, adverse public, professional and House of Commons reactions to certain key proposals in the draft planning bill, which followed the White Paper, resulted in many months of delay and lost opportunity. A new Secretary of State, Michael Gove, was appointed in Autumn 2021 to drive forward the reform programme within the context of an expanded Department, which included the levelling-up strategy, with proposed changes to the planning system being included in a new Levelling-Up and Regeneration Bill (see Chapter 13). One of the perennial problems at ministerial level in English government has been the generally short tenancies of ministers and Secretaries of State in MHCLG, the Department for Transport and the Department of the Environment. These are all departments which require a long-term vision and strategic approach to much of their work. The newly published 25 Year

Conclusions – Learning from the Past and Meeting Future Challenges 345

Environment Plan is a positive step towards a more strategic approach. The short tenancies of key ministers clearly create problems in terms of the commitment and understanding of individuals and the continuity of policy application. The opposition to the original reform proposals from within the governing party, the changes of secretaries of state, and the lack of clear direction has added to the urgency of putting in place a strong and clear legislative and policy framework for planning at the national level in England.

There is widespread agreement that some changes are necessary and the comprehensive Raynsford Review, together with papers produced by the RTPI and TCPA, have advocated changes to the system. The Government's proposals regarding the achievement of enhanced design and environmental outcomes, including the achievement of beauty in development, have been strongly supported. There was a recognition in the Planning White Paper that a whole series of previous piecemeal alterations to the planning system have in fact been harmful to its operational effectiveness, but there is no acknowledgement that politicians have been responsible for most of these changes.

Prime Minister Boris Johnson, when introducing the August 2020 White Paper, would have us believe that the planning system in England is 'a relic from the middle of the twentieth century'. He referred to it as 'our outdated and ineffective planning system' and goes on to lay the blame for the lack of building sufficient homes in the right places wholly at the door of the planning system. Had he read the numerous reports and reviews carried out into this topic over the last two decades he would know that such blame lies largely elsewhere, as I seek to demonstrate in Chapter 11. His blunt and wide of the mark comments made me angry and have spurred me on to meet my objectives in writing this book. Many colleagues in the RTPI share these feelings and the House of Commons Select Committee takes a much more balanced and positive approach to the best way forward.

Two major factors suggest how reforms should be progressed. The first is that the current planning system is far from broken and in need of total replacement. The second is the urgency surrounding the three main challenges that we face and the speed and commitment that will be required to produce the necessary revisions and will help us to fulfil our role as planners in meeting these challenges.

The Raynsford Review (Final Report November 2018) carried out by the TCPA is undoubtedly the best and most comprehensive examination of the current system and the need for change. It puts forward 10 Propositions based on evidence gathered and the expertise of the Task Force. The report goes on to make 24 Recommendations grouped under the 10 Propositions. I agree with all the Propositions and the great majority of the Recommendations. In distilling my own conclusions, on the extent and programming of necessary changes to the planning system and how we should deliver these, I have tried to rationalise what is achievable in the shortest possible time. The overall problem arising from the urgency/time constraints is that we cannot afford to wait for changes at one level to be completed before examining how this will impact on changes at another. A rapid and iterative process therefore needs to be quickly devised, without much argument or dissent, so that we can totally streamline the changes without adversely impacting the ongoing operation of the system. Given the expertise available I believe this is achievable and it will need the full backing and involvement of government and the private sector to provide the necessary impetus.

The first set of changes required in a renewed and rationalised planning system are the review of certain sections of the NPPF and the creation of a single Act of Parliament from the three existing Acts of 1990, 2004 and 2008.[1] The existing NPPF policy on sustainable development is stated as a presumption in favour of policies and developments that meet the three-overarching economic, social, and environmental objectives. It has long been argued by planning and environmental organisations that this should be enshrined in planning law. I can see no reason why national government should not now take this important step and make the achievement of sustainable development mandatory and strengthen the wording in the NPPF accordingly. Other headline changes that should be incorporated into a new planning act and a revised NPPF include the requirement of planning services to meet the three interlocking challenges of climate change, environmental enhancement and the health and wellbeing of the population. These are all components of sustainable development and statutory recognition of all three would give the necessary strength and clarity to policies and programmes aimed at meeting these challenges. The proposal to include the achievement of beauty in developments is now a welcome policy inclusion in the revised NPPF (July 2021 version), under the sustainable development heading in the first instance, and more specifically in section 12: 'Achieving well designed places', with a requirement for net gains in the quality of places, buildings and living environments. The early production of a single Planning Act will require the engagement of skilful legal draughtsmen and an overarching brief from the Government to align statute and policy wherever necessary.

A new set of planning reforms have recently been published by DLUHC as a consultation document ('Levelling- Up and Regeneration Bill: reforms to national planning policy' 22nd December 2022). The consultation period on these more immediate proposals ends on the 2nd March 2023. Further

Conclusions – Learning from the Past and Meeting Future Challenges 347

proposed reforms to the planning system and a subsequent fuller set of changes to the NPPF are outlined in this consultation paper. Most of these further changes are likely to follow the enactment of the Bill. The opportunity for consolidating planning law in England is diminished by the introduction of these further planning reforms into a new Act of Parliament.

The initial consultation proposals include some positive moves which will enhance certain national policies, particularly those requiring beauty in development and place-making and protecting and enhancing the environment. Greater policy alignment with the Environment Act is promised. New National Development Management Policies are proposed, and these are likely to prove controversial given the perception that they will interfere unnecessarily in local decision making. Changes are proposed to speed up local plan-making and to simplify and streamline the content with a key objective of facilitating enhanced public engagement and understanding.

I highlight these new two-stage reforms flowing from the Levelling-Up and Regeneration Bill and its subsequent enactment as the replacements for the earlier set of reforms proposed in the August 2020 White Paper. We will, however, need to wait some months, beyond the publication date of this book, to comprehensively assess the final content and any future performance of this latest set of reform proposals. There is also the prospect of a general election in May 2024. Given these uncertainties I return to my own thoughts on this vital topic.

The NPPF, and arguably a new Planning Act, should incorporate requirements for collaborative working across the main departments of national government and the number of government agencies involved in planning and environmental programmes.

A renewed development plan system should effectively comprise a four-tier system including national, regional, local and neighbourhood dimensions. Some commentators have advocated the production of a national plan. It is most unlikely that the current English Government will support the idea of a national plan. They are however proposing national development management policies in the latest batch of planning reforms, which, if agreed, would be incorporated in future revisions to the NPPF.

I believe that a powerful case exists for producing a national spatial plan that integrates into one multi-layered document the spatial content of existing national infrastructure programmes (energy and transport networks etc), the extent of supra-regional environmental enhancement plans, such as the National Forest, the Nature Recovery Network, and the nationally protected area designations (National Parks, AONBs, National Nature Reserves etc). These are just some of the supra-regional land use designations and requirements that need to be mapped at the national level. In subsequent paragraphs I briefly examine the additional arguments for this type of plan, at the national, regional, and local levels, based on the increasing competition for land use from food production, energy generation, transport and utilities, recreation, housing, and commercial development. A national spatial plan can be produced relatively quickly from existing data sources and should be set up as a 'living plan' that is regularly updated. The NPPF has incorporated, from the outset a broad national policy approach to make the most effective, multiple use of land wherever possible.

I have made a case for the reinstatement of a regional planning tier in Chapter 9. I believe that there are three principal arguments for the production of a new and simplified type of regional plan. The first comprises the many spatial planning issues that spread beyond the border of a single local planning authority and where the extent, interaction and desirable integration of competing land uses must be determined. These include the plans of government agencies such as National Highways, the Environment Agency, and Natural England and those of public utilities and public transport providers. The current loose and often ineffectual duty to cooperate should be changed into a new mandatory requirement to work together in producing new regional plans. This should be incorporated into a new consolidating Planning Act. The establishment of new regional agencies should wherever possible be based on existing successful arrangements, such as the metropolitan mayoralties and combined authorities. No rural areas should be left outside these regional plan arrangements. Existing regional databases should be given new purpose for providing, updating, and distributing information on ecology, landscape, transport, heritage, and other topics required for effective planning. Where such agencies do not exist, new ones should be established. These agencies will use digital mapping techniques to provide information to local and neighbourhood planning authorities and to the public to aid their access to and understanding of plan-making evidence. This level and type of collaboration will assist in speeding up plan making and will be more efficient in terms of the use of resources. Up-to-date regional framework plans are an essential component for determining how and where large-scale strategic development proposals are to be accommodated. It is now arguable that the determination of Nationally Significant Infrastructure Projects should be handled by the established regional government bodies rather than the detached Planning Inspectorate.

Improved and expanded guidance at the national level in the NPPG, coupled with the incorporation of expert advice from the RTPI and the Planning Advisory Service (PAS), will facilitate the more expeditious production of regional and local plans and could also help in the production

Conclusions – Learning from the Past and Meeting Future Challenges 349

of neighbourhood plans, giving the plan-making body greater confidence to progress a plan for their community.

The government's proposals for a new form of local plan that relied on a broadly based three zone system (Growth areas, Renewal areas and Protected areas) was widely criticised. In the comments I submitted to MHCLG on this aspect of the White Paper proposals I stated that these blanket zoning proposals would weaken the planning system and lacked sufficient sophistication to meet the challenges that planning now faces. In this context I stress that sophistication does not equal complication and I accept many of the arguments for streamlining the production of local development plans. The Raynsford Review recommends that the removal of one of the draft stages of plan production would be damaging and would in some way reduce the opportunities for public engagement. I am far from convinced that this would be detrimental, and it could significantly reduce the timescale from commencement to adoption by as much as 12 months. I support a more extensive use of quality graphics, including the spatial mapping of data via GIS to present constraints and opportunities in a sequenced format. This would, as suggested, improve public understanding and engagement with the local plan process. There is also scope for some rationalisation of the number of evidence base documents required to produce local development plans and part of this would be achieved by, in the first instance, using the evidence base produced collaboratively for regional plans and then commissioning any additional work required to fully reflect local conditions. Flood risk, housing need and supply, transport planning and expenditure, minerals and waste management are some of the examples where catchment wide assessment must be qualified by more local studies.

The production of neighbourhood plans has proved to be a valuable addition to the hierarchy of development plans with several benefits arising, including community engagement with the planning system and an improved understanding of the processes involved. There is still considerable scope for improvements in the coverage of these plans, for enhanced community engagement and for greater synchronisation with the production of local plans. This would assist the ability of communities to incorporate their desired specific allocations for development based on the distributed requirement for housing and employment land contained in the local plan. Too many of the currently adopted plans have policy only content.

The Government need to place greater emphasis on achieving public engagement in planning via the NPPF and expanded guidance in the NPPG. The aim in the Planning White Paper to make planning more accessible to the public via improving the presentation of development plans, including computer graphics and digital mapping, should be quickly adopted into the system. Proposals to achieve this are now contained in the latest government consultation on planning reform. Given the number of studies on this topic, going as far back as the landmark Skeffington Report of 1969, it should not be necessary to commission further studies to devise strengthened requirements for public engagement. The great truth that engagement is much wider than staged consultation should be the starting point for early reform of this key part of the planning system. In the concluding section of Chapter 2 I examine the evolution of the relationship between professional planners, the public, and politicians. The need to bring public and planners together into a shared understanding of the issues faced is long overdue.

The concept of planning being in the public interest lies in the DNA of modern planning and is regarded as a key defining feature of the planning profession by the RTPI. What constitutes the public interest has evolved over the years since the introduction of the 1947 planning system and differences of opinion and interpretation remain. The RTPI Ethics and Professional Standards Advice for its members notes that acting in the public interest involves 'the expectations of clients, employers, the local community, and politicians as well as future generations'. There can be little doubt that the health and wellbeing of communities should lie at the heart of any definition of public interest. The need for urgent action on the interlocking challenges of climate change and environmental loss must also be considered as fundamentally in the public interest. Consequently, engaging with the public on development plan policies, site allocations and many individual development proposals must be accompanied by greater clarity of presentation, openness and understanding of the impacted communities. Evidence provided to the House of Commons Select Committee on the Future of the Planning System in England[2] demonstrates the low participation levels in a range of public engagement exercises and the strong bias towards the 55 plus age group as the main participants. There is consequently a need to expand presentations to directly target under-represented age groups. The Committee considered that the involvement of the public is a crucial element of the planning system.

The case for re-resourcing and upskilling local authority planning departments has been made by the RTPI and the House of Commons MHCLG Committee[2] and was supported in general in the Government's White Paper. A 15 per cent reduction in local authority planning staff in England occurred between 2006 and 2016 and this reduction comes on top of earlier losses. The supply of certain skill sets required in the planning process have been particularly affected. Only 26 per cent of LPAs now employ an ecologist, heritage specialists working on conservation and listed buildings are

Conclusions – Learning from the Past and Meeting Future Challenges

in short supply and design and placemaking specialists are urgently needed. The loss of these skill sets is only one part of the problem being faced by English LPAs. Replenishing general planning practitioner posts lost over the preceding 20 years or more is essential. Part of this re-resourcing of staff complements needs to include the rebuilding of morale amongst those planners who have remained in local government departments and coped with a heavy and demanding workload. The Raynsford Review gathered evidence that demonstrated that many planners in local government were feeling a loss of role and purpose in the true positive objectives of planning. This is not a new phenomenon. I experienced this towards the end of the local government part of my career and then subsequently via meetings with planning officers over several years as a consultant. The Raynsford Review findings are backed up by those of RTPI surveys of the feelings of their membership. The final component is the provision of a wide range of up-to-date and focussed training programmes to attract new students and entrants into the profession and to ensure that all practicing planners are equipped with the expanded skill sets they require to positively face the challenges ahead.

How can these resourcing problems be addressed given the wish to improve the system and the quality and effectiveness of its outcomes to meet those key challenges? It is necessary for those in government, public agencies, the private sector, and the public, both as consumer and commentator, to recognise that the planning system is made up of a number of related parts, all of which have impacts on our daily lives and need to contribute to the key challenges we face. The Raynsford Review, in the evidence gathered by the team, identified a tension between a narrow view of planning which they referred to as 'land licensing' and a much more positive view, which regards planning as a largely creative practice of placemaking and working with local communities to achieve all three objectives of sustainable development. My experience over decades is to recognise only partially the narrow 'land licensing' characterisation. In my view this would comprise the worst aspects of development control, the lack of control through ill-thought-out permitted development rights and the achievement of land allocations and designations through the blandest of local plans which are not fit for purpose. I also tend to think that this evidence has in part emerged because local government planners have been highly constrained by lack of resourcing and support in their jobs with limited time devoted to the overtly positive elements of planning. The control or management of development (whichever description you prefer) must also be seen as a positive process that has always involved a skilful balancing exercise, which is now more complex for many development projects as there are more material considerations to consider.

The Planning White Paper acknowledged certain aspects of the wider role definition of planning and some of the valuable work being done by planners with quality outcome examples. The Levelling-Up the United Kingdom White Paper (February 2022) provides further recognition of the important role of planners, but then subsumes proposed reforms within a wider policy context. This becomes even more apparent in the lengthy Levelling-Up and Regeneration Bill, which contains a mixed bag of planning reforms that will not result in an holistic outcome. We need to quickly arrive at a consensus definition of the purpose of town planning that encompasses the delivery of truly sustainable development, the achievement of gains in the quality of all the main services provided by the environment, beauty in built development and protected and restored landscapes, all of which needs to be achieved in the public interest. This should be short and punchy in character with any supporting explanation extending to no more than a single page. This should be quickly agreed and subsequently incorporated into the introductory sections of a new single planning act and the NPPF. Much greater attention must be given to the thorough and helpful work carried out by the House of Commons Select Committee dealing with planning, housing and now levelling-up. The definition and an accompanying acknowledgement of town and country planning's role is a starting point for re-energising and re-resourcing planning and should have a beneficial impact on those already working in public and private practice and in planning education.

There is some evidence of a limited increase in planning recruitment within the UK as we hopefully emerge from the worst impacts of the Covid pandemic. It is clear, however, that there is much work to be done to restaff planning departments to levels that can deal with the challenging workload to the end of this decade and beyond. Staffing levels need to be at least comparable with those which existed in the first decade of this century. The Planning White Paper stated that the Government would 'develop a comprehensive resources and skills strategy for the planning sector to support the implementation of our reforms'. While this is another urgent requirement, any delay in producing a comprehensive strategy should not inhibit immediate funding of new appointments and training initiatives.

There are a number of ways in which the full financial burden of the personnel requirement of re-resourcing all local planning departments can be eased. The sharing of specialist staff between neighbouring councils has been common in the past and will work well in a future system incorporating the proposed regional plan tier. This should work particularly well where there are existing Combined Authorities working at the city-region level, and in existing and future proposed arrangements across rural areas. The use

Conclusions – Learning from the Past and Meeting Future Challenges

of artificial intelligence (AI) has already proved a valuable resource in legal, accountancy and other major consultancies. It provides the local planning authorities and consultancies with the ability to speedily handle some of the basic search and information gathering functions of planning, when preparing evidence for development plans and carrying out background work for the assessment of planning applications. There are, however, risks associated with too great a use of AI in the decision-making process, both in development plan preparation and determining planning applications, where the skills of planners in balancing material considerations and the opinions of all types of consultees must remain paramount. Digital technology and modern databases, mapping systems and graphics can result in better presentations and save time, while also adding to public understanding.

Improved working across local government departments and the sharing of expertise, especially where there are common goals and objectives, such as improving the health and wellbeing of citizens and adapting to climate change, will help to improve efficiency. There needs to be a better understanding of the roles of allied professionals.

Enhanced training is recognised as an important component of re-resourcing in order that planners can refresh and expand their skill base. The RTPI's Research Paper 'Resourcing Public Planning'[3] points out that more needs to be done to make local authorities an attractive place to work to counter the drift of planners to the private sector. Part of the case for re-resourcing is to free up time for existing planners to undergo training. New recruits will also need training programmes. The RTPI paper states the importance of this approach given the difficulties that will be faced in delivering major increases in the number of local authority planners in the short term. My recent experience as an assessor of apprentices on this route to becoming a chartered member of the RTPI confirms the quality of the Institute's training programme.

I have recently reviewed a sample of seven university town planning degree courses and found that they were all up to date and focussed on the key challenges facing planning. To deliver the best comprehensive training and continuing professional development (CPD) programmes the RTPI need to bring the three components of the profession (academia, public service, and private practice) closer together.

The team building and project coordination roles of planners need to be clearly established and enhanced as part of the required upskilling. Assembling, briefing, and coordinating teams to effectively produce development, spatial and specific project plans must take on much greater significance if these plans are to make early and positive contributions to meeting the challenges that lie ahead. This applies to local authority and private sector led projects.

The team leader role of the planner will encompass relationships with the client on delivery objectives, programmes and finance, engagement with the public and all necessary consultees. To achieve delivery of fully sustainable development that truly provides gains, when assessed on the basis of achieving all three objectives, will demand a greater collective team commitment and more effective integration of the skill sets of team members.

There are many management texts available on building, running, and participating in a project team, but there are several key factors that are particularly pertinent to the planning context. These merit emphasis to assist teams in achieving the required high-quality outcomes. The following examples are provided to inform and encourage participants.

1. The scale of the challenges involved in each project should be clearly identified and discussed by the team so that all involved have a clear and shared understanding.
2. The nature of these challenges and the scale and relative complexity of the project may give rise to concerns among team members. Time should be provided at the outset for team and one-to-one discussion to achieve resolution of any concerns, either at the beginning of the project or through adaptive approaches in the early stages.
3. As well as potential concerns there should be excitement at the prospect of what can and needs to be achieved. The team leader will need to communicate this feeling of excitement, if it is not already apparent, and provide a positive approach.
4. A team briefing document incorporating a programme for the work streams should be prepared by the team leader, with all team members being asked for their views on a draft version.
5. A good team leader will require certain personal qualities beyond the necessary planning skills. These will include understanding and empathy; determination and commitment; full recognition of the skills, experience, and all-round level of ability of team members, the contributions which they need to make, and how these should be integrated.
6. The team leader will need to display the appropriate level of professional understanding and negotiating abilities to ensure that full and programmed responses are received from all essential consultees.
7. Team working experience on challenging projects should be part of the in-house experience of all trainee planners.

* * *

Conclusions – Learning from the Past and Meeting Future Challenges

The Value of Planning- Past, Present and Future

The modern origins of our current planning system in the UK were founded on the health and wellbeing of the population and a central belief that planning was in the public interest. In my introduction I set out these twin themes and a core question that aimed to assess whether the planning system, established in 1947, has to date been a force for good with largely benign outcomes, or a restrictive and misused system with negative impacts on people's livelihoods and the environment in which they live. This central question needs to be extended to consider what the experiences of the past have taught us and what are the new lessons we must rapidly learn. Such learning is essential if we are to gain the maximum benefits from a renewed system of town and country planning.

When introduced in 1947, there can be little doubt that the public good and the future health and wellbeing of people and communities were the core purpose behind the establishment of the first modern and holistic planning system. The fact that the 1947 Planning Act, and the associated statutes, were an integral part of the wider welfare system, designed to look after people from 'the cradle to the grave' demonstrate this centrality.

Limited parts of the 1947 system did not have a particularly long life. The betterment tax, which had been introduced at 100 per cent of the increase in land values on the grant of planning permission was removed 6 years later by the 1954 Town and Country Planning Act. The 1946 New Towns Act did not survive much longer due to major legal changes to the Compensation Code which made it much harder for LPAs and development corporations to purchase land at existing use value. For the first two decades of operation of the 1947 system much positive planning was achieved in the form of extensive house building, the protection of quality landscapes in national parks and the early AONBs, together with improvements to public access and management. More than 30 new towns and villages were completed. During this period there were some highly negative urban planning schemes that produced dominant highway networks, often with little respect for the heritage and character of the central and inner areas of several of our major towns and cities.

The first main reform of the planning system came in 1968 following the Planning Advisory Group's (PAG) Report on 'The Future of Development Plans'. The Town and Country Planning Act of that year introduced new types of development plan in the form of a two-tier system of Structure and Local Plans with statutory requirements for publicity and public participation. These were meaningful improvements to the development plan part of the 1947 system. The holistic nature of the revised system was retained.

Subsequent changes to this system were mostly politically motivated, sometimes based on one-off reports that reviewed only a limited part of the total content and resulted in few material improvements. Over the last four decades national politicians have all too often attacked the planning system and local government's operation of the system, blaming it for several problems, including holding back economic progress, building too few houses, and causing significant delays in the determination of planning applications. It is no exaggeration to say that the planning system has survived extensive political criticism and blame over this extended period. This has often been accompanied by adverse public comment usually led by vocal minorities in the retired age groups, who have the time and inclination to defend their local communities against new housing and other types of development. Local politicians and MPs often support these 'action groups' and echo some of the criticisms of their national colleagues. These situations demonstrate the need for enhanced public engagement in the planning system both in terms of the groups who are not being reached and, in the quality, and duration of the methods deployed.

It is not surprising that some local government planners, faced by cuts in resources, these types of criticism and a narrowing of their perceived roles by local and national government, lose their motivation and positivity. Despite these multiple negatives many planners manage to remain generally positive about their chosen profession. There are several examples of great results being achieved despite some of the adverse conditions experienced over the last 40 years. The RTPI and other bodies have rightly celebrated several of these achievements through their national award ceremonies. For the last couple of years, I have acted as a judge on certain of the RTPI award categories and have been impressed by the quality and enthusiasm shown by planners in the public and private sectors in their major contributions to model developments and system innovations.

Towards the end of Chapter 2 I referred to two definitions of what constitutes 'the public interest'. In the first of these, liberal and utilitarian values are the central consideration and individual interests are paramount, and consequently the public interest is either made up of the summation of all individual interests in a community or the greatest good for the greatest number. The second definition of the public interest gives importance to the shared values that are greater than a simple summation of individual preferences and includes the conviction that there are circumstances where market processes are not the best means of determining desirable outcomes. In this second definition, which has increased in significance over the last two decades, while individual interests retain some importance, it is felt that an outside overview is required to ensure that the needs of different groups of the population are considered. Using this second definition as the basis for assessment I have carried out a summary review of the achievements of planning over the last 20 years.

Conclusions – Learning from the Past and Meeting Future Challenges

Several positive outcomes continue to be delivered by the current planning system. Areas of the UK protected for their national landscape and nature conservation value have expanded to include new and enlarged national parks and AONBs. As at July 2020 some 28 per cent of the UK's land area was subject to these protections and had improved management and access systems. The Government's target of 30 per cent coverage by 2030 is now within reach.

Some quality housing schemes are being delivered where good spatial planning, layouts and selection of house designs are produced on a cooperative basis with developers, who aspire to deliver projects that follow local design guidance and make a positive contribution to local character. I have referred to award-winning projects at Marmalade Lane Cambridge, Goldsmith Street Norwich, and the Leeds Climate Innovation District. While there are many other schemes that deserve to be celebrated, collectively these still comprise a minority of all completed projects. Many modern housing developments, particularly those produced by some of the volume housebuilders, are of poor design quality and fail to create an individual sense of place. Many of the recommendations of the Building Better, Building Beautiful Commission, in their report, 'Living with Beauty' (January 2020), being adopted to ensure that all placemaking and community developments result in high quality outcomes. A very positive start has been made by incorporating the requirement for beauty in new developments and places into the NPPF. All LPAs will need to be adequately resourced with the skilled officers to negotiate quality outcomes, which then result in a range of benefits for residents.

In Chapter 6 I review examples of how development management/control operates in the public interest and produces outcomes that are beneficial to the health and wellbeing of individuals and communities. These include the reclamation of derelict land for development and environmental improvement; protecting the best and most versatile land for future food production; the provision of affordable housing and ensuring in specific parts of the country that there is an appropriate community balance between housing for local people and those seeking second homes; avoiding the extraction of minerals where this would result in unacceptable impacts on local communities and protecting heritage assets and valued local environments from adverse development impacts.

Over the years, many material considerations have been added to the planning balance when determining planning applications. These considerations now include impacts on human health and wellbeing. In Chapter 8 I specifically consider the emergence of public concern and the public perception of risk, which can give rise to individual and collective anxiety and fear, which then leads to adverse effects on health and wellbeing. Protecting the public and their health and wellbeing from the potential adverse impacts of specific types of development have become legitimate concerns to be considered. Such developments include waste treatment plants, telecommunications equipment, development on or near unstable land and development that may add to the risk of flooding.

The protection that has been provided to public interests via this part of the UK planning system has increased over the last 30 years. There can be little doubt that this discretionary decision-making approach to individual development applications is more sophisticated and offers greater consideration to public interests than the broad zoning systems deployed in some countries.

The actual and potential values of planning and their contributions to all three parts of sustainable development are much greater than those I have already referred to. Research studies and reports produced by policy institutes and other commentators/political advisors very often ignore the broad economic advantages of good planning. The RTPI has produced a series of research briefings about the value of planning with a focus on the economic advantages.[4] These briefings, produced by leading academics in economics and planning, are a response to the criticisms that planning holds back economic growth and an explanation of the benefits of proactive planning. This economic growth does not need to follow the old capitalism. It does need, amongst other things, to be more ethical, greener, have many more stakeholders and create a fairer society. In the introduction to the first of the four papers, the RTPI state that 'planning is about improving places by helping them to function better economically as well as socially and environmentally'. Good planning through well prepared development plans, policies, spatial plans, and positive development management can shape places and stimulate development activity and investment.

The design of new communities and the reshaping of existing ones via urban renewal projects provides the opportunity to create better connectivity, in terms of public and active transport between homes, workplaces and services, socially through enhanced neighbour connections and meeting places and environmentally by ease of access to habitats and recreation spaces. Policy makers have tended to focus planning on its regulatory development control function and the need to make this more efficient rather than taking a much broader view of planning's capabilities. The second and third papers in this series look at examples of value outcomes from positive and proactive planning in Hamburg, Lille, and Nijmegen and in case studies of Chinese cities. The examples in Western Europe demonstrate that planning interventions can boost labour productivity and values and that these developments tend to also be those that have very high-quality design, good public, and active transport provision and ample and enhancing greenspace, which produce additional values in terms of the health and wellbeing of residents.

The researchers covering the examples in continental Europe concluded that these were well resourced and politically supported and that too often in the UK development is left to the individual entrepreneurial interventions of private sector developers. The UK has produced such positive outcomes, particularly where development corporations and agencies have carried out the strategic plan-making, master planning and provision of essential physical and social infrastructure. However, there have been all too few examples of LPAs being enabled to carry out these types of positive planning due to a lack of adequate resourcing, a concentration on what is perceived as the regulatory role of planning and government's over-reliance on private sector developers who individually rarely have the breadth of vision, land control and commitment required.

Exceptions do exist where there are large landowners or active city councils willing and able to assemble multiple land ownerships to create large mixed use regeneration schemes. This has occurred at King's Cross in London where the landowners, London and Continental Railways and DHL selected Argent as their development partner. The same company replaced Rosehaugh in 1993 as the developer and manager of Brindley Place, a large mixed-use development in the Westside District of Birmingham. These schemes and other examples demonstrate a quality in their urban design, architecture and public realm adding value in terms of attraction and public enjoyment.

It is essential that the full values of planning are realised by both the public and politicians. The verb to realise is used here in every sense as described in the Concise Oxford English Dictionary: 1. to become fully aware of as a fact or understand clearly; 2. to cause to happen or achieve; 3. give actual or physical form to and 4. convert (an asset) into cash. Not all aspects of the value of planning can be monetised, although attempts have been made to do this. In some areas, especially environmental improvements, there are now metrics available to place a financial value on enhanced biodiversity and ecosystem services. Planning fees from applicants for planning permission contribute to the running of planning departments. Developer contributions to physical, social, and environmental infrastructure, through section 106 planning agreements and Community Infrastructure Levy (CIL) contributions, meet some of the costs of new development in terms of its demand on public services. Enhanced land value capture through the planning system would help to provide something approaching the full public service costs of new development.

* * *

The Multiple Challenges Facing the Planning System

On 8 September 2022 Queen Elizabeth II died, ending a remarkable reign spanning 70 years. Only two days earlier she had appointed Liz Truss as prime minister, replacing Boris Johnson following his resignation. The end of the second Elizabethan era marks a major turning point in the history of these islands. After a period of mourning and reflection life has returned to something like normal, yet during the remainder of the month of September 2022 the socio-economic problems facing the country and individual households deepened. This rapid and worrying trend was the result of a cocktail of international (particularly the war in Ukraine) and domestic factors. The new Truss Government rapidly introduced a package of financial measures to encourage economic growth and attempt to reduce the adverse impacts of energy price rises. The package of measures included tax cuts for households, the reversal of previously planned changes to business and employment taxes and a raft of promised deregulation to spur growth. 'Planning reform' featured in the early headlines but only in the form of deregulation. This was an echo of the incoming Government of Margaret Thatcher in 1979. New low tax/low regulation growth zones were being promised. The package of changes in this new economic strategy, to be funded largely by debt, and lacking a parallel independent analysis which usually accompanied budgets, quickly produced damaging economic consequences. Consequently, my review of the values of good planning closes in a period of socio-economic turbulence. This must not be allowed to detract from the need for a strong, positive, and well-resourced planning system. Advocacy for the essential central role of planning, in addressing the many challenges faced, must continue and be given greater prominence in policymaking and decision-taking. Clear governance and direction regarding planning's future roles is urgently needed. Many of the challenges faced require common or integrated policy and spatial planning solutions.

The strong linkages between climate change, environmental and health and wellbeing solutions have been identified along with the planning policies, spatial planning and land use actions required. The 'housing crisis' continues in the UK. The need to shift personal travel from the private car to active and sustainable modes grows in significance and urgency as air pollution and other adverse impacts are not being significantly reduced. Our future approaches to urban and rural planning require regular review and action. The welcome decision to incorporate a requirement for beauty in placemaking and future development presents a new and positive challenge to be addressed by planners, communities, and other stakeholders.

The solutions to these challenges lie within a renewed, re-resourced and holistic planning system. This will rely on an integrated approach to governance

Conclusions – Learning from the Past and Meeting Future Challenges

at national and local levels with cross-departmental solutions. Working in government and departmental silos must be consigned to history.

Health and wellbeing, in the aftermath of the Covid-19 pandemic and entering a period of energy, materials and food price inflation, will continue to grow in importance, and strengthen its position at the core of the planning system and the delivery of sustainable development.

Integrated rural planning is now more essential than ever due to the need to balance, align and suitably mix an array of competing land uses. These include provision for agriculture and future food security, environmental restoration and nature recovery plans, climate change mitigation and adaptation measures, infrastructure provision, widening access for public recreation and some of the required employment and residential developments that cannot be sustainably accommodated within our urban areas. In the early 1980s Marion Shoard and others argued that planning should have a more significant role in protecting the environmental and landscape qualities of the countryside (see Chapter 6). The intensified competition for the use of rural land and the consequences arising from the implementation of the new agriculture and environment Acts demand a greater role for spatial planning and development management.

The RTPI recently commissioned a major study of the issues that planning will need to address in rural areas over, at least, the remaining years of the current decade.[5] The study concludes that a planning framework will be required to balance the needs of the UK's rural settlements and a 'myriad of competing claims in rural space'. Rural communities are now facing a number of additional pressures to the long-standing problems of affordable housing, transport availability and diversification of the rural economy. These include the impacts of Brexit and the changes proposed in the Agriculture Act, climate change and the residual effects of the Covid-19 pandemic. The RTPI will be using the results of this study, carried out in 2021/22 across the UK, to determine and advocate an enhanced approach to policy and plan formulation.

The study defines this new approach to rural planning as essentially being the creation of sustainable development centred around land use planning, the spatial planning of communities and environments, countryside management and environmental protection and enhancement. The global challenges of climate change, biodiversity loss, the security of food and energy supplies must be embraced now and in the future. The focus advocated in the study is the creation of an integrated framework approach to the future planning of rural areas and communities. Good planning is seen as central to the necessary transitions that must be achieved and this needs to be delivered by a broad partnership of community, public, voluntary, and private interests.

The knowledge and skills of local people are recognised as being central to successful 'place-specific planning'.

The 'Land Use in England Committee' was appointed in January 2022 by the House of Lords, and this gave recognition to the need to address changing land use requirements in England, the competition for land use and the need to achieve multiple use and certain outcomes which are beneficial to the public and the environment. In May 2022 the Committee was charged by government to produce a report and recommendations on how the issues surrounding competition for land use are best resolved and how multiple use and greater public benefits can be achieved. The report was published in December 2022 and recognises that radical change is taking place in how we use our land resources.[6] There is already a move away from a landscape which is dominated by food production. There is a recognition that we need to accommodate the challenges and opportunities of moving to a new environment where the restoration of nature and biodiversity, carbon sequestration, new development and infrastructure needs, the inclusion of sustainable energy facilities and greater provision for public access, recreation and wellbeing are given greater priority. These changes do not mean that the food security target, set by DEFRA in 2021, needs to be compromised. The Report recommends the establishment of a Land Commission for England and the production of a land use framework. The format and content of the proposed framework has yet to be clarified, including whether it will incorporate spatial plans.

While farming is clearly important to the future of rural areas and national food security, the planning of rural areas over the coming years will be about much more than food production. The national governments that make up the UK must quickly give recognition to the central role that planning has to play in the essential transition to new mixed rural economies. National and regional spatial plans and policies are necessary to provide 'umbrella' frameworks for local and community spatial plans and policies. This planning vacuum must be urgently addressed via cross department cooperation at national level.

* * *

Planning is good for all of us

My purpose in writing this book has been based around two key themes:

A) That the operation of the town and country planning system is highly beneficial for the nations of the UK and for individuals and communities.

Conclusions – Learning from the Past and Meeting Future Challenges

B) The health and wellbeing of the population was the core concern leading to the origins of modern town planning and the subsequent introduction of the first holistic set of laws governing planning as part of the 1945 Labour Government's welfare state. Current and future concerns for public health and wellbeing are now strongly re-established at the core of national planning policy. All three of the component policy objectives of delivering sustainable development, economic, social, and environmental require positive outcomes that should benefit the health and wellbeing of communities and individuals.

Health and wellbeing considerations are multi-faceted, and the planning system has the proven capabilities to deliver on several fronts. The economic benefits have been summarised in this concluding chapter and the social and environmental benefits are widely demonstrated in the preceding pages and in a world-wide research base. The benefits resulting from good planning include contributions to disease prevention via the provision of access to health and fitness facilities. The provision of access to natural environments and full recreation opportunities are proven to have physical and mental health benefits. Public health plans and programmes increasingly recognise these benefits, including the financial savings that can result from reducing reliance on all tiers of healthcare provision. The lessons learned from the Covid-19 pandemic underline these benefits. The provision of good quality housing and external living environments and beautiful places also bring health and wellbeing benefits. While the benefits of decent and well-designed housing and places are well recognised, they are far from being universally available. Successive UK governments and housebuilders have given differing degrees of recognition to these needs but have generally placed emphasis on the quantity of housing delivered rather than the qualitative aspects of such provision. Some might argue that the provision of a new dwelling brings improvements in quality, but this is not always the case. A particular example is the recent decision to grant permitted development rights to the conversion of some commercial buildings into residential units. This has all too often resulted in inferior internal and external space standards and lack of greenspace provision.

The impacts of the recent energy price crisis and the need for climate mitigation actions in the domestic market have highlighted the widespread inadequacies of the thermal insulation, ventilation, and heating systems in a large proportion of the UK's housing stock. The lack of a retrofitting programme for these dwellings has been frequently highlighted together with the adverse impacts of such inferior living conditions on the physical and mental health of households.

The quality of housing provision and the design of dwellings and places has at last achieved full recognition via the publication of the National Design Guide[7] and subsequently via the incorporation of the need to achieve beauty in the July 2021 amendments to the NPPF. This most welcome amendment makes it clear that the creation of high-quality buildings and places is fundamental to the outcomes of the planning and development process. The National Design Guide emphasises the ways in which good design combine to influence the quality of life (wellbeing). A well-designed living environment is a key contributor to happiness and a feeling of contentment, which when combined with ease of access to nature, recreation opportunities and local services, enhances physical and mental health outcomes. The desire to create quality liveable and healthy environments has existed within planning and allied professions for some considerable time. Planners, architects, landscape architects, urban designers, and the environmental professions now have the tools in the national planning policy and guidance to insist on high qualities of development across all projects, including the regeneration of existing communities. However, as we have seen, they do not have the full resources required, particularly in the public sector, to achieve these necessary outcomes.

In Chapter 8 I examined a number of land use and development examples where planning plays a role in protecting public safety. Individual and community fear and anxiety issues became a material planning consideration, which now must be weighed in the planning balance when determining certain types of planning applications. The safety of the public in such situations is a significant contributor to their health and wellbeing.

There should be little doubt that town and country planning has brought many benefits to individuals and communities. The aims and operation of the planning system have generally been a force for good in our lives.

The planning system must be a major player in meeting the three challenges of climate change, loss of biodiversity and the need to enhance health and wellbeing. By delivering truly sustainable developments, which meet all aspects of the environmental, social, and economic objectives, as defined in the NPPF, the future scope of planning to act in the collective public interest and deliver valued and quality outcomes is considerably enhanced (see illustration 19 the Wheel of Sustainability & the three interlocking challenges). Much of the reason for this enhanced potential is because the challenges are much greater than they have previously been, they are better understood and development in recent years has all too often not achieved comprehensive sustainability.

Conclusions – Learning from the Past and Meeting Future Challenges

To achieve these extensive value benefits full engagement with the public is essential. In this context the public need to be presented with opportunities to contribute to the formulation of ideas, policies and plans and to influence the content and form of projected developments that will shape the future of their neighbourhood. This engagement needs to be a shared experience with the central aim of arriving at agreement in each stage of the planning process. While there is some need for the education of the public to provide them with a good understanding, this is a two-way street where both parties, public and planners, need to gain necessary insights from each other. A teacher-pupil relationship must be avoided.

The upskilling of planners as individuals, and collectively within teams and departments, will strengthen their ability to add value to development activity and community outcomes. The converse also applies in that areas that are poorly planned can result adverse outcomes and cost implications for communities, individuals, and developers.

The next seven years to 2030 will be the most challenging and exciting time to be involved in town and country planning in many countries across the world. This is the focal period when considerable progress must be made in the delivery of those planning actions required to make the greatest possible contribution to the three stated earth challenges. These can be summarised as meeting net zero and no greater than 1.5 degrees of warming by 2050, restoring degraded environments and halting, and wherever possible reversing, the decline in biodiversity, and improving all aspects of people's health and wellbeing. This is an agenda that should inspire those already engaged in the profession, as well as students on existing courses and school children searching for a meaningful future career, where the work involved has positive and tangible outcomes, which benefit their fellow human beings and the planet.

Notes

Chapter 1
1. David Kynaston – *Austerity Britain 1945–51, Volume 1: Tales of a New Jerusalem* – Bloomsbury Publishing 2007.
2. Andrew Marr – *A History of Modern Britain* – Pan Books Edition 2017.
3. Jonathan Schneer – *Ministers at War (Winston Churchill and his War Cabinet 1940-45)* – One World 2016 Edition.
4. Marr – Ibid.
5. Peter Self – *Cities in Flood- The Problems of Urban Growth 2nd Edition* – Faber 1961.
6. Ibid.
7. Planning 2020 – The Raynsford Review of Planning in England, Final Report Nov 2018 – published by the TCPA.
8. Social Insurance and Allied Services – Report by Sir William Beveridge presented to Parliament November 1942, Cmnd 6404.
9. Raynsford – Ibid.
10. White Paper – The Control of Land Use, Cmnd 6537, published by the Coalition Government in 1944.
11. Marr – Ibid.
12. Nicholas Pevsner – *An Outline of European Architecture 7th Edition* – Pelican Books 1963.
13. William Ashworth – *The Genesis of Modern British Town Planning – A Study in Economic and Social History of the Nineteenth and Twentieth Centuries* – part of the International Library of Sociology and Social Reconstruction – Routledge & Kegan Paul – first published in 1954. My references are to the 1972 reprint.
14. Ebenezer Howard – *Garden Cities of To-morrow* – Attic Books 1989 reprint.
15. Ray Thomas – Introduction to *Garden Cities of To-morrow* – Attic Books 1989 reprint.
16. Ibid.
17. Ashworth and Raynsford – Ibid.
18. Raynsford – Ibid.
19. Ashworth – Ibid.
20. Raynsford – Ibid.
21. S.A. de Smith – 'Town & Country Planning Act 1947' – The Modern Law Review Vol 11 No.1 (Jan 1948) pp 72–81.
22. S.A. de Smith – Ibid.
23. Raynsford – Ibid.
24. Roger Suddards – founding partner Last Suddards planning solicitors – drafted planning acts for Mauritius and St Lucia & Nevis.
25. Gerald Manners – author of *The Geography of Energy* and one of the 'Cambridge Six' geographers including Peter Hall who became a leading planning academic and author of books on urban and regional planning.

Chapter 2

1. John Gyford – Lecturer in planning at the Chelmer Institute of Higher Education 1966–69 and subsequently at the Bartlett School of Architecture & Planning, University College London – author and co-author of books and papers on town planning and politics, including – 'Town Planning and the Practice of Politics', Town Planning Discussion Papers No.29 – Bartlett School 1979.
2. David McKie – *A Sadly Mismanaged Affair- A Political History of the Third London Airport* – Croom Helm 1973.
3. Colin Buchanan – author of *Traffic in Towns* – later Sir Colin Buchanan and president of the RTPI.
4. McKie – Ibid.
5. Peter Hall – One of the 'Cambridge Six' referred to at the end of Chapter 2. He was the author of several books on urban and regional geography.
6. Hansard 1968.
7. Routledge, the original publishers of the Skeffington Report in 1969, republished this as part of their 'Studies in International Planning History' in 2014.
8. A series of articles in the journal Planning Theory and Practice – Vol 20 No.5 Dec 2019 (published by Routledge Taylor & Francis Group), reviewing the Skeffington Report 'People and Planning' on public participation in planning, on the 50th anniversary of its publication.
    i) Andy Inch – introduction – 'People and Planning at 50'.
    ii) Francesca Sartorio – 'People and Planning 50 years on – The Never-Ending Struggle for Planning to Engage with People'.
    iii) Jeff Bishop – 'Skeffington – A View from the Coalface'.
    iv) Yasmina Beebeejaun – 'From Participation to Inclusion'.
    v) Katie McClymont – 'Marking the 50th Anniversary of Skeffington: Reflections from a Day of Discussion'.
    vi) Kathryn S Quick – 'An American's Reflections on Skeffington's Relevance at 50'.
9. Dr Victoria Winckler – 'Forty years of Regeneration in the Upper Afan Valley'.
10. Yvonne Rydin – 'Public Participation in Planning'– Chapter 13 in *British Planning – 50 years of Urban and Regional Policy* – Edited by Barry Cullingworth – Athlone Press 1999.
11. RTPI – Practice Advice Paper – 'Probity and the Professional Planner – Exercising your Professional Judgement' – January 2020.
12. Heather Campbell & Robert Marshall – 'Moral Obligations, Planning and the Public Interest – a commentary on current British Practice'.
13. The Nolan Committee on Standards in Public Life – Third Report on Local Government gave particular attention to the operation of the planning system. Lord Nolan was appointed to chair the Committee by Prime Minister John Major in October 1994 and the Third Report was published in 1997.
14. R Bolan – 'The structure of ethical choice in planning practice' – Journal of Planning Education and Research 3, 1983 pp 23–24.
15. Campbell & Marshall – Ibid.
16. John Gyford – *The Politics of Local Socialism* – Allen & Unwin London 1985.
17. The Redcliffe Maud Report (Cmnd 4040) 1969 was published by the Royal Commission on Local Government. The Baines Report – 'The New Local Authorities Management and Structure' – was produced by a study group appointed to examine management principles and structures in local government at both elected member and officer levels – published by the Department for the Environment 1972.
18. Gyford – Ibid.

19. Campbell & Marshall – Ibid.
20. Gyford – Ibid.
21. RTPI – 'Probity and the Professional Planner- Exercising your independent professional judgement'. Practice Advice Jan 2020.
22. RTPI – 'Code of Professional Conduct' – the RTPI professional standards as last amended by the Board of Trustees – 10 Feb 2016.
23. Professor Cliff Hague – 'Transforming Planning: Transforming the Planners' – Paper presented at the 50th Anniversary Conference of the Department of Town & Country Planning, University of Newcastle-upon-Tyne 1996.
24. V Nadin & S Jones – 'A Profile of the Profession' – Journal of the RTPI 73 1990 pp 13–24.
25. Campbell & Marshall Ibid.

Chapter 3
1. Robin McKie – 'Why is life expectancy faltering' – Special Report in The Observer – 23 June 2019.
2. 'The Social Determinants of Health – The Solid Facts' – World Health Organisation Europe with the International Centre for Health & Society, University College London 1998.
3. 'What Makes us Healthy' – The Health Foundation – Infographic 29 June 2017.
4. 'Plan the World We Need – the contribution of planning to a sustainable, resilient and inclusive recovery'. RTPI Research Paper June 2020.
5. 'Planning for Post Pandemic Recovery' RTPI webinars and video debates 29 June – 3 July 2020 including the following:
    i) 'The risks exposed: What Covid-19 is telling us about the way we plan and build our living environment' – A discussion with a neuroscientist and a planner focussing on the key themes of health, inequality, environment, and the need to plan for more resilient homes and neighbourhoods.
    ii) 'Transport for Green Recovery – What are the long-term impacts for active and mass transport and the impact and implications for transport infrastructure planning in the future?'
    iii) 'New priorities for health and wellbeing – a function of place?'
    iv) 'The future of commercial centres'.
    v) 'How will Covid-19 change the way we plan our towns and cities?' Public health has always been a central concern of planners and disease outbreaks have at various points in history had fundamental and enduring impacts on our towns and cities.
    vi) 'Climate action and environmental impact – the role of the local planning authority' – many commentators note the impact of fewer journeys on the climate – what is the role of the local planning authority in capitalising on this?
    vii) 'Planning for Post Covid-19 economic and social recovery and planning's contribution'. This is a unique time in history and planners will be central to the recovery of our economies and communities. What will planners contribute and what will future generations thank the planners of now for?
    viii) 'Inclusive economies'. The pandemic is catalysing the greatest economic downturn for centuries. What can planners do to help facilitate the recovery and bring in more inclusive economic growth?
6. 'Measuring Wellbeing: A guide for practitioners'. New Economics Foundation London 2012.
7. 'Concepts of Health and Wellbeing'– Health Knowledge – This is a UK based learning resource for anyone working in health, social care, and wellbeing.
8. Anna Cronin de Chavez, Kathryn Backett-Milburn, Odette Parry, and Stephen Platt- 'Understanding and Researching Wellbeing: Its usage in different disciplines and potential

for health research and health promotion.' Health Education Journal 64 (1) 2005 pp 70–87.
9. 'Promoting Health and Wellbeing in Wales.' Health Promotion Wales 2003 – https://www.hpw.wales.gov.uk.
10. 'Wellbeing Policy and Analysis: An Update of Wellbeing Work across Whitehall.' ONS June 2013. In this paper the ONS chart how they established the National Wellbeing Programme to monitor and report UK progress by producing accepted and trusted measures of the wellbeing of the nation.
11. Ibid.
12. 'Securing the Future – delivering the UK Sustainable Development Strategy.' The UK Government Sustainable Development Strategy Cmnd 6467 – March 2005.
13. United Nations Conference on the Environment and Development 1992 held in Rio de Janeiro. At this conference the international community adopted Agenda 21, an unprecedented plan of action for sustainable development.
14. 'Saving Lives – Our Healthier Nation' – White Paper published by the Labour Government 5 July 1999 – Cmnd 4386.
15. 'Healthy and Safe Communities' – National Planning Practice Guidance (NPPG) – first published 6 March 2014 and last updated 1 Nov 2019 – Ref ID: 53-001 to 012- 20190722.
16. Health and Wellbeing Boards (HWBs) explained – The Kings Fund – 22 June 2016.
17. 'Born in Bradford' news item – https://web.archive.org/web/20131029212458/https:www.borninbradford.nhs.uk/news/ – 28 October 2013.
18. 'Wellbeing of Future Generations (Wales) Act – Future Policy.org – an online database of sustainable policy solutions.
19. 'Wellbeing of Future Generations (Wales) Act 2015' [Acts of the National Assembly for Wales] – https://www.legislation.gov.uk/anaw/2015/2/contents/enacted .
20. 'Wellbeing of Future Generations (Wales) Act 2015 – The Essentials – Welsh Government (www.gov.wales)
21. 'The Future Generations Report 2020' – Future Generations Commissioner for Wales May 2020 (www.futuregenerations.wales)
22. Town and Country Planning Act 1990, s336 (1) – see Introduction to this Act in the Encyclopedia of Planning Law and Practice – Thompson: Sweet & Maxwell.
23. Planning & Compulsory Purchase Act 2004 – s39 – Sustainable Development – Encyclopedia of Planning Law and Practice – general note ref 2-4524.3 – Thompson Sweet & Maxwell.
24. Russell Shorto – *Amsterdam – A History of the World's Most Liberal City* – Chapter 1 – Abacus 2014 paperback edition.
25. 'The Amsterdam City Doughnut – A Tool for Transformative Action'. The 'City Portrait for Amsterdam' is known as the 'Amsterdam City Doughnut'. This was created by Doughnut Economics Action Lab (DEAL) in collaboration with Biomimicry 3.8, Circle Economy, and C40 Cities – published April 2020.
26. Kate Raworth – 'Introducing the Amsterdam City Doughnut' – https://www.kateraworth.com/2020/04/08/amsterdam-city-doughnut/ .
27. Kate Raworth – 'A Safe and Just Space for Humanity: can we live within the doughnut?' Oxfam Discussion Paper 2012 – Oxfam International Oxford.
28. Kate Raworth – 'Doughnut Economics- Seven Ways to Think Like a 21st Century Economist' – Random House Business Books 2018.
29. Ibid. – 27.
30. Ibid. – 27.
31. George Monbiot – *The Guardian* April 2020.
32. Malcolm Sayers & Katherine Trebeck – 'The UK Doughnut – A Framework for environmental sustainability and social justice'. Oxfam Research Report February 2015.

Planning for Good or Ill

33. Goldsmith Street Norwich housing development wins the prestigious UK Stirling Prize for Architecture in 2019 – *The Guardian* 8 October 2019/RIBA Architecture.com.
34. Ibid.
35. 'Homes and Neighbourhoods – A Guide to Designing in Bradford' – Tibbalds Planning and Urban Design and Urban Design Skills – adopted by Bradford Metropolitan District Council as a Supplementary Planning Document 4 February 2020.

Chapter 4

1. David Gosling – *Gordon Cullen – Visions of Urban Design* Academy Editions London 1996.
2. Section 106 Agreements were first provided for and defined in the 1990 Town and Country Planning Act though the history of such developer contribution agreements goes back through several previous planning acts including the 1971 Act (included as s52 of that Act), the 1962 Act, the 1947 Act, the 1943 Act and the 1932 Act – see Encyclopedia of Planning Law and Practice volume 2. The 1990 Act has often been referred to as the Principal Act. It should be noted that S106 of the !990 Act was repealed and replaced by widened powers in the 1991 Planning and Compensation Act referred to as planning obligations. These obligations differ from the earlier agreements in that they can be either bilateral or unilateral.
3. John Rose – Obituary of Robert Turley – Place North West 20 May 2016.
4. Professor Otto Saumarez Smith – Oxford Dictionary of National Biography February 2019.
5. Professor Otto Saumarez Smith – *Boom Cities – Architect Planners and the Politics of Radical Urban Renewal in 1960s Britain* – Oxford University Press April 2019.
6. Smith – Ibid. – 5.
7. James Fennell – 'A short history of planning consultancy' – based around the career of Geoff Smith. Planning Matters Litchfield's UK planning blog.
8. 'The UK Planning Profession in 2019' – statistics on the size and make-up of the planning profession in the UK- RTPI Research Paper June 2019.
9. 'Investing in Planning Delivery: How we can respond to the pressures on local authority planning'. RTPI Research Report No. 10.
10. 'Planning for the Future' White Paper August 2020 – Ministry of Housing, Communities and Local Government.
11. Gavin Parker, Emma Street and Matthew Wargent – 'The Rise of the Private Sector in Fragmentary Planning in England'. Planning Theory & Practice 2018, volume 19 No.5 pp 734–750 – Routledge Taylor & Francis Group.
12. Orly Linovski – 'Firm of the Future: Planning Practice in Publicly Traded Companies'. Planning Theory & Practice volume 21 No.3 July 2020.
13. Parker, Street and Wargent – Ibid.

Chapter 5

1. Ministry of Housing Communities and Local Government (MHCLG) – 'Planning For the Future' White Paper August 2020.
2. Report of the Building Better Building Beautiful Commission – 'Living with Beauty – Promoting health, wellbeing and sustainable growth' January 2020. The Commission was established by the Government and its terms of reference are set out in the appendices to the Report.
3. Dame Fiona Reynolds – 'The Fight for Beauty – Our Path to a Better Future'– Oneworld Publications 2017.
4. Ibid.
5. Eric Carlson – 'Why Beauty is important for our lives and cities' – Modern City 4 March 2019.

6. Ibid, 'Living with Beauty'.
7. Helicon – 'Is Beauty a Basic Need and Right' – December 2019 – Https://Heliconcollab.net/feed/
8. Theaster Gates – Ted Talk- 'Why Beauty Matters' – 14 July 2015 – https://ideas.ted.com/why-beauth-matters/
9. Ibid.
10. Rainer Zerbst – *Gaudi – The Complete Buildings* – Taschen 2005.
11. John McKean & Colin Baxter – *Charles Rennie Macintosh: Architect, Artist, Icon* – Lomond.
12. Cesar Manrique – *In his own words* – Fundacion Cesar Manrique 1997.
13. McKeen & Baxter – Ibid.
14. Trewin Copplestone – *Twentieth-Century World Architecture* – Brian Trodd Publishing House Limited 1991.
15. Manrique – Ibid.
16. David Gosling – *Gordon Cullen – Visions of Urban Design* – Academy Editions 1996. This is a decade-by-decade anthology of Cullen's work, which is beautifully illustrated by his drawings, plans and paintings.
17. John Punter in 'British Planning – 50 Years of Urban & Regional Policy'– edited by Barry Cullingworth – Athlone Press 1999.
18. Broadway Malyan – 'Bradford Growth Assessment'– part of the evidence base for producing successive versions of the Council's local development plan.

Chapter 6
1. Barry Cullingworth & Vincent Nadin – 'Town & Country Planning in Britain' – Routledge – currently in its 15th edition 2015.
2. Malcolm Grant – *Urban Planning Law* – First edition 1982 Sweet & Maxwell.
3. Peter Morgan & Susan Nott – *Development Control-Law, Policy, and Practice* – Butterworths a division of Reed Elsevier (UK) Ltd – Second edition 1995.
4. National Planning Policy Framework (NPPF) Section 15 – Conserving & Enhancing the Natural Environment- paragraphs 170–172.
5. National Planning Practice Guidance (NPPG) – Minerals – Guidance on the planning for mineral extraction in plan-making and the application process – page 2 of 75 pages of guidance – 17 October 2014.
6. Hansard – 8 January 2002, provides a record of the debate.
7. Secretary of State's decision letter 19 November 2002 with the inspector's report and conclusions attached. Appeal ref – APP/D34350/A/1071002.
8. NPPF Section 17 – Facilitating the sustainable use of minerals.
9. Jeremy Burchardt, Joe Doak and Peter Gavin – 'Review of Key Trends and Issues in UK Rural Land Use' – Final Report to the Royal Society – University of Reading August 2020.
10. Burchardt et al – Ibid. – para. 2.3.3.
11. Marion Shoard – *The Theft of the Countryside* – www.marionshoard.co.uk/Books/The-Theft-Of-The-Countryside/Introduction.php
12. Marion Shoard – *The Planner* 1982.
13. John Bowers & Paul Cheshire – A condensed contribution to *The Changing Countryside* – The Open University & Croom Helm Ltd 1985 and taken from their book – *Agriculture, the Countryside & Land Use: An Economic Critique* pp 135–6 & 142–3 Methuen 1983.
14. Ministry of Agriculture Fisheries and Food (MAFF) – Technical Report 11– Agricultural Land Classification for England & Wales.
15. Natural England (part of DEFRA) – Guide to Assessing development proposals on Agricultural Land – 16 January 2018.
16. NPPF Section 15 – Conserving and Enhancing the Natural Environment – paragraph 170 a) & b).

17. NPPG – Guidance for the Natural Environment 21 July 2019 para 002 – Ref ID8-001/002-20190721.
18. George Bangham – *Game of Homes – The rise of multiple property ownership in Great Britain* – Resolution Foundation June 2019.
19. Department for the Environment, Food and Rural Affairs (DEFRA): 'English National Parks and the Broads' UK Government Vision and Circular March 2010.
20. DEFRA – Ibid. – paras 76 & 77.
21. The Taylor Review of the Rural Economy and Affordable Housing – July 2008.
22. 'England's Rural Areas – Steps to Release their economic potential' – Report of the Rural Advocate June 2008.
23. Second and Holiday Homes: Housing Evidence Base – Cornwall Local Plan Briefing Note 11 Version 2 December 2015.
24. Cornwall Local Plan Briefing Note – Ibid.
25. RLT Built Environment Ltd v Cornwall County Council [2016] EWHC 2817(Admin) – judgement of Hickinbottom J.
26. *Encyclopedia of Planning Law & Practice* – Sweet & Maxwell.
27. FH Stephen – *The Economics of the Law* – Wheatsheaf Books London 1988.
28. Cooke J in Stringer v MHLG [1971] AUER 65 at 77.
29. Westminster City Council v Great Portland Estates plc [1985] AC 661.
30. R Wright v Resilient Energy Severndale Ltd and Forest of Dean Council [2019] UKSC 53.
31. Mark Lowe QC & Estelle Dehon (barrister) – 'Heads up on Material Considerations: You really can't buy a planning permission says the Supreme Court'. Cornerstone Barristers 21 November 2019.
32. George Dobry QC – 'Review of the Development Control System' – Final Report February 1975.
33. Department of the Environment (DOE) Circular 22/80 – 'Development Control- Policy and Practice' – 28 November 1980.
34. Encyclopedia of Planning Law & Practice – commentary leading up to the Planning & Compulsory Purchase Act 2004.
35. RTPI – 'The Top Five Planning Myths' – 6 September 2011.
36. 'Development Management – Designing Buildings' – Wiki – last edited 25 September 2020.
37. Sir Michael Lyons – 'Place-shaping: a shared ambition for the future of local government' – March 2007.
38. Richard Pritchard – 'Development Management' – 10 March 2008 – An unofficial blog of the Planning Advisory Service (PAS) team – originally published in *The Planner*.
39. Citiesmode, Arup and PAS – 'Good Development Management' – June 2019 – produced for the Local Government Association (LGA) and MHCLG.

Chapter 7
1. NPPF (as revised) July 2021 Section 11 – 'Making Effective Use of Land'.
2. The Old Testament Book of Numbers – Chapter 5 v 1-5.
3. For further information on the roles of Patrick Geddes & Raymond Unwin in the 'Movement for Town Planning' see William Ashworth's book *The Genesis of Modern British Town Planning* as referred to in Chapter 1 of this book.
4. Martin J Elson – *Green Belts – Conflict Mediation in the Urban Fringe* Heinemann London 1986 – with reference to Raymond Unwin's role in promoting Green Belts.
5. Peter Hall – *Urban & Regional Planning* – 3rd edition Routledge 1992.
6. Circular 42/55 MHLG – 'Green Belts' 3 May 1955 – The Minister's statement of 26 April is annexed to the Circular.

7. Elson – Ibid.
8. Peter Self – *Cities in Flood – The Problems of Urban Growth* – Faber 2nd edition 1961.
9. Green Belt Statistics – additional research briefing produced by the UK Parliament Library – 20 November 2020 – figures reproduced from MHCLG local authority Green Belt statistics for England 2019/20 – September 2020.
10. The Land Cover Atlas of the UK – University of Sheffield Department of Urban Studies and Planning.
11. The Joint Nature Conservation Committee (advisor to government on nature conservation) – UK Biodiversity Indicators Protected Areas 15 October 2020. NB these areas are calculated on a basis which removes the land area overlaps of different designations.
12. NPPF July 2021 Section 13 – 'Protecting Green Belt land' paragraph 137.
13. Planning Policy Guidance PPG2 – Green Belts January 1995.
14. *The Times* newspaper 5 October 2018.
15. Local Planning Authority Green Belt: England 2018/19 – Planning Statistical Release 18 October 2019 MHCLG.
16. 'State of Brownfield 2019 – An updated analysis on the potential of brownfield land for housing' – Council for the Preservation of Rural England (CPRE) March 2019.
17. Cecilia Wong & Baing Schultze – *Brownfield residential redevelopment in England- What happens to the most deprived neighbourhoods?* Joseph Rowntree Foundation June 2010.
18. TCPA – Green Belts Policy Statement May 2002. This policy statement was produced by the TCPA's Housing & New Communities Task Team. This team notably included Martin Elson and Peter Hall.
19. 'The Green Noose: An analysis of Green Belts and Proposals for Reform' – The Adam Smith Institute 14 January 2015.
20. 'Delivering Change: Building Homes where we need them' – Centre for Cities 2015.
21. Gabrielle Garton Grimwood & Cassie Barton – 'Green Belt' – Briefing Paper No. 00934-House of Commons Library 20 November 2020.
22. OECD – 'Economic Survey March 2011 – United Kingdom Overview' – 16 March 2011.
23. 'Green Belt Policy' – analysis of the results of a consultation by the Landscape Institute August 2016.
24. The Landscape Institute – 'Green Belt Policy Briefing- A New Vision' – April 2018.
25. Sir Dieter Helm – 'Green and Prosperous Land – A Blueprint for Rescuing the British Countryside'. William Collins 2019. The author is Professor of Economic Policy at the University of Oxford and Chair of the Natural Capital Committee, which advises government.

Chapter 8

1. The International Commission referred to in paragraph 118 of the July 2021 version of the NPPF is the International Commission on Non-Ionising Radiation Protection (ICNIRP). This organisation was founded in 1992 by the International Radiation Protection Association. ICNIRP's activities are of a scientific nature and deal only with health risk assessment (source: Wikipedia).
2. National Planning Policy for Waste – Department of Communities and Local Government October 2014.
3. National Planning Practice Guidance – Land Stability – paragraph 001 Ref ID: 45-001-20190722.
4. Health & Safety Laboratory – 'Review of the Public Perception of Risk and Stakeholder Engagement' – HSL/2005/16.
5. Chris Hilson – 'Planning law and public perceptions of risk: evidence of concern or concern based on evidence?' – *Journal of Planning & Environmental Law* December 2004 pp 1638–1648.

6. Andrew Piatt – 'Public concern – a material consideration?' – *Journal of Planning & Environmental Law* May 1997 pp 397–400.
7. Neil Stanley – 'Public concern: the decision-makers dilemma' – *Journal of Planning & Environmental Law* October 1998 pp 919–934.
8. Gateshead MBC v SoSE [1995] Env.L.R 37; Newport CBC v SoS Wales [1998] JPL 377; Envirocor Waste Holdings v SoSE & others [1996] Env.L.R 49.
9. Case commentary in the JPL April 1998 pp 377–387 – considers the various types of fear based on the Newport and Gateshead cases quoted above.
10. Planning Inspectorate England appeal decision ref APP/X4725/A/99/1035502.
11. Friends of the Earth – 'The Incineration Campaign Guide' December 1997.
12. Planning Policy Guidance Note 8 – Telecommunications – August 2001. This version of PPG 8 replaces that of December 1992 and introduces the first advice on taking account of health considerations in planning decisions on telecoms infrastructure.
13. Public Health England – 'Guidance Mobile Phone Base Stations: Radio waves and health' – updated 24 September 2020.
14. Eileen O'Connor – Director Radiation Research Trust UK- letter of 18 March 2010 addressed to members of Ofcom (the government appointed regulator for all communications services).
15. Phillips v First Secretary of State [2003] EWHC Admin 2415 – 22 October 2003.
16. Chris Hilson – Ibid.
17. Trevett v Secretary of State for Transport, Local Government & the Regions, and others [2002] EWHC Admin 2696, 25 November 2002.
18. Chris Hilson – Ibid.

Chapter 9
1. Peter Hall – 'The Regional Dimension' (Chapter 6) in 'British Planning – 50 years of Urban and Regional Policy' – Edited by Barry Cullingworth – The Athlone Press 1999.
2. Hall – Ibid.
3. Peter Self – 'Cities in Flood – The problems of Urban Growth' Faber Second edition 1961.
4. Ian Scargill – *Urban France* – Croom Helm Ltd 1983.
5. 'Planning Control in Westen Europe' – HMSO London 1989. This Report was commissioned by the Planning & Land Use Policy Directorate of the Department of the Environment and produced by the Joint Centre for Land Development Studies (College of Estate Management)/University of Reading.
6. 'Levelling-Up the United Kingdom' – White Paper presented to Parliament by the Secretary of State for Levelling-Up, Housing and Communities on 22 February 2022. Michael Gove became the first Secretary of State for this expanded Government department which replaced the former MHCLG.
7. Letter from the Chief Planner at the Department.
8. House of Commons – Communities and Local Government Committee – 'Abolition of Regional Spatial Strategies: a planning vacuum' – Second Report of Session 2010–2011 published 17 March 2011 – HMSO.
9. House of Commons – Ibid. – Evidence item 105.
10. House of Commons – Ibid. – Evidence item 3937 – http://www.rtpi.org.uk
11. 'Decentralisation and the Localism Bill – an essential guide' – HM Government December 2010.
12. RTPI – 'Localism Bill' – 'Living' Brief 21 December 2010.
13. Geoff Walker – 'The South West: current planning issues, major projects and professional planning activity'. This paper was written to brief the President of the RTPI for her 2020 regional visits. Geoff is a member of the RTPI Regional Activities Committee South West.
14. Geoff Walker – Ibid.

15. West of England Combined Authority – West of England Strategic Planning – Autumn 2020.
16. Local Plans Expert Group (LPEG) – 'Local Plans – Report to the Communities Secretary and to the Minister of Housing & Planning' – March 2016.
17. 'Fixing our broken housing market' – Housing White Paper 7 February 2017 MHCLG.
18. The National Infrastructure Commission (NIC) Final Report – Partnering for Prosperity – A new deal for the Cambridge – Milton Keynes – Oxford Arc' – 2017.
19. Ed Cox & Katie Schmuecker – 'Beyond Banks and Big Government – strategies for local authorities to promote investment' – Institute for Public Policy Research – IPPR North – March 2013.
20. Cox & Schmuecker – Ibid.
21. One Powerhouse Consortium/Royal Society of Arts (RSA): 'A One Powerhouse framework for national convergence and prosperity – A Vision for Britain. Planned.' January 2021.
22. One Powerhouse Consortium/RSA/Atkins – 'One Powerhouse Towards a Spatial Blueprint – The North' 2020.

Chapter 10
1. Source BBC News article 2017.
2. Society of Motor Manufacturers & Traders – March/April 2020.
3. Otto Saumarez Smith- Addition to the Oxford Dictionary of National Biography 2019.
4. Otto Saumarez Smith – 'Boom Cities, Architect Planners and the Politics of Radical Urban Renewal in 1960s Britain' – Oxford University Press April 2019.
5. Colin Buchanan – 'Traffic in Towns' – Ministry of Transport November 1963.
6. Ministry of Transport & Ministry of Housing and Local Government – Circular 1/64 – 'Buchanan Report on traffic in towns' – HMSO 1964.
7. Michael J Bruton – *Introduction to Transportation Planning, 3rd Edition* – University College London Press 1992.
8. Peter Hall – *Great Planning Disasters of our Time* – Weidenfeld & Nicholson London 1980.
9. Clive A Brook – 'The Implementation of City Transportation Strategies – a planning perspective' – paper delivered at the First International Conference on Urban Transport and the Environment – Southampton June 1995 – published by Computational Mechanics Publications – Southampton & Boston edited by LJ Sucharov.
10. 'A New Deal for Transport- Better for Everyone' – White Paper on Transport HMSO July 1998.
11. NPPF July 2021 – Section 9 – Promoting Sustainable Transport.
12. Alastair Baldwin & Kelly Shuttleworth – 'How governments use evidence to make transport policy' – Institute for Government February 2021.
13. Department for Transport – Road Investment Strategy (RIS2) – 11 March 2020. This comprises a £27 billion road building programme to 2025.
14. (A) Royal College of Physicians Report – 'Every Breath We Take: The Lifelong Impact of Air Pollution' 2016
    (B) DEFRA Report on nitrogen dioxide exposure – 2015.
15. Tim Smedley – article in *The Sunday Times* 20 December 2020. He is the author of *Clearing the Air – The Beginning & the End of Air Pollution* Bloomsbury.
16. Government Office of Science – 'A time of unprecedented change in the transport system' – part of The Future Mobility Programme – January 2019.
17. Government Office for Science – Future of Mobility: Evidence Review – 'Governance of UK Transport Infrastructures' – produced by Professor Greg Marsden (Institute for Transport Studies University of Leeds) & Professor Iain Docherty (Adam Smith Business School University of Glasgow).

376   Planning for Good or Ill

18. Chartered Institute of Highways & Transportation: 'Better planning, better transport, better places'– August 2019.
19. Serena Ralston – 'Revolutions in the Arc' – Tech-Landscape, article in *The Planner* April 2018.
20. 'Connecting Leeds Transport Strategy'– Leeds City Council – produced as a draft for public consultation closing on 11 April 2021.

Chapter 11

1. Andrew Gimson – 'How Macmillan built 300,000 houses a year' – Conservative Home – 17 October 2013 (https://www.conservativehome.com)
2. Gimson – Ibid.
3. www.law.ox.ac.uk/housing-after-grenfell/blog/2019/02/human-right-housing.
4. MHCLG – 'Fixing our broken housing market' Cmnd 9352 February 2017.
5. MHCLG – 'Planning for the Future' White Paper August 2020.
6. Shelter – 'Denied the right to a safe home – Exposing the housing emergency' 2021.
7. Kate Barker – 'Review of Housing Supply- Delivering Stability: Securing our Future Housing Needs'– prepared for HM Treasury and the Office of the Deputy Prime Minister – HMSO March 2004.
8. Ian Mulheirn – 'Tackling the UK housing crisis: is supply the answer?' – UK Collaborative Centre for Housing Evidence – August 2019.
9. Wendy Wilson & Cassie Barton – 'Tackling the under-supply of housing in England' – House of Commons Library Briefing Paper No. 07671 – 14 January 2021.
10. Alan Holmans: 'New estimates of housing demand and need in England 2011 to 2031 –Town & Country Planning Tomorrow series paper 16.
11. Holmans – Ibid.
12. Wendy Wilson & Cassie Barton – Ibid.
13. Barker – Ibid.
14. KPMG & Shelter – 'Building the Homes we need: A programme for the 2015 government' – 2014.
15. 'The Lyons Housing Review – Mobilising across the nation to build the homes our children need'– 2014. The Review was chaired by Sir Michael Lyons.
16. Rt Hon Sir Oliver Letwin MP – 'Independent Review of Build Out' – Final Report presented to Parliament by the Secretary of State for Housing, Communities and Local Government – October 2018.
17. Wilson & Barton – Ibid.
18. 'Forecasting UK House Prices and Home Ownership' – Oxford Economics November 2016.
19. Mulheirn – Ibid.
20. MCHLG – 'Affordable Housing Supply – April 2019 – March 2020 England.
21. Glen Bramley – 'Housing Supply requirements across Great Britain: for low-income households and homeless people' – produced for Crisis and the National Housing Federation November 2018.
22. MHCLG Parliamentary Select Committee – 'Land Value Capture'.
23. RICS – 'Land value capture: Attitudes from the housebuilding industry on alternative mechanisms' – August 2020.
24. Pat McAllister – 'The Taxing Problems of Land Value Capture, Planning Obligations and Viability Tests: Some Reasonable Models? – *Town Planning Review*, 90(4) pp 429–451 – University of Reading.
25. NPPG – Viability – Gov.uk. – latest partial update 1 September 2019.
26. Adam Tinson & Amy Clair – 'Better Housing is crucial for our health and the Covid-19 recovery'– The Health Foundation 28 December 2020.

Chapter 12

1. Crispin Tickell in his foreword to James Lovelock's book – *The Revenge of Gaia* – Penguin Books 2007.
2. NPPF – July 2021 Section 14 'Meeting the Challenge of Climate Change, Flooding and Coastal Change'.
3. NPPG – Climate Change – Gov.uk. Last updated 15 March 2019.
4. TCPA/RTP – 'The Climate Crisis – A Guide for Local Authorities on Planning for Climate Change – 3rd edition October 2021.
5. House of Commons MHCLG Committee – 'The future of the planning system in England' – First Report of Session 2021–22, 27 May 2021.
6. RTPI – response to the Committee on Climate Change progress report to Parliament – July 2020.
7. Climate Change Committee (CCC) – 'The Sixth Carbon Budget – The UK's path to Net Zero' December 2020 – CCC copyright.
8. CCC – 'Independent Assessment of UK Climate Risk'. Advice to Government for the UK's third Climate Change Risk Assessment (CCRA3) – June 2021 – CCC copyright.
9. NPPG – Climate Change – Paragraph: 004 Ref ID: 004-20140612.
10. RTPI – 'Plan the World We Need – The contribution of planning to a sustainable, resilient and inclusive recovery' – Research Paper June 2020.
11. RTPI – 'Five Reasons for Climate Justice in Spatial Planning' – Position Paper January 2020.
12. RTPI – 'Strategic Planning for Climate Resilience' – recommendations to the Liverpool City Region Combined Authority – Research Paper November 2020.
13. RTPI – 'Place-Based Approaches to Climate Change – Opportunities for collaboration in local authorities' – Research Paper March 2021.
14. CCC – 'Local Authorities and the Sixth Carbon Budget' – December 2020.
15. TCPA & The Centre for Sustainable Energy: 'Why the Planning System needs to be at the heart of delivering the UK's Climate Change targets' – response to the Planning White Paper Consultation October 2020.
16. Dr Emma Lawrance, Rhiannon Thompson, Gianluca Fontana, Dr Neil Jennings – 'The impact of climate change on mental health and emotional wellbeing: current evidence and implications for policy and practice' – The Grantham Institute for Climate Change and the Environment – Imperial College London – Briefing Paper No. 36 May 2021.
17. IPCC – The UN Intergovernmental Panel on Climate Change – 'The Physical Science Basis of Climate Change'. This comprises part 1 of a major three-part report series from working groups, which together make up the Sixth Assessment Report (AR6) of the IPCC. Part 1 was published on 9 August 2021. Parts 2 & 3 – 'Impacts, Adaptation and Vulnerability' and 'Mitigation of Climate Change', were published in February and March 2022 with a Synthesis Report expected in October 2022.

Chapter 13

1. Professor Alasdair Rae – 'The Land Cover Atlas of the UK' – University of Sheffield Department of Urban Studies and Planning – published 8 November 2017 and compiled from Corine (Co-ordination of Information on the Environment) Land Cover data from 2012, which includes 44 different types of land use classification. Corine is now one of the most widely used land classification systems and provides comprehensive European coverage.
2. Government Office for Science: Trend Deck 28 June 2021 – includes MHCLG data on Green Belt land designation.
3. MHCLG – 'Land Use in England 2017' – published 31 May 2019 & 'Land Use Change Statistics in England 2017–18'.

4. 'The Glover Review' – commissioned by DEFRA – May 2018. The independent panel chaired by Julian Glover published their report – 'The Landscapes Review' in September 2019. Their proposals for new and expanded Areas of Outstanding Natural Beauty (AONBs) are being progressed by Natural England – see statement by Secretary of State DEFRA 24 June 2021.
5. Gov.uk – press release regarding the new Nature Recovery Network (NRN) Delivery Partnership – 5 November 2020.
6. Joint Nature Conservation Committee (JNCC) – 'UK Biodiversity Indicators – C1 Protected areas – 15 October 2020.
7. IUCN National Committee United Kingdom – 'Putting Nature on the Map- Identifying Protected Areas in the UK' – February 2012.
8. Thomas Starnes, Alison E Beresford, Graeme M Buchanan, Matthew Lewis, Adrian Hughes, and Richard D Gregory: 'The extent and effectiveness of protected areas in the UK' – Global Ecology and Conservation 2021.
9. Assaf Shwartz, Zoe G Davies, Nicholas A Macgregor, Humphrey Q P Crick & 9 others: 'Scaling up from protected areas in England: The value of establishing large conservation areas' – *Biological Conservation* August 2017.
10. Catherine Early – 'Rewilding' – *The Planner* September 2019.
11. David Woodfall (Editor) – *Rewilding-Real Life Stories of Returning British and Irish Wildlife to Balance* – William Collins 2019.
12. Patrick Barkham (Editor) – *The Wild Isles – An anthology of the best British & Irish nature writing* – Head of Zeus 2021.
13. MHCLG Parliamentary Select Committee May 2021.
14. NPPF – Section 15 – 'Conserving and enhancing the natural environment' – July 2021.
15. NPPG – 'Natural Environment' – including updates to July 2021.
16. Landscape Institute – 'Ecosystem Services' – Technical Information Note – February 2016.
17. HM Government: 'A Green Future: Our 25-year Plan to Improve the Environment' – 2018.
18. Gov.uk – 'The Natural Choice: Securing the Value of Nature' – White Paper 2011.
19. Natural Capital Committee: Advice to government on net environmental gain.
20. NPPG Paragraph 028, Ref ID: 8-028-20190721.
21. Natural England – 'Biodiversity Metric 3.0: Auditing and accounting for biodiversity – User Guide July 2021. Separate documents published alongside the User Guide include a calculation tool and technical report.
22. RTPI Research Paper – 'Green Growth Boards – Communicate, Collaborate, Innovate' – November 2021.
23. The Institute for European Environmental Policy – 'The Health and Social Benefits of Nature and Biodiversity Protection' – Final Report 28 April 2016.

Chapter 14

1. – The Town & Country Planning Act 1990.
   – The Planning & Compulsory purchase Act 2004.
   – The Planning Act 2008 – This Act introduced legislation to speed up the handling of Nationally Significant Infrastructure Projects by a new Infrastructure Planning Commission. The Commission was subsequently abolished in 2011 with its functions being transferred to the Planning Inspectorate.
   – Other Acts with relevant planning content include the Climate Change Act 2008; The Localism Act 2011, which established the ability of communities to produce Neighbourhood Plans; the Housing & Planning Act 2016; The Neighbourhood Planning Act 2017.

2. House of Commons MHCLG Select Committee – 'The Future of the Planning System in England' – First Report of Session 2021–22 – 10 June 2021.
3. RTPI – 'Resourcing Public Planning' Research Paper July 2019.
4. RTPI – Research Briefings:
    'The Value of Planning' – Research Briefing No. 5 – June 2014 – based on research conducted by the Universities of Glasgow & Sheffield.
    'Planning as market maker: How planning stimulates development in continental Europe' – Research Briefing No.12 November 2015 – based on research conducted by the University of Liverpool.
    'Planning of China's future: How planners contribute to growth and development' – Research Briefing No.13 December 2015 – research conducted for the RTPI by Professor Fulong Wu, Dr Fangzhu Zhang & Zheng Wang from the Bartlett School of Planning University College London.
    'Delivering the value of planning' – Research Briefing No. 20 August 2016 – based on research conducted by the Universities of Glasgow & Sheffield and the RTPI.
5. 'Rural Planning in the 2020s' – RTPI Research Paper July 2022 – produced by a project team led by Professor John Sturzaker from the University of Hertfordshire.
6. 'Making the most out of England's land' – House of Lords, Land Use in England Committee – 13th December 2022.
7. MHCLG: National Design Guide: Planning Practice Guidance for beautiful, enduring, and successful places – January 2021.

Select Bibliography

Modern town and country planning has a wide and challenging coverage with sustainable development and health and wellbeing at its heart. The natural and built environments together with urban and rural communities are its focus. Its key pursuit of sustainable development has three components – economic, social, and environmental. Today's planning systems throughout the world face the three interlocking challenges of climate change, environmental loss and issues surrounding the health and wellbeing of communities and individuals. As a result, the literature on these multiple related topics is vast. My sources for this book cover only a minute proportion of this vast and expanding literature. My selection for this bibliography is limited by many factors, in particular my own knowledge and research base, the size of my library, the scope of my book and the time available to ensure I get my message out there. My selection includes dated but foundational texts, which I have collected over the years including works by Ebenezer Howard, Lewis Keeble, and Peter Self. I have also included books that have inspired me along the way to becoming a first-time author and to continue my involvement and interest in the planning profession.

Ashworth, William – *The Genesis of Modern British Town Planning* (London: Routledge & Kegan Paul, 1972 reprint)

Attenborough, David – *A Life on Our Planet: My Witness Statement and a Vision for the Future* (London: Witness Books, 2020)

Barkham, Patrick (Editor) – *The Wild Isles: An anthology of the best British & Irish nature writing* (London: Head of Zeus, 2021)

Berners-Lee, Mike – *There is no Planet B: A Handbook for the Make or Break Years* (Cambridge: Cambridge University Press, 2019)

Berry, Jim, McGreal, Stanley & Deddis, Bill (Editors) – *Urban Regeneration, Property Investment and Development* (London: E & FN Spon, 1993)

Blowers, Andrew (Editor) – *Planning for a sustainable environment: A report by the Town & Country Planning Association* (London: Earthscan Publications, 1993)

Brand, Stewart – *Whole Earth Discipline* (London: Atlantic Books, 2010)

Bruton, Michael J – *Introduction to Transportation Planning* (London: UCL Press 1992, 3rd edition)

Bunting, Madeleine – *The Plot: A Biography of an English Acre* (London: Granta Publications, 2009)

Carney, Mark – *Value(s): Building a Better World For All* (London: Harper Collins 2021)

Cowen, Rob – *Common Ground* (London: Hutchinson, 2015)

Crane, Nicholas – *The Making of the British Landscape: From the Ice Age to the Present* (London: Weidenfeld & Nicholson, 2016)

Cullingworth, Barry (Editor) – *British Planning: Fifty Years of Urban and Regional Policy* (London: Athlone Press, 1999)

Cullingworth, Barry & Nadin, Vincent – *Town and Country Planning in Britain* (London: Routledge, 2014, 15th edition)

Curry, Nigel – *Countryside Recreation, Access and Land Use Planning* (London: E & FN Spon, 1994)
Elson, Martin J – *Green Belts: Conflict Mediation in the Urban Fringe* (London: William Heinemann, 1986)
Environment, Department of – *Planning Control in Western Europe* (London: HMSO, 1989)
Fairbrother, Nan – *New Lives New Landscapes* (London: Penguin Books, 1972)
Glasson, John – An Introduction to Regional Planning: Concepts, Theory & Practice (London: Hutchinson, 1975)
Gore, Al – *An inconvenient truth: The Planetary Emergency of Global Warming and what we can do about it* (London: Bloomsbury, 2006)
Gore, Al – *Our Choice: A Plan to Solve the Climate Crisis* (London: Bloomsbury, 2009)
Gosling, David – *Gordon Cullen: Visions of Urban Design* (London: Academy Editions, 1996)
Grant, Malcolm – *Urban Planning Law* (London: Sweet & Maxwell, 1982)
Hall, Peter – *Urban and Regional Planning* (London: Routledge, 1992, 3rd edition)
Hawken, Paul (Editor) – *Drawdown: The Most Comprehensive Plan Ever Proposed to Reverse Global Warming* (London: Penguin Books, 2018)
Healey, Patsy & others (Editors) – *Rebuilding the City: Property-led urban regeneration* (London: E & FN Spon, 1992)
Helm, Dieter – *Green and Prosperous Land: A Blueprint for Rescuing the British Countryside* (London: William Collins, 2019)
Hoskins, W G – *The Making of the English Landscape* (London: Guild Publishing, 1988)
Howard, Ebenezer – *Garden Cities of To-morrow* (Builth Wells, Wales: Attic Books – new revised edition 1985)
Juniper, Tony – *Saving Planet Earth* (London: Collins, 2007)
Kynaston, David – *Austerity Britain 1945–51* (London: Bloomsbury, 2007)
Keeble, Lewis – *Principles and Practice of Town and Country Planning* (London: The Estates Gazette, 1969, 4th edition)
Lovelock, James – *The Revenge of Gaia* (London: Penguin Books, 2007)
Lovelock, James – *The Vanishing Face of Gaia: A Final Warning* (London: Penguin Group, 2009)
Lovelock, James – *A Rough Ride to the Future* (London: Penguin Group, 2014)
Macfarlane, Robert – *The Wild Places* (London: Granta Books, 2008)
Macfarlane, Robert – *Landmarks* (London: Hamish Hamilton, 2015)
Manrique, Cesar – *In his own words* (Lanzarote: Fundacion Cesar Manrique, 1995)
Marr, Andrew – *A History of Modern Britain* (London: Pan Macmillan, 2017)
McKean, John & Baxter, Collin – *Charles Rennie Mackintosh: Architect, Artist, Icon* (Edinburgh: Lomond Books, 2000)
McKie, David – *A Sadly Mismanaged Affair: A Political History of the Third London Airport* (London: Croom Helm, 1973)
Meadows, Donella H, Meadows, Denis L, Randers, Jorgen, Behrens, William W – *The Limits to Growth: A Report for the Club of Rome's project on the predicament of mankind* (London: Pan Books, 1974)
Morgan, Peter & Nott, Susan – *Development Control: Law, Policy and Practice* (London: Butterworth, 1995)
Open University in association with the Countryside Commission – *The Changing Countryside* (London: Croom Helm, 1985)
Porritt, Jonathan – *The World We Made: Alex McKay's Story from 2050* (London: Phaidon Press, 2013)
Rackham, Oliver – *History of the Countryside* (London: BCA illustrated version, 1994)
Raworth, Kate – *Doughnut Economics* (London: Penguin Random House, 2018)
Raynsford, Rt Hon Nicholas (Chair) – *Raynsford Review of Planning in England: Planning 2020* (London: TCPA, 2018)

Reynolds, Fiona – *The Fight For Beauty: Our Path to a Better Future* (London: Oneworld Publications, 2017)
Richards, Brian – *Transport in Cities* (London: Architecture, Design & Technology Press, 1990)
Rydin, Yvonne – *The British Planning System: An Introduction* (London: Macmillan Press, 1993)
Scargill, Ian – *Urban France* (London: Croom Helm, 1988)
Schneer, Jonathan – *Ministers at War: Winston Churchill and his War Cabinet* (London: Oneworld Publications, 2016)
Scruton, Sir Roger (Chairman of the Building Better Building Beautiful Commission) – *Living with Beauty: Promoting Health, Wellbeing and Sustainable Growth* (January 2020)
Self, Peter – *Cities in Flood: The Problems of Urban Growth* (London: Faber & Faber, 1961, 2nd edition)
Shoard, Marion – *The Theft of the Countryside* (London: Maurice Temple Smith, 1980)
Shrubsole, Guy – *Who owns England: How we lost our Green & Pleasant Land & How to Take It Back* (London: William Collins, 2019)
Singleton, Ross, Castle, Pamela, & Short David – *Environmental Assessment* (London: Thomas Telford Publishing, 1999)
Stern, Nicholas – *The Economics of Climate Change: The Stern Review* (Cambridge: Cambridge University Press, 2007)
Stern, Nicholas – *A Blueprint for a Safer Planet* (London: The Bodley Head, 2009)
Sucharov L J (Editor) – *Urban Transport and the Environment for the 21st Century: First International Conference Southampton* (Southampton: Computational Mechanics Publications, 1995)
Sucharov L J & Baldasano, Recio (Editors) – *Urban Transport and the Environment: Second international Conference, Barcelona* (Southampton: Computational Mechanics Publications, 1996)

Townroe, Peter & Martin, Ron (Editors) – *Regional Development in the 1990s: The British Isles in Transition* (London: Jessica Kingsley Publications & the Regional Studies Association, 1992)
Tree, Isabella – *Wilding: The return of nature to a British farm* (London: Picador, 2018)
Truelove, Paul – *Decision making in Transport Planning* (Harlow, Essex: Longman Scientific & Technical, 1992)
Walmsley, David & Perrett, Ken – *The Effects of Rapid Transit on public Transport and Urban Development* (London: HMSO, 1992)
Ward, Barbara & Dubos Rene – *Only One Earth: The Care and Maintenance of a Small Planet – An unofficial report commissioned by the Secretary General of the United Nations Conference on the Human Environment, prepared with the assistance of a 152-member Committee of Corresponding Consultants in 58 countries* (Middlesex, England: Penguin Books, 1972)
Woodfall, David (Editor) – *Rewilding: Real Life Stories of Returning British & Irish Wildlife to Balance* (London: William Collins, 2019)
Zerbst, Rainer – *Gaudi – The Complete Buildings* (Cologne: Taschen, 2005)

Acknowledgements

As a first-time author producing a book based on my chosen profession and my experiences over more than five decades, I owe debts of gratitude to many people who have helped me along the way.

First and foremost, my greatest thanks must go to my wife, Jill, who has supported me in so many ways since our marriage in 1967. For the first couple of years, we spent many weekday evenings working around the dining table of our rented ground floor flat in Chelmsford, Jill marking books and preparing lessons in her new job as a languages teacher at a large comprehensive school and me preparing for exams and carrying out project work. Special support was provided during my period of career change from local government to a one-man band consultancy and subsequently in 1989 when I persuaded her to take on the management of finance, personnel and other support services at Clive Brook Associates, a role she fulfilled for eleven years, despite one or two close friends saying they gave this new relationship a matter of months. The end product of my last three years of writing would not have been possible without her continued help.

Our children, Claire, Paul, and Richard, have all pursued careers related to the urban and rural environments. Claire is an environmental and planning law specialist and now main board director of a leading law firm; Paul is an interior architect/designer and Richard a landscape architect/designer. This has provided me with some useful insights and information, at times robust discussions, and added to my inner circle of inspiration without which my book would not have been completed.

Our four grandchildren, Alex, Hayden, Florence, and Rory were particularly in my mind when writing the last three chapters, for it is their generation that will face the consequences of any failures to meet the three interlocking challenges.

As I have grown older and more experienced in life in general, the start provided by my parents has taken on greater significance. While they both left school at the age of 13 they provided me with considerable help and encouragement in my school years.

Dale Robinson and John Gyford, my lecturers/tutors at the Chelmer Institute of Higher Education (later Anglia Ruskin University) provided a unique and intensive style sandwich course in town planning. They demonstrated considerable dedication, knowledge and good humour in their work and were truly inspirational.

In my local government career, I was fortunate to work under the direction of a number of inspiring and talented leaders, including Jack Lowe and Colin Slater at Nottinghamshire, Graham King at West Glamorgan, and John Finney at Leeds.

In consultancy I have worked with some amazing professionals who have all in various ways contributed to my personal learning curve and many friendships. My thanks go to Graham Connell and Sue Ansbro, the initial Directors working alongside me at Clive Brook Associates (CBA). Maureen Pollard was my secretary for 20 years at CBA and Turley Associates and was frequently mentioned by clients as an excellent communicator on behalf of the business. Andy Rollinson and Richard Baxter were subsequently appointed as directors and joined me and our team at Turley Associates. I feel privileged to have worked with Rob Turley, an excellent and visionary planner who demonstrated the value of Christian ethics in business. We had established our companies in Manchester and Leeds in 1983/84 and wondered what might have happened had we met at that foundation stage. Jonathan Isles, MD at Dacre Son & Hartley, had been a friend who introduced several clients to CBA throughout its 16 years and then welcomed me as leader of a reconstituted planning team at Dacres in 2004. Jenny Hanbidge joined Dacres shortly after my arrival as my secretary and subsequently gained a planning degree and membership of the RTPI. She was a great support in developing the work of the planning team. Mark Johnson arrived as my successor at Dacres some four years later and we developed a close working relationship leading to an invitation to join him in establishing Johnson Brook as a new planning consultancy in Leeds in 2013. The success of this new company and our excellent working relationship, backed by a great team, provided inspiration to continue my full-time role for a few more years.

Working in consultancy teams both large and small has introduced me to a wide range of talented individuals and I have learned a great deal from these experiences. So many clients and consultants have helped me as an individual and to grow the companies I have helped to lead. The following deserve mention for their friendship and long support: clients including Sir Ken Morrison (who provided me with my first project), Gwen Fuller; Tony Clegg, Geoff Goodwill and Rob Stansfield at Mountleigh (for a period one of the leading commercial developers in Western Europe); Peter Gilman, Graham and Paul Gilliatt; and consultants including Paul Winter, Richard Fletcher, Peter Batty, Ron Watson and Robert Tregay.

Andrew Williamson, planning lawyer, chartered town planner and barrister, provided great support in my early years as a planning consultant in the form of project and client introductions and sound advice. We quickly developed a strong mutual understanding and an open and discursive approach to the projects, planning appeals and public inquiries on which we worked together. This proved to be a major contribution to my development and confidence as a consultant.

Friends who have given me support and encouragement across the years are many especially Lee and Britt Heyworth, Stephen and Pat Shaw, Stephen and Margaret Nixon, Jim Kelly, John and Carol Rushfirth, and Derek and Rhoda Twine.

Finally, after rejections from several publishers and literary agents I feel especially fortunate to have discovered The Self-Publishing Partnership Ltd. They provided a welcoming and helpful approach from the outset. I am extremely grateful to Frances Prior-Reeves my managing editor, Douglas Walker, proof-reader Chris Walkley and cover designer Kevin Rylands.

Index

Illustrations are referenced in italics (see full list at page viii)

Abercrombie, Patrick 174, 183, 227, 249
Afan Forest Park 39
Afan Valley (Upper) 38–39
affordable housing 156, 290
Agriculture Act 2020 152, 326
Agricultural Land Classification 150
agricultural land:
   control of development 148–152
AI (artificial intelligence) 353
air pollution 263–264
Amsterdam 85
Approach 99
   City Portrait 101
Area of Outstanding Natural Beauty (AONB) 320–323
Ashworth, William 8–14, 114
Attenborough, Sir David 302–304
Attlee, Clement 3, 6
Attlee Government (1945–1950) xiii
'Austerity Britain' 3

Barker, Dame Kate 279
Barker Review of Housing Supply 279–280
Barlow Report 4
beauty 115
   Covid 19 and 134
Benchmark Land Value (BLV) 293–294
betterment 12–13, 40, 291, 291–294
Bevan, Aneurin 270, 271
Beveridge, Sir William 5, 6
Beveridge Report (1942) 5, 61, 342
Biodiversity Net Gain (BNG) 331
Bourne Leisure 101–102
Bournville 10–11
Bradford and Airedale Health and Wellbeing Board 78

Bradford city centre highway plan 250
Bradford Council:
   approach to health and wellbeing 75–79
   Homes and Neighbourhoods Guide 89
   'Born in Bradford' Study 79
Brand, Stewart 301
Buchanan, Sir Colin 25, 250–251, 253–254
Building Better Building Beautiful Commission 115, 117, 335

Cadbury, George xii, 11
Cambridge- Oxford Arc (CAMKOX) 240, 266–267
Churchill, Sir Winston 3, 13–14, 270
Clark, Greg 70, 239
Clegg, Tony 95–98
climate change
   adaptation and mitigation 310
   health and wellbeing 317
   NPPF 305–306
   NPPG 305–306
   recommended reading 299–304
   RTPI and 306–307, 310–314
   spatial planning and 314–316
   TCPA and 306–307, 310, 313, 314
   transport and 269
   UK risks 308-309
climate justice 312
Clive Brook Associates 90–103
Commissioner for Sustainable Futures (Wales) 80–83
Committee for Climate Change (CCC) 306, 307–310, 313–314, 342
compensation and betterment 40
control of land use 142–143
Control of Land Use (1944 White Paper) 5–6, 61, 148

Covid 19 Pandemic 61–62, 116
   housing and 289, 294, 317
   transport and 247, 249, 263
Cubitt, Thomas 7
Cullen, Gordon 97–98, 124–127

Dacre Son & Hartley 104–106
DEFRA (Department for Environment, Food and Rural Affairs) 321, 331, 333
Department for Levelling-Up, Housing and Regeneration 327
design:
   beauty and 115–118
   personal involvement 118–121, 122–124, 132–133
   personal reflections 121–122, 124–127
   skill requirements 135–136
development control 138–141, 152–153, 171
   agriculture 148–152
   minerals 144–147
   second homes 153–155, 157–158
   tall buildings 158–159
development – definition of 139
development management 164–168, 170
Doughnut Economics:
   Amsterdam 86–87
   UK 86–87

East Leeds Urban Extension 99, 133
Enterprise 5 Bradford 90–92
Environment Act 2021 326, 327, 331
Environmental Land Management Schemes (ELMS) 326
Environmental Net Gain 329, 330
Environment Plan 333
environment and spatial planning 334–335
environment and master plans 335–336
Essex County Council 21, 24, 27
Essex Design Guide 119

Finney, John 41, 43, 120
Fletcher, Richard 97
Freiburg 269
Fuller, Gwen 100

Garden Cities Movement 11–12
Garden City Association 11–12

Gates, Theaster 118
Gaudi, Antonio 121
Glover Review 321
Goldsmith Street, Norwich 88
Gore, Al 300
Gove, Michael 327, 328
Gower AONB 33, *3*
Green Belts:
   case for and against change 184–197
   enabling development 173
   five purposes of 180–181
   land coverage 177–178, 184–186
   multiple land use potential 172–173
   openness 181–182
   very special circumstances 184
Gyford, John 22, 24

Hall, David 11
Hall, Peter 19, 26, 186
Hausmann, Baron Georges-Eugene 7
Hawken, Paul 304
Health:
   life expectancy 62
   obesity 62
   social determinants of 63–65
   trends 62, 64–65
Health Impact Assessment 76
Health and Social Care Act 2012 76
health and wellbeing:
   access to nature 338, 340
   access to recreation 338, 340
   'Born in Bradford' 79
   Bradford policy approach 75–79
   Covid 19 and 62
   definitions 66–69
   design and 74–75, 88–89, 114
   environment and 339–340
   housing and 289, 290, 294–295
   increasing significance of 61
   leading approaches, Wales and Amsterdam 79–87
   mental health 62–63
   NPPF 71–72, 74
   NPPG 72–74
   planning's positive role 65–66
   RTPI approach 87–88
   transport and 248, 262–264, 267–268
Health and Wellbeing Boards 78

Heap, Sir Desmond 17, 176
Helm, Sir Dieter 195–196
Heseltine, Michael 71
Howard, Ebenezer 11–14, 174, 179, 183, 193, 197, 227, 341–342
Huddersfield 17–19, 107

Ilkley, West Yorkshire:
  All Saints Church 120, 122–124, *11*, *12*, *13*, *14*
  Area Plan 132–133, 194–195, *5*, *6*, *7*, *8*, *9*, *10*
  International Panel on Climate Change (IPCC) 303, 314, 317–318, 342
  International Union for the Conservation of Nature (IUCN) 323
Isles, Jonathan 105

Johnson, Boris 327, 345
Johnson Brook 106
Joint Nature Conservation Committee 302

Kenyon, Stan 41–42
Kynaston, David 3, 50

Land Cover Atlas of the UK 320, 322
Land Use in England Committee 362
Land Value Capture (LVC) 291–292
Leeds-Bradford Airport Inquiry 45–46
Leeds City Council 39–50
  city transportation strategies 252–256, 267–268
  Climate Innovation District 289
  Health and Wellbeing Strategy 268
Letchworth Garden City 12
Levelling-up the UK White Paper (2022) 352
Levelling-Up and Regeneration Bill 328, 344, 346, 347
Lever, William (1st Viscount Leverhulme) 11
Litchfields 108–109
Litchfield, Nathaniel xv
Living with Beauty Report 115, 117, 130–131, 134–135
Localism Act (2011) 113
Local Environment Improvement Plans 339

Local Nature Recovery Strategies 333
London – third airport site selection 25–26
London – National Park City 338–339
Lovelock, James 300
Lowe, Jack 31
Lower Swansea Valley Project 34–35

Macmillan, Harold 270–271
Major, John 271
Manners, Gerald 19
Manrique, Cesar 122
Marmalade Lane, Cambridge 88, *16*
Marr, Andrew 3
material considerations 159–163
  anxiety and fear 169, 217
  financial 161
  private interests 161
  safety 161
  residential development 161–162
McKie, David 25–26
McKintosh, Charles Rennie 121
mineral extraction and development 144–147
  fracking 146–147
Monbiot, George 86
Morrison, Sir Ken 90–92
Mountleigh Developments 95–98
movement for town planning 8–9, 14, 61

Nash, John 7
National Design Guide 137
  health and wellbeing 74
national parks 320–323
National Parks and Access to the Countryside Act (1949) 15
National Nature Reserves (NNR) 321
national spatial planning 347–348
Natural Capital 329, 330
Natural Capital Committee (NCC) 329, 330
Natural England 329, 331, 332
Nature Recovery Networks (NNR) 321, 325, 332
neighbourhood plans 113, 349
Nidderdale AONB 144, *4*
Nolan Committee – third report 56–59
Nottinghamshire County Council 31–33

obesity 62

planners – role, definition of 60
   role and ethics 21, 36, 55–58
planning applications- determination of 140–141
planning balance sheet xv
planning:
   consultancy 106, 110–112
   and design 118–121, 127–132
   material considerations 159–163
   and politics 50, 54
   and project co-ordination 353–354
   reform of 343–347, 352
   resourcing of 351
   team building 353–354
   and rural areas 361–362
Planning & Compulsory Purchase Act 2004 84, 138–139, 230
Planning for the Future White Paper (2020) 110, 273
Porritt, Jonathan 302
Port Sunlight 11
public consultation 28-31, 51–52
public engagement 349–350
public good (interest), planning for 52–55, 59, 69, 74, 350, 355, 356–359, 362–365

Raworth, Kate 86–87
Raynsford Review- 'Planning 2020' 4, 17, 113, 291
regional planning 226–246
   French approach 228–229
   Regional Planning Bodies 230
   reinstatement of 348
   Regional Spatial Strategies (RSS) 230–233
   levelling-up and 242–243
rewilding 324
Rewilding Britain 324–326
Reynolds, Dame Fiona 115
risk management and planning 210–225
Robinson, Dale 22–24, 119
Roskill Commission 25–26
Rowntree, Joseph xii
Royal Town Planning Institute (RTPI) 14,
   Articles of Association 14, 52
   Code of Professional Conduct 58
   climate change 306–307, 310–314
   education 22
   membership 108–109
   national awards 356
   and rural planning 361–362
Rydin, Yvonne 51

Saltaire – World Heritage Site 10, *1, 2*
Salt, Titus xii, 10
Saumarez Smith, Professor Otto 107, 249–250
de Smith, S A 15–16
Scott Report 4, 5
Scruton, Sir Roger 115–116
second homes 153–155, 157–158
Self, Peter 5, 19, 176–177
Shields, Mike 41, 44
Shoard, Marion 149
Simon, Sir John 9
Skeffington Report (1969) 28–31
Slater, Colin 32
South and North Pennines SPAs/SACs 132–133, *8*
South West England-regional planning 235–238
Springswood new town project 96–98
Stern, Nicholas 301–302
sustainable development:
   definition of xiii, xiv,
   NPPF 3 objectives xiv, 74
Suddards, Roger 17
Swansea City Council 33–37
Sustainability and Transformation Partnerships 73, 79

Telecommunications infrastructure 219–223
the three interlocking challenges 343, *17*
Thatcher, Margaret 271–272
Thomas, Ray 12
Town & Country Planning Act 1947 4,6, 15–17, 31, 40, 51, 57, 138, 342
Town & Country Planning Act 1968 29
Town & Country Planning Act 1990 84, 138, 139
TCPA 11, 113, 189–190, 306–307, 310, 313, 314

transport:
    car usage 248–249, 260–261
    city transportation strategies 252–256
    and health and wellbeing 248, 261–264
    integration with planning 259–261, 264–265
    land-use transportation studies 251
Turley, Robert 119–120
Turley Associates 102–104

urbanisation 319-320
Uthwatt Report 4, 5

Wardley, Stanley Gordon 250
welfare state xiii

Wellbeing of Future Generations (Wales) Act 2015 80–83, 332
Welwyn Garden City 12
West Glamorgan County Council 37–39
Wheel of Sustainable Development *17*
Williamson, Andrew 95–96, 181, 215
Winter, Paul 99, 102
Wordsworth, William 115
World War II (1939-1945) xii, 1–3, 4

Yorkshire Dales National Park 154–156, 321, *4*